T0255547

Lecture Notes in Artificial Intelligence 13404

Subseries of Lecture Notes in Computer Science

More information about this subseries at https://link.springer.com/bookseries/1244

Ralph Bergmann · Lukas Malburg ·
Stephanie C. Rodermund · Ingo J. Timm (Eds.)

KI 2022: Advances in Artificial Intelligence

45th German Conference on AI
Trier, Germany, September 19–23, 2022
Proceedings

Editors
Ralph Bergmann
University of Trier
Trier, Rheinland-Pfalz, Germany

Lukas Malburg
University of Trier
Trier, Germany

Stephanie C. Rodermund
University of Trier
Trier, Germany

Ingo J. Timm
University of Trier
Trier, Germany

ISSN 0302-9743 ISSN 1611-3349 (electronic)
Lecture Notes in Artificial Intelligence
ISBN 978-3-031-15790-5 ISBN 978-3-031-15791-2 (eBook)
https://doi.org/10.1007/978-3-031-15791-2

LNCS Sublibrary: SL7 – Artificial Intelligence

This Springer imprint is published by the registered company Springer Nature Switzerland AG
The registered company address is: Gewerbestrasse 11, 6330 Cham, Switzerland

Preface

The proceedings volume contains the papers presented at the 45th German Conference on Artificial Intelligence (KI 2022), held as a virtual edition of this conference series during September 19–23, 2022, and hosted by the University of Trier, Germany.

KI 2022 was the 45th German Conference on Artificial Intelligence organized in cooperation with the German Section on Artificial Intelligence of the German Informatics Society (GI-FBKI, Fachbereich Künstliche Intelligenz der Gesellschaft für Informatik (GI) e.V.). The German AI Conference basically started 47 years ago with the first meeting of the national special interest group on AI within the GI on October 7, 1975. KI is one of the major European AI conferences and traditionally brings together academic and industrial researchers from all areas of AI, providing an ideal place for exchanging news and research results on theory and applications. While KI is primarily attended by researchers from Germany and neighboring countries, it warmly welcomes international participation.

The technical program of KI 2022 comprised paper as well a tutorial, a doctoral consortium, and workshops. Overall KI 2022 received about 47 submissions from authors in 18 countries, which were reviewed by three Program Committee members each. The Program Committee, comprising 58 experts from seven countries, accepted 12 full papers and five technical communications. As a highlight of this year's edition of the KI conference, the GI-FBKI and German's Platform for Artificial Intelligence PLS (Plattform Lernende Systeme) jointly organized a half-day event on privacy and data use, consisting of a keynote talk by Ahmad-Reza Sadeghi as well as a panel discussion. We were honored that very prominent researchers kindly agreed to give very interesting keynote talks (alphabetical order, see also the abstracts below):

- Bruce Edmonds, Manchester Metropolitan University, UK
- Eyke Hüllermeier, LMU Munich, Germany
- Sven Körner, thingsTHINKING GmbH, Germany
- Ahmad-Reza Sadeghi, TU Darmstadt, Germany
- Manuela Veloso, J. P. Morgan Chase AI Research, USA

As Program Committee (PC) chairs, we would like to thank our speakers for their interesting and inspirational talks, the Workshop Chair Dorothea Koert, the Doctoral Consortium Chair Mirjam Minor, the Industry Chair Stefan Wess, and our Local Chairs Stephanie Rodermund and Lukas Malburg. Our special gratitude goes to the Program Committee, whose sophisticated and conscientious judgement ensures the high quality of the KI conference. Without their substantial voluntary work, this conference would not have been possible.

In addition the following tutorial and workshops took place:

- Stephan Sahm: Tutorial on Universal Differential Equations in Julia (UDE 2022)
- Christoph Beierle, Marco Ragni, Kai Sauerwald, Frieder Stolzenburg, and Matthias Thimm: 8th Workshop on Formal and Cognitive Reasoning (FCR 2022)
- Ulrich Furbach, Alexandra Kirsch, Michael Sioutis, and Diedrich Wolter: Robust AI for High-Stakes Applications (RAIHS 2022)
- Martin Atzmueller, Tomáš Kliegr, and Ute Schmid: Explainable and Interpretable Machine Learning (XI-ML 2022)
- Mirko Lenz, Lorik Dumani, Philipp Heinrich, Nathan Dykes, Merlin Humml, Alexander Bondarenko, Shahbaz Syed, Adrian Ulges, Stephanie Evert, Lutz Schröder, Achim Rettinger, and Martin Vogt: Text Mining and Generation (TMG 2022)
- Sylvia Melzer, Stefan Thiemann, and Hagen Peukert: 2nd Workshop on Humanities-Centred AI (CHAI 2022)
- Lars Schaupeter, Felix Theusch, Achim Guldner, and Benjamin Weyers: AI and Cyber-Physical Process Systems Workshop 2022 (AI-CPPS 2022)
- Falco Nogatz and Mario Wenzel: 36th Workshop on (Constraint) Logic Programming (WLP 2022)
- Petra Gospodnetić, Claudia Redenbach, Niklas Rottmayer, and Katja Schladitz: Generating synthetic image data for AI (GSID-AI 2022)
- Jürgen Sauer and Stefan Edelkamp: 33rd Workshop Planning, Scheduling, Design and Configuration (PuK 2022)

Furthermore, we would like to thank our sponsors:

- Aimpulse Intelligent Systems GmbH (https://www.aimpulse.com)
- Advancis Software & Services GmbH (https://advancis.net)
- Dedalus HealthCare GmbH (https://www.dedalus.com)
- Empolis Information Management GmbH (https://www.empolis.com)
- Encoway GmbH (https://www.encoway.de)
- German Research Center for Artificial Intelligence (https://www.dfki.de)
- IOS Press (https://www.iospress.com)
- jolin.io consulting (https://www.jolin.io/de)
- Klinikum Mutterhaus der Borromäerinnen gGmbH (https://www.mutterhaus.de)
- Livereader GmbH (https://livereader.com)
- Plattform Lernende Systeme (https://www.plattform-lernende-systeme.de)
- Springer (https://www.springer.com)
- SWT Stadtwerke Trier Versorgungs-GmbH (https://www.SWT.de)
- Verband für Sicherheitstechnik e.V. (https://www.vfs-hh.de)

Last but not least, many thanks go to Silke Kruft for her extensive support with the organization of the accompanying program as well as to Felix Theusch and Benedikt Lüken-Winkels for their support for web conferencing technology. Additionally, our thanks go to Daniel Krupka and Alexander Scheibe from GI for providing extensive support in the organization of the conference. We would also like to thank EasyChair for their support in handling submissions and Springer for their support in making these proceedings possible. Our institutions, the University of Trier (Germany) and the German Research Center for Artificial Intelligence (Germany), also provided support, for which we are grateful.

July 2022

Ralph Bergmann
Lukas Malburg
Stephanie Rodermund
Ingo Timm

Organization

Program Committee

Martin Aleksandrov	FU Berlin, Germany
Klaus-Dieter Althoff	German Research Center for Artificial Intelligence (DFKI)/University of Hildesheim, Germany
Martin Atzmueller	Osnabrück University, Germany
Mirjam Augstein	University of Applied Sciences Upper Austria, Austria
Franz Baader	TU Dresden, Germany
Kerstin Bach	Norwegian University of Science and Technology, Norway
Joachim Baumeister	denkbares GmbH, Germany
Christoph Benzmüller	FU Berlin, Germany
Ralph Bergmann (PC Chair)	University of Trier/German Research Center for Artificial Intelligence (DFKI), Germany
Tarek Richard Besold	TU Eindhoven, The Netherlands
Tanya Braun	University of Münster, Germany
Ulf Brefeld	Leuphana Universität Lüneburg, Germany
Philipp Cimiano	Bielefeld University, Germany
Stefan Edelkamp	CTU Prague, Czech Republic
Manfred Eppe	University of Hamburg, Germany
Ulrich Furbach	University of Koblenz, Germany
Johannes Fähndrich	Hochschule für Polizei Baden-Württemberg, Germany
Johannes Fürnkranz	Johannes Kepler University Linz, Austria
Andreas Hotho	University of Würzburg, Germany
Steffen Hölldobler	TU Dresden, Germany
Eyke Hüllermeier	LMU Munich, Germany
Gabriele Kern-Isberner	TU Dortmund, Germany
Margret Keuper	University of Mannheim, Germany
Matthias Klusch	German Research Center for Artificial Intelligence (DFKI), Germany
Franziska Klügl	Örebro University, Sweden
Dorothea Koert (Workshop Chair)	TU Darmstadt, Germany
Stefan Kopp	Bielefeld University, Germany
Ralf Krestel	ZBW - Leibniz Information Centre for Economics/Kiel University, Germany

Antonio Krueger	German Research Center for Artificial Intelligence (DFKI), Germany
Fabian Lorig	Malmö University, Sweden
Bernd Ludwig	University of Regensburg, Germany
Thomas Lukasiewicz	University of Oxford, UK
Lukas Malburg (Local Chair)	University of Trier/German Research Center for Artificial Intelligence (DFKI), Germany
Mirjam Minor (DC Chair)	Goethe University Frankfurt, Germany
Till Mossakowski	University of Magdeburg, Germany
Ralf Möller	University of Lübeck, Germany
Falco Nogatz	German Research Center for Artificial Intelligence (DFKI), Germany
Özgür Lütfü Özcep	University of Lübeck, Germany
Heiko Paulheim	University of Mannheim, Germany
Stephanie Rodermund (Local Chair)	University of Trier/German Research Center for Artificial Intelligence (DFKI), Germany
Elmar Rueckert	Montanuniversität Leoben, Austria
Jürgen Sauer	University of Oldenburg, Germany
Ute Schmid	University of Bamberg, Germany
Lars Schmidt-Thieme	University of Hildesheim, Germany
Claudia Schon	University of Koblenz and Landau, Germany
Lutz Schröder	Friedrich-Alexander-Universität Erlangen-Nürnberg, Germany
René Schumann	HES-SO University of Applied Sciences Western Switzerland, Switzerland
Dietmar Seipel	University of Würzburg, Germany
Daniel Sonntag	German Research Center for Artificial Intelligence (DFKI), Germany
Myra Spiliopoulou	Otto-von-Guericke-University Magdeburg, Germany
Steffen Staab	University of Stuttgart, Germany/University of Southampton, UK
Alexander Steen	University of Greifswald, Germany
Frieder Stolzenburg	Harz University of Applied Sciences, Germany
Heiner Stuckenschmidt	University of Mannheim, Germany
Matthias Thimm	FernUniversität in Hagen, Germany
Ingo J. Timm (PC Chair)	University of Trier/German Research Center for Artificial Intelligence (DFKI), Germany
Diedrich Wolter	University of Bamberg, Germany
Stefan Wrobel	Fraunhofer IAIS/University of Bonn, Germany

Additional Reviewers

Finzel, Bettina
Fischer, Elisabeth
Frank, Daniel
Hartmann, Mareike
Iurshina, Anastasiia
Kobs, Konstantin
Kuhlmann, Isabelle
Liu, Jing
Meilicke, Christian
Memariani, Adel
Muschalik, Maximilian
Rechenberger, Sascha
Scheele, Stephan
Schlör, Daniel
Slany, Emanuel
Solopova, Veronika
Vestrucci, Andrea
Weidner, Daniel
Wilken, Nils

Abstracts of Invited Talks

Prospects for Using Context to Integrate Reasoning and Learning

Bruce Edmonds

Centre for Policy Modelling, Manchester Metropolitan University, UK

Whilst the AI and ML communities are no longer completely separate (as they were for 3 decades), principled ways of integrating them are still not common. I suggest that a kind of context-dependent cognition, that is suggested by human cognitive abilities could play this role. This approach is sketched, after briefly making clear what I mean by context. This move would also: make reasoning more feasible, belief revision more feasible, and provide principled strategies for dealing with the cases with over- or under-determined conclusions.

Representation and Quantification of Uncertainty in Machine Learning

Eyke Hüllermeier

Institute for Informatics, LMU Munich, Germany

Due to the steadily increasing relevance of machine learning for practical applications, many of which are coming with safety requirements, the notion of uncertainty has received increasing attention in machine learning research in the recent past. This talk will address questions regarding the representation and adequate handling of (predictive) uncertainty in (supervised) machine learning. A specific focus will be put on the distinction between two important types of uncertainty, often referred to as aleatoric and epistemic, and how to quantify these uncertainties in terms of suitable numerical measures. Roughly speaking, while aleatoric uncertainty is due to randomness inherent in the data generating process, epistemic uncertainty is caused by the learner's ignorance about the true underlying model. Going beyond purely conceptual considerations, the use of ensemble learning methods will be discussed as a concrete approach to uncertainty quantification in machine learning.

The First Rule of AI: Hard Things are Easy, Easy Things are Hard

Sven Körner

thingsTHINKING GmbH, Karlsruhe

Artificial intelligence is not only relevant for high-tech large corporations, but can be a game changer for different companies of all sizes. Nevertheless, smaller companies in particular do not use artificial intelligence in their value chain, and effective use tends to be rare, especially in the midmarket. Why is that? Often, there is a lack of appropriate know-how on how and in which processes the technology can be used at all. In my talk, I will discuss, how academia and industry can grow together, need each other, and should cooperate in making AI the pervasive technology that it already is.

Federated Learning: Promises, Opportunities and Security Challenges

Ahmad-Reza Sadeghi

Head of System Security Lab, TU Darmstadt, Germany

Federated Learning (FL) is a collaborative machine learning approach allowing the involved participants to jointly train a model without having to mutually share their private, potentially sensitive local datasets. As an enabling technology FL can benefit a variety of sensitive distributed applications in practice. However, despite its benefits, FL is shown to be susceptible to so-called backdoor attacks, in which an adversary injects manipulated model updates into the federated model aggregation process so that the resulting model provides targeted predictions for specific adversary-chosen inputs. In this talk, we present our research and experiences, also with industrial partners, in utilizing FL to enhance the security of large scale systems and applications, as well as in building FL systems that are resilient to backdoor attacks.

AI in Robotics and AI in Finance: Challenges, Contributions, and Discussion

Manuela Veloso

J. P. Morgan Chase AI Research, USA

My talk will follow up on my many years of research in AI and robotics and my few recent years of research in AI in finance. I will present challenges and solutions on the two areas, in data processing, reasoning, including planning and learning, and execution. I will conclude with a discussion of the future towards a lasting human-AI seamless interaction.

Contents

An Implementation of Nonmonotonic Reasoning with System W 1
Christoph Beierle, Jonas Haldimann, Daniel Kollar, Kai Sauerwald, and Leon Schwarzer

Leveraging Implicit Gaze-Based User Feedback for Interactive Machine Learning 9
Omair Bhatti, Michael Barz, and Daniel Sonntag

The Randomness of Input Data Spaces is an A Priori Predictor for Generalization 17
Martin Briesch, Dominik Sobania, and Franz Rothlauf

Communicating Safety of Planned Paths via Optimally-Simple Explanations 31
Noel Brindise and Cedric Langbort

Assessing the Performance Gain on Retail Article Categorization at the Expense of Explainability and Resource Efficiency 45
Eduardo Brito, Vishwani Gupta, Eric Hahn, and Sven Giesselbach

Enabling Supervised Machine Learning Through Data Pooling: A Case Study with Small and Medium-Sized Enterprises in the Service Industry 53
Leonhard Czarnetzki, Fabian Kainz, Fabian Lächler, Catherine Laflamme, and Daniel Bachlechner

Unsupervised Alignment of Distributional Word Embeddings 60
Aïssatou Diallo and Johannes Fürnkranz

NeuralPDE: Modelling Dynamical Systems from Data 75
Andrzej Dulny, Andreas Hotho, and Anna Krause

Deep Neural Networks for Geometric Shape Deformation 90
Aida Farahani, Julien Vitay, and Fred H. Hamker

Dynamically Self-adjusting Gaussian Processes for Data Stream Modelling 96
Jan David Hüwel, Florian Haselbeck, Dominik G. Grimm, and Christian Beecks

Optimal Fixed-Premise Repairs of $\mathcal{EL}$ TBoxes 115
Francesco Kriegel

Health and Habit: An Agent-based Approach 131
Veronika Kurchyna, Stephanie Rodermund, Jan Ole Berndt, Heike Spaderna, and Ingo J. Timm

Knowledge Graph Embeddings with Ontologies: Reification for Representing Arbitrary Relations 146
Mena Leemhuis, Özgür L. Özçep, and Diedrich Wolter

Solving the Traveling Salesperson Problem with Precedence Constraints by Deep Reinforcement Learning 160
Christian Löwens, Inaam Ashraf, Alexander Gembus, Genesis Cuizon, Jonas K. Falkner, and Lars Schmidt-Thieme

HanKA: Enriched Knowledge Used by an Adaptive Cooking Assistant 173
Nils Neumann and Sven Wachsmuth

Automated Kantian Ethics: A Faithful Implementation 187
Lavanya Singh

PEBAM: A Profile-Based Evaluation Method for Bias Assessment on Mixed Datasets 209
Mieke Wilms, Giovanni Sileno, and Hinda Haned

Author Index 225

An Implementation of Nonmonotonic Reasoning with System W

Christoph Beierle(✉), Jonas Haldimann, Daniel Kollar, Kai Sauerwald, and Leon Schwarzer

Faculty of Mathematics and Computer Science, Knowledge-Based Systems, FernUniversität in Hagen, 58084 Hagen, Germany
christoph.beierle@fernuni-hagen.de

Abstract. System W is a recently introduced reasoning method for conditional belief bases. While system W exhibits various desirable properties for nonmonotonic reasoning like extending rational closure and fully complying with syntax splitting, an implementation of it has been missing so far. In this paper, we present a first implementation of system W. The implementation is accessible via an extension of an online platform supporting a variety of nonmonotonic reasoning approaches.

1 Introduction

A conditional, denoted as $(B|A)$, formalizes a defeasible rule "If A then usually B" for logical formulas A, B. Two well known inference methods for conditional belief bases consisting of such conditionals are p-entailment that is characterized by the axioms of system P [1,20] and system Z [11,26]. Newer approaches include model-based inference with single c-representations [15,16], (skeptical) c-inference taking all c-representations into account [2,5,6], and the more recently introduced system W [18,19]. Notable properties of system W include capturing and properly going beyond p-entailment, system Z, and c-inference, and avoiding the drowning problem [9,26]. Unlike c-inference, system W extends rational closure [24], and unlike system Z, it fully complies with syntax splitting [12,17,25].

While for all other reasoning approaches cited above implementations are available, for instance in the InfOCF system [3] or in the online reasoning platform InfOCF-Web [22], so far an implementation of system W has been missing. In this paper, we present a first implementation of system W. The implementation is realized as an extension of the Java library InfOCF-Lib [21], and it has also been integrated into InfOCF-Web, thus providing easy online access for carrying out experiments and supporting direct comparisons to the other reasoning methods provided by InfOCF-Web.

After briefly recalling the necessary basics of conditional logic in Section 2, we introduce system W in Section 3. The implementation of system W, some first evaluation results, and the created online interface are presented in Section 4. Section 5 concludes and points out further work.

R. Bergmann et al. (Eds.): KI 2022, LNAI 13404, pp. 1–8, 2022.
https://doi.org/10.1007/978-3-031-15791-2_1

2 Background on Conditional Logic

A *(propositional) signature* is a finite set Σ of propositional variables. The propositional language over Σ is denoted by $\mathcal{L}_\Sigma$. Usually, we denote elements of the signatures with lowercase letters $a, b, c, \ldots$ and formulas with uppercase letters $A, B, C, \ldots$. We may denote a conjunction $A \wedge B$ by AB and a negation $\neg A$ by $\overline{A}$ for brevity of notation. The set of interpretations over Σ, also called *worlds*, is denoted as Ω_Σ. An interpretation $\omega \in \Omega_\Sigma$ is a *model* of a formula $A \in \mathcal{L}_\Sigma$ if A holds in ω, denoted as $\omega \models A$, and the set of models of A is $\Omega_A = \{\omega \in \Omega_\Sigma \mid \omega \models A\}$. A formula A *entails* a formula B, denoted by $A \models B$, if $\Omega_A \subseteq \Omega_B$.

A *conditional* $(B|A)$ connects two formulas A, B and represents the rule "If A then usually B". The conditional language over a signature Σ is denoted as $(\mathcal{L}|\mathcal{L})_\Sigma = \{(B|A) \mid A, B \in \mathcal{L}_\Sigma\}$. A finite set Δ of conditionals is called a *(conditional) belief base*. A belief base Δ is called *consistent* if there is a ranking model for Δ [11,27].

We use a three-valued semantics of conditionals in this paper [10]. For a world ω a conditional $(B|A)$ is either *verified* by ω if $\omega \models AB$, *falsified* by ω if $\omega \models A\overline{B}$, or *not applicable* to ω if $\omega \models \overline{A}$.

Reasoning with conditionals is often modelled by inference relations. An *inference relation* is a binary relation $\mid\!\sim$ on formulas over an underlying signature Σ with the intuition that $A \mid\!\sim B$ means that A (plausibly) entails B. (Non-monotonic) inference is closely related to conditionals: an inference relation $\mid\!\sim$ can also be seen as a set of conditionals $\{(B|A) \mid A, B \in \mathcal{L}_\Sigma, A \mid\!\sim B\}$. An *inductive inference operator* [17] is a function that maps each belief base to an inference relation. Well-known examples for inductive inference operators are p-entailment [1], denoted by $\mid\!\sim^p$, and system Z [26], denoted by $\mid\!\sim^z$.

3 System W

Recently, system W has been introduced as a new inductive inference operator [18, 19]. System W takes into account the tolerance information expressed by the ordered partition of Δ that can be used for checking the consistency of Δ [11] and that underlies the definition of system Z [26].

Definition 1 (inclusion maximal tolerance partition [26]**).** *A conditional $(B|A)$ is* tolerated *by Δ if there exists a world $\omega \in \Omega_\Sigma$ such that ω verifies $(B|A)$ and ω does not falsify any conditional in Δ. The inclusion maximal* tolerance partition $OP(\Delta) = (\Delta^0, \ldots, \Delta^k)$ *of a consistent belief base Δ is defined as follows. The first set Δ^0 in the tolerance partitioning contains all conditionals from Δ that are tolerated by Δ. Analogously, Δ^i contains all conditionals from $\Delta \backslash (\bigcup_{j<i} \Delta^j)$ which are tolerated by $\Delta \backslash (\bigcup_{j<i} \Delta^j)$, until $\Delta \backslash (\bigcup_{j<k+1} \Delta^j) = \emptyset$.*

It is well-known that $OP(\Delta)$ exists iff Δ is consistent [26]. In addition to the tolerance partition, system W also takes into account the structural information about which conditionals are falsified by a world, yielding the preferred structure on worlds.

Definition 2 **(ξ^j, ξ, preferred structure $<^{\mathsf{w}}_{\Delta}$ on worlds** [19]**).** *Consider a consistent belief base $\Delta = \{r_i = (B_i|A_i) \mid i \in \{1,\dots,n\}\}$ with $OP(\Delta) = (\Delta^0,\dots,\Delta^k)$. For $j = 0,\dots,k$, the functions ξ^j and ξ are the* functions mapping worlds to the set of falsified conditionals *from the set Δ^j in the tolerance partition and from Δ, respectively, given by*

$$\xi^j(\omega) := \{r_i \in \Delta^j \mid \omega \models A_i\overline{B_i}\}, \tag{1}$$

$$\xi(\omega) := \{r_i \in \Delta \mid \omega \models A_i\overline{B_i}\}. \tag{2}$$

The preferred structure on worlds *is given by the binary relation $<^{\mathsf{w}}_{\Delta} \subseteq \Omega \times \Omega$ defined by, for any $\omega, \omega' \in \Omega$,*

$$\omega <^{\mathsf{w}}_{\Delta} \omega' \text{ iff there exists } m \in \{0, \dots, k\} \text{ such that}$$
$$\xi^i(\omega) = \xi^i(\omega') \quad \forall i \in \{m+1, \dots, k\}, \text{ and}$$
$$\xi^m(\omega) \subsetneqq \xi^m(\omega'). \tag{3}$$

Thus, $\omega <^{\mathsf{w}}_{\Delta} \omega'$ if and only if ω falsifies strictly fewer conditionals than ω' in the partition with the biggest index m where the conditionals falsified by ω and ω' differ. Note that $<^{\mathsf{w}}_{\Delta}$ is a strict partial order. The inductive inference operator system W is based on $<^{\mathsf{w}}_{\Delta}$ and is defined as follows.

Definition 3 **(system W, $\mid\!\sim^{\mathsf{w}}_{\Delta}$** [19]**).** *Let Δ be a belief base and A, B be formulas. Then B is a* system W inference from A (in the context of Δ), *denoted $A \mid\!\sim^{\mathsf{w}}_{\Delta} B$ if for every $\omega' \in \Omega_{A\overline{B}}$ there is an $\omega \in \Omega_{AB}$ such that $\omega <^{\mathsf{w}}_{\Delta} \omega'$.*

System W extends system Z and c-inference and enjoys further desirable properties for nonmonotonic reasoning like avoiding the drowning problem. For more information on system W we refer to [12, 13, 19]. We illustrate system W with an example.

Example 1. Consider the belief base $\Delta = \{(b|a), (\overline{a}\overline{b}|\overline{a} \vee \overline{b}), (c|\top)\}$ over the signature $\Sigma = \{a, b, c\}$. Because every conditional in Δ is tolerated by Δ, the ordered partition of Δ is trivial, i.e., $OP(\Delta) = \{\Delta\}$. The preferred structure $<^{\mathsf{w}}_{\Delta}$ on worlds is given in Fig. 1; note that $<^{\mathsf{w}}_{\Delta}$ is not a total preorder, and thus, it cannot be expressed by system Z nor by any other ranking model of Δ.

Let us consider the question whether from $\overline{a}b \vee a\overline{b}$ we can infer $\overline{a}b$ in the context of Δ. This inference can not be obtained with p-entailment and neither with system Z. However, using the preferred structure $<^{\mathsf{w}}_{\Delta}$ given in Fig. 1, it is straightforward to verify that for each world ω' with $\omega' \models a\overline{b}$ there is a world ω with $\omega \models \overline{a}b$ such that $\omega <^{\mathsf{w}}_{\Delta} \omega'$. Thus, with system W we obtain the inference $\overline{a}b \vee a\overline{b} \mid\!\sim^{\mathsf{w}}_{\Delta} \overline{a}b$.

4 Implementation and System Walkthrough

InfOCF-Lib [21] is a Java library for representing conditional beliefs and for nonmonotonic reasoning from conditional belief bases. Initially focussed on reasoning with ranking functions (OCFs), it now contains, among other things, implementations of the inductive inference operators system P, system Z, and different kinds of c-inference.

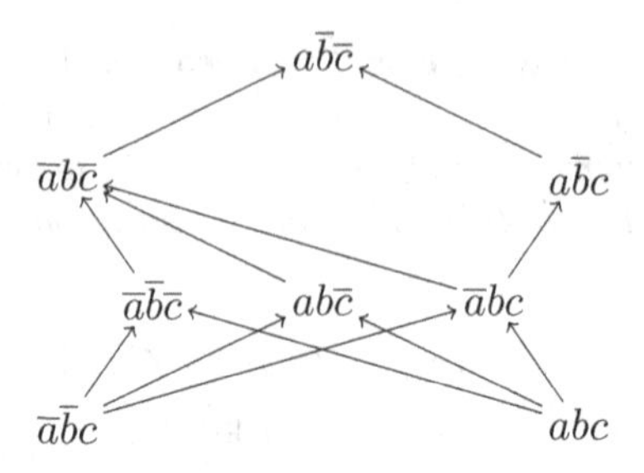

Fig. 1. The preferred structure on worlds induced by the belief base Δ from Example 1. Edges that can be obtained from transitivity are omitted.

Implementation of System W. The implementation of system W inference extends the library InfOCF-Lib. Answering a query q with system W for a given belief base Δ is done in three steps: First, the inclusion maximal ordered partition $OP(\Delta)$ is calculated. This is done by using the consistency test algorithm given in [11] which is already implemented in *InfOCF-Lib*. Then, based on $OP(\Delta)$ the preferred structure $<^{\mathsf{w}}_{\Delta}$ on worlds is computed and stored as a directed graph using the graph library *JGraphT*[1]. Finally, $<^{\mathsf{w}}_{\Delta}$ is used to answer the query q.

To compute $<^{\mathsf{w}}_{\Delta}$ the implementation first calculates the functions ξ^j for each $j \in \{0,\ldots,k\}$ from the previously generated ordered partition $OP(\Delta) = \{\Delta^0,\ldots,\Delta^k\}$. This is done by iterating over all possible worlds $\omega \in \Omega_\Sigma$ and all partitions $\Delta^j \in OP(\Delta)$ and checking for each combination what conditionals in Δ_j are falsified by ω. To represent the functions ξ^j, a hash map is created for each $\omega \in \Omega$ mapping $j \in \{0,\ldots,k\}$ to $\xi^j(\omega)$. Then a new empty instance of a undirected JGraphT graph is created and the edges in $<^{\mathsf{w}}_{\Delta}$ are added. For each tuple $(\omega,\omega') \in \Omega_\Sigma \times \Omega_\Sigma$ it is checked if the edge ω,ω' is in $<^{\mathsf{w}}_{\Delta}$. This is done by comparing $\xi^m(\omega),\xi^m(\omega')$ for every $m \in \{k,\ldots,0\}$ in descending order until $\xi^m(\omega) \neq \xi^m(\omega')$. For the last considered m, we have $\omega <^{\mathsf{w}}_{\Delta} \omega'$ iff $\xi^m(\omega) \subsetneqq \xi^m(\omega')$ in which case an edge is added to the directed graph representing the preferred structure on worlds. This approach computes a directed graph representing the relation $<^{\mathsf{w}}_{\Delta}$. To draw illustrations of this graph, the library *JGraphX*[2] is used.

To answer a query "Does A entail B?" with $A, B \in \mathcal{L}_\Sigma$ the set of worlds Ω_{AB} verifying and the set of worlds $\Omega_{A\overline{B}}$ falsifying the conditional $(B|A)$ are computed first. After that the system checks if for every world $\omega \in \Omega_{A\overline{B}}$ there is a world $\omega' \in \Omega_{AB}$ such that $\omega' <^{\mathsf{w}}_{\Delta} \omega$. This is done by iterating over the set $\Omega_{A\overline{B}}$ of falsifying worlds and checking if one of its predecessors in the directed graph is an element of the set Ω_{AB} of the verifying worlds. If for each of the falsifying worlds $\omega \in \Omega_{A\overline{B}}$ this is the case, the query is answered with *true*, otherwise, it is answered with *false*. Note that this covers the special cases that $\Omega_{A\overline{B}}$ is empty, in which case the result is *true*, and also the case in which a world in $\Omega_{A\overline{B}}$ is minimal and thus has no predecessor at all, in which case the result is *false*.

[1] https://jgrapht.org/.

[2] https://github.com/jgraph/jgraphx.

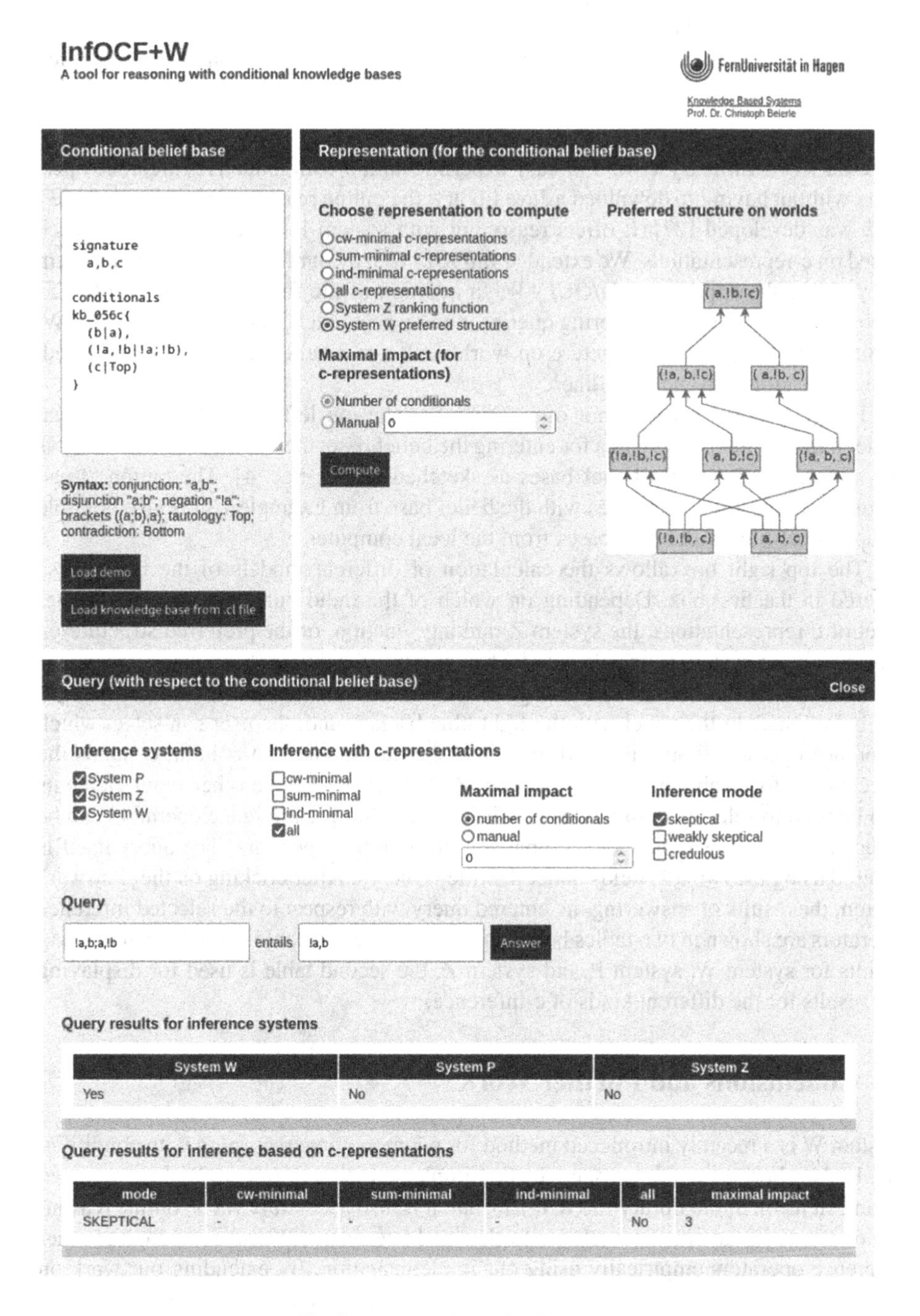

Fig. 2. User interface of InfOCF+W.

We have successfully applied this implementation to the real life examples for belief bases presented in [14] as well as to several thousands of automatically generated belief bases for which complete induced system W inference relations were computed [7,8,23].

InfOCF+W Online System. For easy experimentation with inductive inference operators without having to download a Java library, the online reasoning platform *InfOCF-Web* was developed [22]. It offers reasoning with system P, system Z, and inference based on c-representations. We extended InfOCF-Web by our implementation of system W, yielding the application *InfOCF+W*. In addition of the functionalities of InfOCF-Web, InfOCF+W offers answering queries with respect to a belief base with system W; additionally the preferred structure on worlds $<^{\mathsf{w}}_{\Delta}$ for a belief base can be displayed. The application is available online[3].

Figure 2 shows a screenshot of InfOCF+W. The top left box titled "Conditional belief base" contains a text area for entering the belief base; the syntax used here follows the syntax of .cl-files for belief bases as sketched in [21, Sec. 4]. The button "Load Demo" will pre-fill the text area with the belief base from Example 1. It is also possible to upload .cl-files with belief bases from the local computer.

The top right box allows the calculation of different models of the belief base entered in the first box. Depending on which of the radio buttons is selected, either a set of c-representations, the system Z ranking function, or the preferred structure on worlds $<^{\mathsf{w}}_{\Delta}$ for system W is calculated when clicking the button "Compute".

The box on the bottom of the page is used for answering queries with respect to the belief base. In the panel with the light blue background, the user can select which inference operator should be used to answer the query. The first column contains the checkboxes for enabling system P, system Z, and system W; the other input elements can be used to select certain modes of c-inference. Multiple inference operators can be selected at once to simplify the comparison of inference operators. The query itself is entered using the two text fields under the title "Query". After clicking on the "Answer" button, the results of answering the entered query with respect to the selected inference operators are shown in two tables below the button. The first table is used for displaying results for system W, system P, and system Z; the second table is used for displaying the results for the different kinds of c-inference.

5 Conclusions and Further Work

System W is a recently introduced method for nonmonotonic reasoning with conditionals. In this short paper, we presented a first implementation of system W. Its is realized as an extension of the library InfOCF-Lib, and it is also accessible via an online reasoning platform. In our current work, we further evaluate system W and compare it to other inference operators empirically using our implementation. By extending our work on employing SAT solvers for nonmonotonic inference [4], we are also working on using state-of-the-art SAT solvers for implementing reasoning with system W.

[3] http://wbs2.fernuni-hagen.de:18081/systemWTest/InfOCF.jsp.

References

1. Adams, E.: The logic of conditionals. Inquiry **8**(1–4), 166–197 (1965)
2. Beierle, C., Eichhorn, C., Kern-Isberner, G.: Skeptical inference based on c-representations and its characterization as a constraint satisfaction problem. In: Gyssens, M., Simari, G. (eds.) FoIKS 2016. LNCS, vol. 9616, pp. 65–82. Springer, Cham (2016). https://doi.org/10.1007/978-3-319-30024-5_4
3. Beierle, C., Eichhorn, C., Kutsch, S.: A practical comparison of qualitative inferences with preferred ranking models. KI - Künstliche Intelligenz **31**(1), 41–52 (2017)
4. Beierle, C., von Berg, M., Sanin, A.: Realization of skeptical c-inference as a SAT problem. In: Keshtkar, F., Franklin, M. (eds.) Proceedings of the Thirty-Fifth International Florida Artificial Intelligence Research Society Conference (FLAIRS), Hutchinson Island, Florida, USA, 15–18 May 2022 (2022)
5. Beierle, C., Eichhorn, C., Kern-Isberner, G., Kutsch, S.: Properties of skeptical c-inference for conditional knowledge bases and its realization as a constraint satisfaction problem. Ann. Math. Artif. Intell. **83**(3–4), 247–275 (2018)
6. Beierle, C., Eichhorn, C., Kern-Isberner, G., Kutsch, S.: Properties and interrelationships of skeptical, weakly skeptical, and credulous inference induced by classes of minimal models. Artif. Intell. **297**, 103489 (2021)
7. Beierle, C., Haldimann, J.: Normal forms of conditional belief bases respecting inductive inference. In: Keshtkar, F., Franklin, M. (eds.) Proceedings of the Thirty-Fifth International Florida Artificial Intelligence Research Society Conference (FLAIRS), Hutchinson Island, Florida, USA, 15–18 May 2022 (2022)
8. Beierle, C., Haldimann, J.: Normal forms of conditional knowledge bases respecting system P-entailments and signature renamings. Ann. Math. Artif. Intell. **90**(2), 149–179 (2022)
9. Benferhat, S., Cayrol, C., Dubois, D., Lang, J., Prade, H.: Inconsistency management and prioritized syntax-based entailment. In: Proceedings of the Thirteenth International Joint Conference on Artificial Intelligence (IJCAI 1993), vol. 1, pp. 640–647. Morgan Kaufmann Publishers, San Francisco (1993)
10. de Finetti, B.: La prévision, ses lois logiques et ses sources subjectives. Ann. Inst. H. Poincaré **7**(1), 1–68 (1937). Engl. Transl. Theory Probab. (1974)
11. Goldszmidt, M., Pearl, J.: Qualitative probabilities for default reasoning, belief revision, and causal modeling. Artif. Intell. **84**(1–2), 57–112 (1996)
12. Haldimann, J., Beierle, C.: Inference with system W satisfies syntax splitting. In: Kern-Isberner, G., Lakemeyer, G., Meyer, T. (eds.) Proceedings of the 19th International Conference on Principles of Knowledge Representation and Reasoning, KR 2022, Haifa, Israel, 31 July–5 August 2022, pp. 405–409 (2022)
13. Haldimann, J., Beierle, C.: Properties of system W and its relationships to other inductive inference operators. In: Varzinczak, I. (ed.) Foundations of Information and Knowledge Systems - 12th International Symposium, FoIKS 2022. LNCS, pp. 206–225. Springer, Cham (2022). https://doi.org/10.1007/978-3-031-11321-5_12
14. Haldimann, J.P., Osiak, A., Beierle, C.: Modelling and reasoning in biomedical applications with qualitative conditional logic. In: Schmid, U., Klügl, F., Wolter, D. (eds.) KI 2020. LNCS (LNAI), vol. 12325, pp. 283–289. Springer, Cham (2020). https://doi.org/10.1007/978-3-030-58285-2_24
15. Kern-Isberner, G.: A thorough axiomatization of a principle of conditional preservation in belief revision. Ann. Math. Artif. Intell. **40**(1–2), 127–164 (2004)
16. Kern-Isberner, G.: Conditionals in Nonmonotonic Reasoning and Belief Revision. LNCS (LNAI), vol. 2087. Springer, Heidelberg (2001). https://doi.org/10.1007/3-540-44600-1

17. Kern-Isberner, G., Beierle, C., Brewka, G.: Syntax splitting = relevance + independence: new postulates for nonmonotonic reasoning from conditional belief bases. In: Calvanese, D., Erdem, E., Thielscher, M. (eds.) Principles of Knowledge Representation and Reasoning: Proceedings of the 17th International Conference, KR 2020, pp. 560–571. IJCAI Organization (2020)
18. Komo, C., Beierle, C.: Nonmonotonic inferences with qualitative conditionals based on preferred structures on worlds. In: Schmid, U., Klügl, F., Wolter, D. (eds.) KI 2020. LNCS (LNAI), vol. 12325, pp. 102–115. Springer, Cham (2020). https://doi.org/10.1007/978-3-030-58285-2_8
19. Komo, C., Beierle, C.: Nonmonotonic reasoning from conditional knowledge bases with system W. Ann. Math. Artif. Intell. **90**(1), 107–144 (2021). https://doi.org/10.1007/s10472-021-09777-9
20. Kraus, S., Lehmann, D.J., Magidor, M.: Nonmonotonic reasoning, preferential models and cumulative logics. Artif. Intell. **44**(1–2), 167–207 (1990)
21. Kutsch, S.: InfOCF-Lib: a Java library for OCF-based conditional inference. In: Proceedings DKB/KIK-2019. CEUR Workshop Proceedings, vol. 2445, pp. 47–58. CEUR-WS.org (2019)
22. Kutsch, S., Beierle, C.: InfOCF-web: an online tool for nonmonotonic reasoning with conditionals and ranking functions. In: Zhou, Z. (ed.) Proceedings of the Thirtieth International Joint Conference on Artificial Intelligence, IJCAI 2021, Virtual Event/Montreal, Canada, 19–27 August 2021, pp. 4996–4999 (2021). ijcai.org
23. Kutsch, S., Beierle, C.: Semantic classification of qualitative conditionals and calculating closures of nonmonotonic inference relations. Int. J. Approx. Reason. **130**, 297–313 (2021). https://doi.org/10.1016/j.ijar.2020.12.020
24. Lehmann, D., Magidor, M.: What does a conditional knowledge base entail? Artif. Intell. **55**, 1–60 (1992)
25. Parikh, R.: Beliefs, belief revision, and splitting languages. Logic Lang. Comput. **2**, 266–278 (1999)
26. Pearl, J.: System Z: a natural ordering of defaults with tractable applications to nonmonotonic reasoning. In: Proceedings of the 3rd Conference on Theoretical Aspects of Reasoning About Knowledge (TARK 1990), pp. 121–135. Morgan Kaufmann Publishers Inc., San Francisco (1990)
27. Spohn, W.: Ordinal conditional functions: a dynamic theory of epistemic states. In: Harper, W., Skyrms, B. (eds.) Causation in Decision, Belief Change, and Statistics, II, pp. 105–134. Kluwer Academic Publishers (1988)

Leveraging Implicit Gaze-Based User Feedback for Interactive Machine Learning

Omair Bhatti[1(✉)], Michael Barz[1,2], and Daniel Sonntag[1,2]

[1] German Research Center for Artificial Intelligence (DFKI), Saarbrücken, Germany
shahzad.bhatti@dfki.de

[2] Oldenburg University, Oldenburg, Germany

Abstract. Interactive Machine Learning (IML) systems incorporate humans into the learning process to enable iterative and continuous model improvements. The interactive process can be designed to leverage the expertise of domain experts with no background in machine learning, for instance, through repeated user feedback requests. However, excessive requests can be perceived as annoying and cumbersome and could reduce user trust. Hence, it is mandatory to establish an efficient dialog between a user and a machine learning system. We aim to detect when a domain expert disagrees with the output of a machine learning system by observing its eye movements and facial expressions. In this paper, we describe our approach for modelling user disagreement and discuss how such a model could be used for triggering user feedback requests in the context of interactive machine learning.

Keywords: Interactive machine learning · Eye tracking · Gaze · Confusion detection · Emotion detection · User disagreement

1 Introduction

Applying machine learning to a new problem or a new domain usually requires a machine learning practitioner to collect a large amount of labelled samples, select representative/discriminating features, and choose an appropriate learning algorithm to model the concepts at hand. In contrast, interactive machine learning enables users, also without a background in machine learning to train a model in a fast-paced, incremental manner [1]. A user can steer the behaviour of the machine learning model by continuously providing feedback, e.g., upon requests from the system. However, repeated feedback queries, such as trivial yes/no questions, can be perceived as frustrating and annoying [6].

This may lead to reduced user trust in model outputs and deteriorate a user's impression of a model's accuracy [12]. Previous research discussed guidelines and rules for developing IML systems and their interfaces to avoid such problems

Supported by organization x.

R. Bergmann et al. (Eds.): KI 2022, LNAI 13404, pp. 9–16, 2022.
https://doi.org/10.1007/978-3-031-15791-2_2

[8,11,24]. Dudley and Kristensson [8] propose to reduce the number of interactions by triggering feedback requests for questions of high relevance to the system.

We propose to limit feedback requests to situations in which a user disagrees with the output of a machine learning models by observing the eye movements and facial expression of that user. In this work, we specify what user disagreement with a machine learning model means, we describe our planned user study and approach for modelling user disagreement based on gaze and facial expressions, and discuss how interactive machine learning systems may benefit from such a model.

2 Background

We hypothesise that user disagreement stems from negative affective states such as frustration, confusion or disappointment. Therefore, we examine the previous literature on how to detect these affective states using implicit user feedback and how they relate to user disagreement. Previous research has shown that human gaze and facial expressions can be used for affect recognition [16,25] and generally are sources for implicit user feedback [3,4]. Lallé et al. [15] introduce predictors for the state of *confusion* leveraging gaze from a user. According to D'Mello and Graesser [9], confusion "is hypothesized to occur when there is a mismatch between incoming information and prior knowledge [...], thereby initiating cognitive disequilibrium" (p. 292). Therefore we hypothesize that user confusion can be an indicator for a user's disagreement with the output of a model. Pollak et al. [20] use facial emotion recognition to detect user satisfaction and dissatisfaction where positive emotional feedback corresponds to satisfaction and negative to dissatisfaction. To detect user disagreement, we look for situations in which the user is *confused* or *dissatisfied* by the model output. We plan an experiment to where we push the user to disagree with the model's output while his gaze and facial expression are recorded.

2.1 Confusion Detection

User confusion occurs when a mismatch exists between prior user knowledge and incoming information [9]. Early research on confusion detection origins from the field of educational computing [5,7], where predictors for confusion leverage facial expression of students, the posture of students or students interface actions and their studying behaviour. Pachman et al. [18] propose the usage of gaze data for confusion prediction in digital learning. In their study, the participants are presented with a puzzle, and while solving it, their gaze is recorded. The authors aim to detect the buildup of confusion during the problem-solving process. On the other hand, we focus on the immediate affective state of confusion resulting from the user processing the information of the model's output. Detecting this type of *immediate* confusion is especially relevant in the field of Human-Computer-Interaction (HCI) since user experience and user satisfaction decreases when the user is in such a confused state [17]. Pentel [19] introduces a predictor for confusion, using mouse movement recorded when playing a simple game. The author show that an SVM model trained on mouse movements can successfully predict user confusion, but it is restricted to the generated game and thus difficult to generalise. [21] create a predictor for confusion based on gaze data on their persona information visualisation tool. The visualisation contains multiple areas of interest (AOIs) with different types of information about a persona. Their model achieves an accuracy of 80% of confusion predictions using the number of fixation, the length of transition paths between AOIs, and the user's demographic data as features. In a follow-up work, Salminen et al. [22] train a model using gaze-based data only, achieving an accuracy of 70%. The accuracy increases to 99% when the model includes demographic data as features. This indicates that the demographic data correlates with confusion in their recorded dataset. Including demographic data as features leads to a significant improvement of the model to 99% in accuracy. The authors state that most instances of confusion occur for non-experienced, old males which indicates that trust correlates with age and gender (demographic features). However, this suggests that demographic features can be used to model how often confusion appears in different user groups but not for real-time monitoring of confusion.

Lallé et al. [15] created a predictor for confusion during interaction with their interactive data visualisation tool ValueChart. The tool's goal is to assist users to make the best suitable decision (finding rental property) based on their preference. In a study with 136 participants, the authors collect gaze and mouse movement data while a user performs tasks on ValueChart. The user can report confusion by clicking a button on the top right corner of the visualisation tool. Their confusion prediction model achieves a accuracy of 61% using a Random Forest Classifier. A more recent contribution from the same group [23] uses deep learning based on eye movements to predict confusion on the same dataset as [15]. Instead of using features calculated from the eye-tracking data, the authors suggest using the raw sequential gaze data and feeding it to a Recurrent Neural Network (RNN), allowing the RNN to pick up discriminators for classifying confusion that would otherwise be lost when using calculated features. Using deep learning, their model outperforms their previous work (61% vs 82%), and their results suggest that deep learning in combination with raw sequential gaze data is a feasible option for affect recognition. A possible limitation is their self-report button for reporting instances of confusion. It can influence the user's gaze because of its placement in the interface. Therefore it is important to provide a non-distracting way to self-report confusion. A trigger placed in the hand of the participants could be a solution.

2.2 Leveraging Emotion Detection for ML

Using implicit emotional feedback for artificial agents is a recent idea, and only a few publications have explored it. Pollak et al. [20] investigated whether emotional feedback from a user can serve as the reward function for a reinforcement learning agent. The reward function corresponds to the user's level of satisfaction inferred from facial emotion recognition. The emotions are classified as negative ('angry', 'disgust', 'fear', 'sad'), positive ('happy') or neutral ('neutral', 'surprise') [10]. In their experiment, the user controls a drone's movement, which then, based on the emotional feedback, learns whether it took the correct corresponding action. Their initial finding suggests that incorporating emotional feedback into the reward function of a reinforcement learning agent can be used to teach an agent. The author indicate that there is a considerable individual difference between participants' strength of emotional feedback, which makes it harder to differentiate between positive and negative feedback. Therefore, we plan to gather facial expressions and gaze data in our study to have a multimodal solution for user disagreement detection.

Krause and Vossen [14] suggest the use of implicit triggers based on user confusion or uncertainty for explanations in human-agent interaction. They argue that explanations should not only be provided when the user explicitly asks for it but also when the agent detects that the user is uncertain or confused. Their work also lists other possible implicit triggers such as conflicts between a user's beliefs and the agent or misunderstanding its output. These triggers are similar to those we propose to detect since they also describe user disagreement with the model's output, but instead of triggering an explanation, we query the user for feedback.

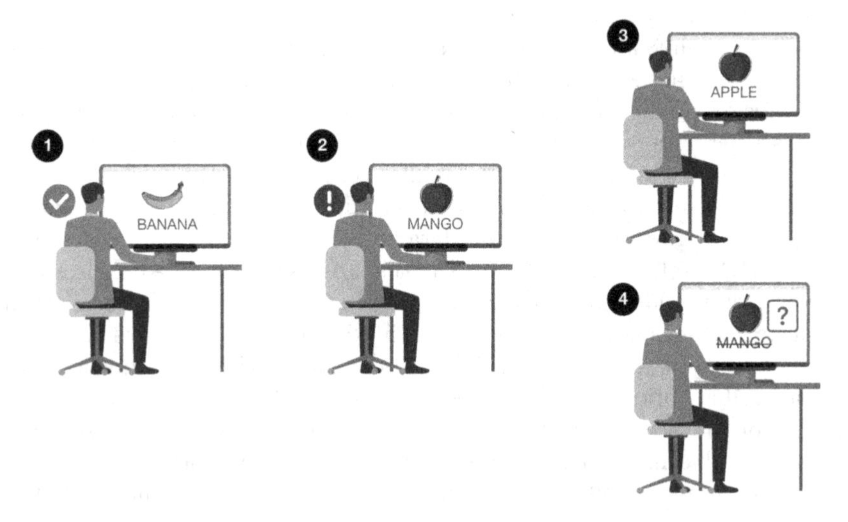

Fig. 1. (1) User interacts with an IML system; (2) our predictor picks up that the user disagrees with the output; (3) the IML system reacts with returning alternative solution or (4) triggers a feedback request

3 Method

In this section we describe our approach, how we plan to collect data for user disagreement, how we plan to create a predictor for user disagreement and how the predictor can be leveraged in IML.

3.1 Data Collection

We plan to conduct an experiment to collect the data necessary to create an effective user disagreement predictor. The experiment will collect the users' gaze data using an eye-tracker and simultaneously record their facial expressions using a video camera. We plan to show a series of images containing an object and its corresponding label, simulating an output of an object detection model. We will randomly include images containing objects with wrong labels. These instances lead to user disagreement. We want to minimise influences on gaze and facial expression; therefore, the participant uses a trigger placed in his hand to confirm user disagreement. The participant will see one object-label pair at a time for a certain amount of time. If the user presses the trigger in his hand, we stop the image sequence and confirm that he disagrees with the output. A possible extension of the study is to show an image depicting a scene and a caption describing it. To robustly record and synchronize the data we intend to use our multisensor-pipeline (MSP) framework for prototyping multimodal-multisensor interfaces based on real-time sensor input [2].

3.2 Disagreement Detection Model

The features for our planned detection model will be sourced from the eye-tracker and the video camera recording the user. Based on previous research, we list the features we hypothesise to be relevant for user disagreement detection (see Table 1).

3.3 Application in IML

User feedback is essential for interactive machine learning. It helps IML systems to become 'lifelong' learners. Hence the importance to enable users to provide feedback by creating effective interfaces and human-agent interactions. A crucial aspect is **when** to trigger feedback requests since repeatedly asking for feedback can be perceived as frustrating [6] and also reduces trust and impression of model accuracy [12]. Therefore, we try to provide implicit user feedback to the IML system with our proposed user disagreement predictor. The feedback from our predictor consists of a confidence value of detecting user disagreement and the gaze scan path leading to his affective state. The IML system then can react either by showing an alternative solution or triggering a request asking the user for explicit feedback. Figure 1 depicts an example of such a pipeline with a IML system for image captioning. When our predictor detects confusion, the IML

Table 1. List of features collected from previous research relevant for user disagreement detection

	Feature	Source
Eye-tracker	Number of fixations	[15,21]
	Fixation durations	[15,21]
	Length of the transition paths between AOIs (image and label)	[18,21,22]
	Image of the scan path visualising the last seconds of eye movement before the user self-reports disagreement	[23]
	Raw sequential gaze data as time-series	[23]
Video camera	Emotion detection based on (FER2013)	[7,9,13]
	Body posture/movement	[5,7]

system gets notified that the user disagrees with the captioning provided for the image. Further, it also receives the previous scan path leading to disagreement. The IML system can return an alternative captioning or explicitly ask the user for correction.

3.4 Limitations

We intend to use a remote eye tracking system for gaze estimation in our disagreement detection system. For this, the interaction screen must be instrumented with an additional piece of hardware that requires a user-specific calibration. Also, individual differences of users' eye movements when expressing disagreement need to be considered. They could have a negative impact on the generalizability of our approach.

4 Conclusion

We have shown the motivation and need for detecting when to ask a user for feedback. The following steps will be to conduct the planned study, collect the dataset, and create a user disagreement detection model using the features we collected from previous works. Further, we will use the detection model as a trigger for querying feedback by integrating it into an IML system.

Acknowledgements. This work was funded by the *German Federal Ministry of Education and Research* (BMBF) under grant number *01JD1811C* (GeAR).

References

1. Amershi, S., Cakmak, M., Knox, W.B., Kulesza, T.: Power to the people: the role of humans in interactive machine learning. AI Mag. **35**(4), 105–120 (2014). https://doi.org/10.1609/aimag.v35i4.2513
2. Barz, M., Bhatti, O.S., Lüers, B., Prange, A., Sonntag, D.: Multisensor-pipeline: a lightweight, flexible, and extensible framework for building multimodal-multisensor interfaces. In: Companion Publication of the 2021 International Conference on Multimodal Interaction, ICMI 2021 Companion, pp. 13–18. Association for Computing Machinery, New York (2021). https://doi.org/10.1145/3461615.3485432. ISBN 9781450384711
3. Barz, M., Bhatti, O.S., Sonntag, D.: Implicit estimation of paragraph relevance from eye movements. Front. Comput. Sci. **3**, 808507 (2021). https://doi.org/10.3389/fcomp.2021.808507
4. Barz, M., Stauden, S., Sonntag, D.: Visual search target inference in natural interaction settings with machine learning. In: Bulling, A., Huckauf, A., Jain, E., Radach, R., Weiskopf, D. (eds.) ETRA 2020: 2020 Symposium on Eye Tracking Research and Applications, Stuttgart, Germany, 2–5 June 2020, pp. 1:1–1:8. ACM (2020). https://doi.org/10.1145/3379155.3391314
5. Bosch, N., Chen, Y., D'Mello, S.: It's written on your face: detecting affective states from facial expressions while learning computer programming. In: Trausan-Matu, S., Boyer, K.E., Crosby, M., Panourgia, K. (eds.) Intelligent Tutoring Systems, pp. 39–44. Springer, Cham (2014). https://doi.org/10.1007/978-3-319-07221-0_5. ISBN 978-3-319-07221-0
6. Cakmak, M., Chao, C., Thomaz, A.L.: Designing interactions for robot active learners. IEEE Trans. Auton. Ment. Dev. **2**(2), 108–118 (2010). https://doi.org/10.1109/TAMD.2010.2051030
7. D'Mello, S.K., Craig, S.D., Graesser, A.C.: Multimethod assessment of affective experience and expression during deep learning. Int. J. Learn. Technol. **4**(3/4), 165–187 (2009). https://doi.org/10.1504/IJLT.2009.028805. ISSN 1477-8386
8. Dudley, J.J., Kristensson, P.O.: A review of user interface design for interactive machine learning. ACM Trans. Interact. Intell. Syst. **8**(2) (2018). https://doi.org/10.1145/3185517. ISSN 2160-6455
9. D'Mello, S.K., Graesser, A.C.: Confusion. In: International Handbook of Emotions in Education, pp. 299–320. Routledge (2014)
10. Ekman, P., et al.: Universals and cultural differences in the judgments of facial expressions of emotion. J. Pers. Soc. Psychol. **53**(4), 712 (1987)
11. Ghajargar, M., Persson, J., Bardzell, J., Holmberg, L., Tegen, A.: The UX of interactive machine learning. Association for Computing Machinery, New York (2020). https://doi.org/10.1145/3419249.3421236. ISBN 9781450375795
12. Honeycutt, D., Nourani, M., Ragan, E.: Soliciting human-in-the-loop user feedback for interactive machine learning reduces user trust and impressions of model accuracy. In: Proceedings of the AAAI Conference on Human Computation and Crowdsourcing, vol. 8, no. 1, pp. 63–72, October 2020. https://ojs.aaai.org/index.php/HCOMP/article/view/7464
13. Khaireddin, Y., Chen, Z.: Facial emotion recognition: state of the art performance on FER2013. arXiv preprint arXiv:2105.03588 (2021)
14. Krause, L., Vossen, P.: When to explain: identifying explanation triggers in human-agent interaction. In: 2nd Workshop on Interactive Natural Language Technology for Explainable Artificial Intelligence, pp. 55–60 (2020)

15. Lallé, S., Conati, C., Carenini, G.: Predicting confusion in information visualization from eye tracking and interaction data. In: Proceedings of the Twenty-Fifth International Joint Conference on Artificial Intelligence, IJCAI 2016, pp. 2529–2535. AAAI Press (2016). ISBN 9781577357704
16. Lim, J.Z., Mountstephens, J., Teo, J.: Emotion recognition using eye-tracking: taxonomy, review and current challenges. Sensors **20**(8) (2020). https://doi.org/10.3390/s20082384. ISSN 1424-8220. https://www.mdpi.com/1424-8220/20/8/2384
17. Nadkarni, S., Gupta, R.: A task-based model of perceived website complexity. MIS Q. **31**(3), 501–524 (2007). ISSN 02767783. https://www.jstor.org/stable/25148805
18. Pachman, M., Arguel, A., Lockyer, L., Kennedy, G., Lodge, J.: Eye tracking and early detection of confusion in digital learning environments: proof of concept. Australas. J. Educ. Technol. **32**(6) (2016). https://doi.org/10.14742/ajet.3060. https://ajet.org.au/index.php/AJET/article/view/3060
19. Pentel, A.: Patterns of confusion: using mouse logs to predict user's emotional state. In: Cristea, A.I., Masthoff, J., Said, A., Tintarev, N. (eds.) Posters, Demos, Late-Breaking Results and Workshop Proceedings of the 23rd Conference on User Modeling, Adaptation, and Personalization (UMAP 2015), Dublin, Ireland, 29 June–3 July 2015, CEUR Workshop Proceedings, vol. 1388. CEUR-WS.org (2015). https://ceur-ws.org/Vol-1388/PALE2015-paper5.pdf
20. Pollak, M., Salfinger, A., Hummel, K.A.: Teaching drones on the fly: can emotional feedback serve as learning signal for training artificial agents? arXiv preprint arXiv:2202.09634 (2022)
21. Salminen, J., Jansen, B.J., An, J., Jung, S.G., Nielsen, L., Kwak, H.: Fixation and confusion: investigating eye-tracking participants' exposure to information in personas. In: Proceedings of the 2018 Conference on Human Information Interaction & Retrieval, CHIIR 2018, pp. 110–119. Association for Computing Machinery, New York (2018). https://doi.org/10.1145/3176349.3176391. ISBN 9781450349253
22. Salminen, J., Nagpal, M., Kwak, H., An, J., Jung, S.g., Jansen, B.J.: Confusion prediction from eye-tracking data: experiments with machine learning. In: Proceedings of the 9th International Conference on Information Systems and Technologies, ICIST 2019. Association for Computing Machinery, New York (2019). https://doi.org/10.1145/3361570.3361577. ISBN 9781450362924
23. Sims, S.D., Conati, C.: A neural architecture for detecting user confusion in eye-tracking data. In: Proceedings of the 2020 International Conference on Multimodal Interaction, ICMI 2020, pp. 15–23. Association for Computing Machinery, New York (2020). https://doi.org/10.1145/3382507.3418828. ISBN 9781450375818
24. Zacharias, J., Barz, M., Sonntag, D.: A survey on deep learning toolkits and libraries for intelligent user interfaces (2018)
25. Zeng, Z., Pantic, M., Roisman, G.I., Huang, T.S.: A survey of affect recognition methods: audio, visual, and spontaneous expressions. IEEE Trans. Pattern Anal. Mach. Intell. **31**(1), 39–58 (2009). https://doi.org/10.1109/TPAMI.2008.52

The Randomness of Input Data Spaces is an A Priori Predictor for Generalization

Martin Briesch(✉), Dominik Sobania, and Franz Rothlauf

Johannes Gutenberg-University, Mainz, Germany
{briesch,dsobania,rothlauf}@uni-mainz.de

Abstract. Over-parameterized models can perfectly learn various types of data distributions, however, generalization error is usually lower for real data in comparison to artificial data. This suggests that the properties of data distributions have an impact on generalization capability. This work focuses on the search space defined by the input data and assumes that the correlation between labels of neighboring input values influences generalization. If correlation is low, the randomness of the input data space is high leading to high generalization error. We suggest to measure the randomness of an input data space using Maurer's universal. Results for synthetic classification tasks and common image classification benchmarks (MNIST, CIFAR10, and Microsoft's cats vs. dogs data set) find a high correlation between the randomness of input data spaces and the generalization error of deep neural networks for binary classification problems.

Keywords: Deep learning · Label landscape · Generalization

1 Introduction

While deep neural networks (DNN) have gained much attention in many machine learning tasks [29], there is still only limited theory explaining the success of DNN. Especially the generalization abilities of DNNs have challenged classical learning theory as standard approaches like VC-dimension [43], Rademacher complexity [7], or uniform stability [10] fail to explain the generalization behavior of over-parameterized DNNs [50]. Most of the existing theory approaches look at the hypothesis space of the model and the properties of the learning algorithm; properties of the data distribution (as well as the machine learning tasks) are addressed to a much lower extent.

Focusing on the data distribution, [50] observed a lower generalization capability of DNNs when randomizing natural data. Arpit et al. [5] find that learning on real data behaves differently than learning on randomized data. DNNs seem to work content-aware and learn certain data points first. Thus, there is evidence that the properties of the input data distribution have an influence on the generalization capabilities of DNNs and natural data has properties that enable DNNs

R. Bergmann et al. (Eds.): KI 2022, LNAI 13404, pp. 17–30, 2022.
https://doi.org/10.1007/978-3-031-15791-2_3

to perform well. This raises the question why DNNs perform well on supervised learning tasks with natural data signals.

This paper studies how the properties of training data influence the generalization capability of DNNs. We assume a label landscape $(X, f, \mathcal{N})$ with the set of training data X, the labeling function $f : X \to Y$ that assigns a label $y \in Y$ to each training instance $x \in X$, and a neighborhood mapping $\mathcal{N} : X \to 2^X$ which assigns to each input x a set of neighboring inputs. We suggest that the properties of the label landscape formed by the training data influences the generalization behavior of DNNs.

To measure the properties of the training data, we perform a random walk through the label landscape $(X, f, \mathcal{N})$. A random walk with N steps iteratively selects a neighboring training instance x_i (based on a distance metric) and returns the corresponding label y_i. Thus, it creates a sequence of labels y^N. We expect that the randomness of y^N (for example measured by Maurer's universal test) influences the generalization capability of DNNs. If Maurer's universal test indicates that y^N is a random sequence, then generalization is expected to be low; in contrast, if y^N is non-random (which means the per-bit entropy of the sequence is low), DNNs are expected to be able to learn well and show high generalization capability for this particular data distribution. Thus, we suggest that the randomness of a sequence of binary labels generated by a random walk through the input data space is a good predictor for the expected generalization capability of DNNs.

We present evidence and experimental results for four types of problems. First, we follow the approach suggested by [50] and systematically randomize the labeling function $f : X \to Y$ by assigning the label y independently at random with probability v. With stronger randomization of the labels, the resulting sequence y^N created by a random walk has higher randomness according to Maurer's universal test and generalization decreases. We present results for different binary instances of synthetic test problems where we know the decision boundaries (an XOR type problem, a majority vote problem, and a parity function problem). Second, we study binary instances of MNIST [30] and CIFAR10 [28] using the same randomization method as in the previous experiments and extend the results with experiments where we randomize the training instances $x \in X$. For the extension, we consider four different variants. We either perform a random permutation $\pi : x \to x$ of all input variables of the training data (*PermutGlobal*), perform a random permutation of all variables for all training instances (*PermutInd*), draw each input value randomly from a Gaussian distribution matching the original distribution of the input values (*GaussianInd*), or draw each input value from a white noise distribution (*NoiseInd*). The results indicate that Maurer's universal test applied to the sequence y^N is a good predictor for the expected generalization capability of a DNN. Finally, we focus on binary instances of the more complex cats vs. dogs data set [15] and distinguish between training instances that are either easy or difficult to learn by a DNN. Experimental results confirm that the randomness of y^N is a good indicator of the expected generalization.

In Sect. 2, we describe preliminaries and present Maurer's universal as a novel measure for the randomness of data sets and related supervised learning tasks. Sect. 3 describes the experimental setting and presents the results. In Sect. 4, we give an overview of related work before concluding the paper in Sect. 5. Sect. 6 describes the limitations and future research directions.

2 Randomness of Data Spaces

Consider a data set $\mathcal{D}$ consisting of a finite number m of pairs (x_i, y_i) where $x \in X$ and $y \in Y$. x_{ij} denotes the value of the j-th input variable of the vector x_i; y_i denotes the corresponding label. All pairs are drawn i.i.d. from the population distribution P_{XY}. The goal of a machine learning model in a supervised classification task is to find a function h^* from a hypothesis space $\mathcal{H}$ given a loss function l that minimizes the population risk $R(h)$:

$$R(h) = \mathbb{E}[l(h(X), Y)]$$

$$h^* = \arg\min_{h \in \mathcal{H}} R(h)$$

Usually, the model does not have access to the complete distribution P_{XY} but rather only to the data set $\mathcal{D}$. Therefore, a common approach in machine learning is to minimize the empirical risk $R_{emp}(h)$ on the given data $\mathcal{D}$:

$$R_{emp}(h) = \frac{1}{m} \sum_{i=1}^{m} l(h(x_i), y_i)$$

$$\hat{h} = \arg\min_{h \in \mathcal{H}} R_{emp}(h)$$

Unfortunately, the empirical risk can be significantly different from the population risk. This makes bounding the gap between $R(h)$ and $R_{emp}(h)$, also called *generalization*, a central challenge in machine learning [42].

In theory, given a sufficient amount of parameters and training time, a multilayer neural network can approximate any function h arbitrarily well [13,22]. Thus, any data set $\mathcal{D}$ can be learned by a large enough model. This is confirmed by empirical studies where complex DNN models can fit both data from natural signals as well as random data [50]. Learning arbitrary h can be achieved by standard DNN models without changing any hyperparameters, neither for the model nor for the used learning algorithm. When fitting DNN models to either natural signals or random data, [50] as well as [5] observed differences in the generalization error. For natural signals, usually the generalization error is low; for random or randomized data, generalization error is high.

We believe that the differences in generalization error g_{err} between different data sets can be explained by the properties of the label landscape defined on the data set $\mathcal{D}$. Analogously to fitness landscapes known in other domains, we define a label landscape $(X, f, \mathcal{N})$, where the labeling function $f : X \rightarrow Y$ assigns a label $y \in Y$ to each training instance $x \in X$ and a neighborhood mapping

$\mathcal{N} : X \to 2^X$ assigns to each input $x \in X$ a set of neighboring inputs. The labeling function f is defined by the input data; the neighborhood mapping $\mathcal{N}$ is usually problem-specific and defines which input/training data is similar to each other [20,49]. Instead of defining $\mathcal{N}$ on the raw input data, we can also define $\mathcal{N}$ on an underlying manifold representing the data.

Using a label landscape defined on the input data, we can calculate relevant properties like the correlation between neighboring data points. Such measures are relevant for combinatorial optimization problems as problems, where the objective values of neighboring solutions are uncorrelated, are difficult to solve [24,39]. If fitness values (labels) of neighbors in the input space are uncorrelated, the no free lunch theorem holds [45–48] and optimization methods can not beat random search. The situation is similar for non-parametric machine learning methods like kernel machines which rely on the smoothness prior $h(x) \approx h(x+\epsilon)$. The smoothness prior assumes that the properties of neighboring inputs (either measured in time or in space) are similar and do not abruptly change. Consequently, kernel machines have problems to learn non-local functions with low smoothness [9], although deep learning is able to learn some variants of non-local functions [23].

Algorithm 1. Random walk

1: Select random start point x_0
2: Initialize $y^N = [y_0]$
3: **for** $z = 1, 2, \ldots, N$ **do**
4: Select x_z randomly from the neighborhood $\mathcal{N}(x_{z-1})$
5: Append y_z to y^N
6: **end for**

We suggest to capture the correlation between labels of neighboring input values (taken from the given data set $\mathcal{D}$) by performing a random walk through $(X, f, \mathcal{N})$ and analyzing the resulting sequence y^N of labels. Algorithm 1 shows the random walk as pseudo-code. We initialize y^N with the label y_0 of a random start point x_0 (lines 1–2) and perform N times a step of the random walk appending the label y_z of a randomly selected x_z from the neighborhood $\mathcal{N}(x_{z-1})$ (lines 3–6).

We expect that the randomness of y^N influences the generalization ability of DNNs trying to learn the properties of $\mathcal{D}$. For example, we assume a binary classification problem that can easily be learned and linearly separated (see Fig. 1a). When performing a random walk through the space of input values, the value of the corresponding label y_i rarely changes and the randomness of the resulting sequence y^N is low. Situation is different, if we assign random labels to the input data points (Fig. 1c). Then, the resulting sequence y^N is random. In contrast, Fig. 1b shows the landscape of the parity problem, which can be well learned using DNN [23] but is a non-local problem. When performing a random walk through such a landscape, the resulting sequence y^N is non-random but highly structured as the labels of neighboring input data points are

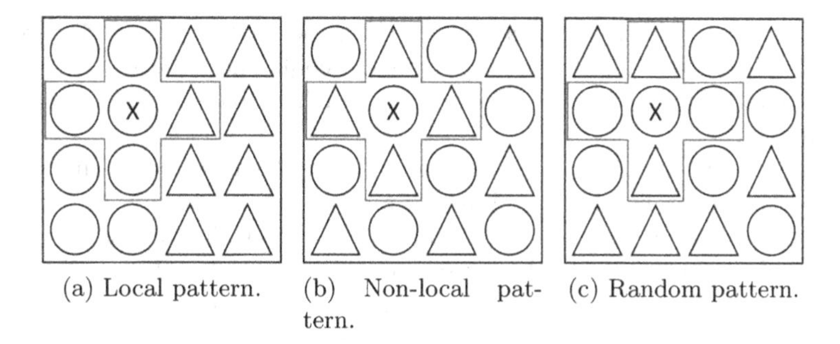

(a) Local pattern. (b) Non-local pattern. (c) Random pattern.

Fig. 1. Resulting landscapes for different example binary classification problems. Each input data has four neighbors. (a) easy problem with high correlation between labels of neighboring input data (b) non-local, but easy problem with low randomness in y^N (c) non-local and difficult problem, where each input data has a randomly chosen label.

always different. This property of the classification problem can be learned by an appropriate model.

To measure the statistical randomness of a binary sequence y^N, we suggest using Maurer's universal test T_U [12,33]. The purpose of Maurer's universal test is to measure the entropy in the sequence y^N. Other possibilities to measure the statistical randomness of a sequence are the Wald-Wolfowitz runs test [44], which measures the number of label changes, or autocorrelation tests [11]. We choose Maurer's universal test as it is able to detect also high-order as well as non-linear dependencies in a sequence.

We use the statistical test Maurer's universal T_U to test if the source process of the sequence is random [12,33]. Maurer's universal takes the sequence y^N of binary labels y (from $B = \{0,1\}$) as input. The test has three parameters $\{L, Q, K\}$. It partitions the sequence in blocks of length L with Q blocks used for initializing the test and K blocks to perform the test. Thus, $N = (Q+K)L$ and $b_n(y^N) = [y_{L(n-1)+1}, \dots, y_{Ln}]$. The test function $f_{T_U} : B^N \rightarrow \mathbb{R}$ measures the per-bit entropy and is defined as

$$f_{T_U}(y^N) = \frac{1}{K} \sum_{n=Q+1}^{Q+K} \log_2 A_n(y^N),$$

where

$$A_n(y^N) = \begin{cases} n \text{ , if } \forall a < n, b_{n-a}(y^N) \neq b_n(y^N) \\ \min\{a : a \geq 1, b_n(y^N) = b_{n-1}(y^N)\} \text{ , otherwise.} \end{cases}$$

This test function can be used to compute the $p \in [0,1]$ value

$$p = \operatorname{erfc}\left(\left|\frac{f_{T_U} - \text{expectedValue}(L)}{\sqrt{2}\sigma}\right|\right),$$

where erfc is the complementary error function. expectedValue(L) and σ are precomputed values [33]. The p value measures the confidence whether the process is non-random. Thus, low values of p indicate a high probability that the process is non-random.

If Maurer's universal test indicates that y^N is a random sequence (high values of p), then the generalization capability of a DNN applied to this data set $\mathcal{D}$ is expected to be low; in contrast, if y^N is non-random (which means the per-bit entropy of the sequence is low), DNNs are expected to be able to learn well the structure of $\mathcal{D}$ and show high generalization capability. Thus, we suggest that the randomness of a sequence of binary labels generated by a random walk through the input data space is a good predictor for the expected generalization capability of DNNs learning the input data.

3 Experiments and Discussion

To study how the properties of input data influences the generalization capability of DNNs, we randomize all studied data sets to different degrees as suggested by [50] and perform random walks through the label landscapes $(X, f, \mathcal{N})$ as described in Algorithm 1. For all considered data sets, we perform 30 random walks with $N = 1,000,000$ steps and calculate the confidence p for the resulting sequence y^N of labels. As data sets, we use synthetic classification tasks as well as on the common classification benchmarks MNIST [30], CIFAR10 [28], and the cats vs. dogs data set [15]. For each test problem, the input data is split into 80% train and 20% test data.

For the synthetic classification tasks as well as MNIST, we train a multilayer perceptron (MLP) consisting of two hidden layers with 4,096 neurons each and ReLU activation functions. For CIFAR10 and the cats vs. dogs data set, we use a small convolutional network (CNN) with three convolutional layers with 32/64/64 filters of kernel size 3×3 followed by a dense layer with 256 hidden neurons. After each convolutional layer we use 2×2 MaxPooling and all layers use ReLU activation functions. The models are trained with the Adam optimizer [26] until convergence to 100% accuracy on the train data. Thus, test error is identical to the generalization error g_{err}.

All experiments were conducted on a workstation using an AMD Ryzen Threadripper 3990X 64×2.90 GHz, an NVIDIA GeForce TITAN RTX and 128 GB DDR4 RAM. The DNNs were implemented using Tensorflow 2 [1].

3.1 Synthetic Classification Problems with Known Decision Boundaries

To analyze whether the suggested measure p properly captures the randomness of a problem for both, local and non-local patterns, we first study problems where we already know the classification problem's decision boundaries. We select three synthetic d-bit binary classification problems. The first one is a XOR type problem with binary input vectors x_i ($x_{ij} \in \{0, 1\}$). The label of each vector x_i

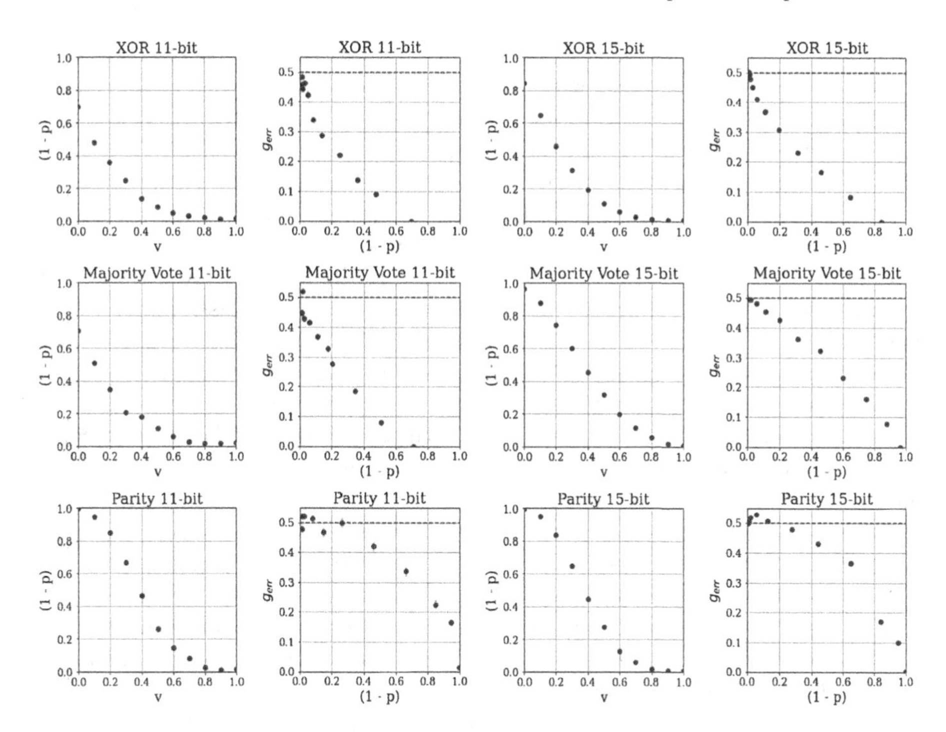

Fig. 2. $(1-p)$ over the randomization level v and generalization error g_{err} over $(1-p)$ for all studied synthetic classification problems (XOR, majority vote, and parity) for $d = 11$ and $d = 15$. The dashed line indicates performance of random guessing. All results are averaged over 30 runs.

depends on the first two input variables while the remaining features hold no explanatory power:

$$y_i = \begin{cases} 1 \text{ for } x_{i,1} = x_{i,2} \\ 0 \text{ for } x_{i,1} \neq x_{i,2} \end{cases}$$

The second test problem uses the same binary input vectors x_i. The label is determined by the majority vote over the elements $x_{i,j}$:

$$y_i = \begin{cases} 1 \text{ for } \sum_{j=1}^{d}(x_{i,j}) \geq \frac{d+1}{2} \\ 0 \text{ for } \sum_{j=1}^{d}(x_{i,j}) < \frac{d+1}{2} \end{cases}$$

The third test problem also uses binary input vectors x_i. The label of each vector is determined by the parity function:

$$y_i = \begin{cases} 1 \text{ if} & \sum_{j=1}^{d}(x_{i,j}) \text{ is even} \\ 0 \text{ otherwise} & \end{cases}$$

For all synthetic classification tasks, we study instances of different size $d \in \{11, 15\}$ and corrupt the labeling processes by changing each label y with probability v to a random class in the training and test set (see [50]) to construct different instances of the tasks with varying degrees of structure. The used data set $\mathcal{D}$ consists of all possible input vectors, as we assume that $X = \mathcal{D}$. The neighborhood function $\mathcal{N}(x)$ maps each input $x_i \in X$ to a set of inputs $x \in X$ that are different from x_i in one position $x_{i,j}$. We measure the randomness of y^N (constructed by the random walk) using Maurer's universal and compare it to the generalization performance of the MLP/CNN.

Figure 2 plots the measure $(1-p)$ over the randomization level v and the generalization error g_{err} over $(1-p)$ for all studied synthetic classification problems for $d = 11$ and $d = 15$. For comparison, the dashed line indicates the performance of random guessing. All results are averaged over 30 runs.

We expect that for higher values of v (which leads to a higher randomness of y^N and a lower correlation between neighboring inputs) the inherent structure of the classification problem sets declines which leads to lower generalization. The results confirm this expectation, as we can observe lower values of $(1-p)$ for larger values of v as well as a lower generalization error g_{err} for high values of $(1-p)$. For the considered test problems, the measure $(1-p)$ is a good predictor for generalization as Pearson's r correlation coefficient between generalization error g_{err} and $(1-p)$ is lower than -0.94 for all studied problem instances. This holds not only for small problems ($d = 11$) but also for larger problem instances ($d = 15$). Furthermore and contrary to the smoothness prior, the measure $(1-p)$ correctly detects structure (non-randomness) not only in local (XOR, majority vote) but also in non-local (parity) patterns.

3.2 Natural Data with Unknown Decision Boundaries

To verify whether our findings also hold on natural data, we extend our experiments to the MNIST and CIFAR10 data sets. We consider a binary classification version of those problems and (as before) corrupt the labeling function f by randomizing each label y with probability v. Again, we study the randomness of y^N (created by a random walk) and compare it to the generalization capability of MLP/CNN. However, since the true decision variables for the MNIST and CIFAR10 problems do not lie in the raw input matrix but rather are represented by latent variables in an underlying manifold [18], we first approximate such manifold by reducing the dimension of the input data with a variational autoencoder [21,27]. Consequently, we define the neighborhood $\mathcal{N}(x)$ on $\mathcal{D}$ as the set of k nearest data points measured by Euclidean distance inside this manifold. In our experiments, we chose $k = 10$.

Figure 3 plots the measure $(1-p)$ over the randomization probability v and the generalization error g_{err} over $(1-p)$ for the binary versions of MNIST and CIFAR10. The dashed line indicates the generalization error g_{err} of random guessing. Again, all results are averaged over 30 runs.

As expected, we also find a strong correlation between p and g_{err} for natural signals. Again, we observe lower values of $(1-p)$ for larger values of v and a lower generalization error g_{err} for high values of $(1-p)$. The Pearson's r correlation coefficient between generalization error g_{err} and $(1-p)$ is lower than -0.97 for

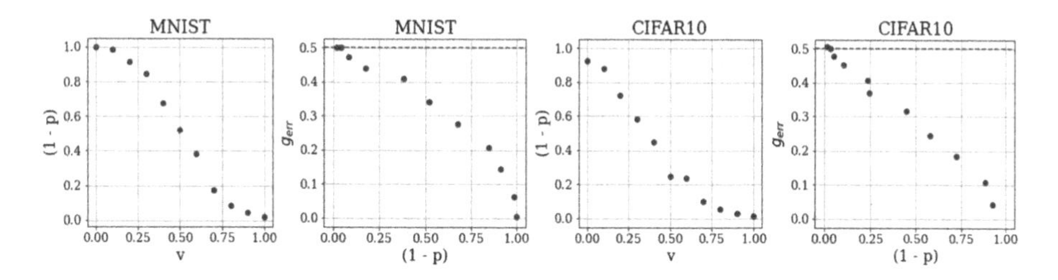

Fig. 3. $(1-p)$ over the randomization level v and the generalization error g_{err} over $(1-p)$ for the binary versions of MNIST and CIFAR10.

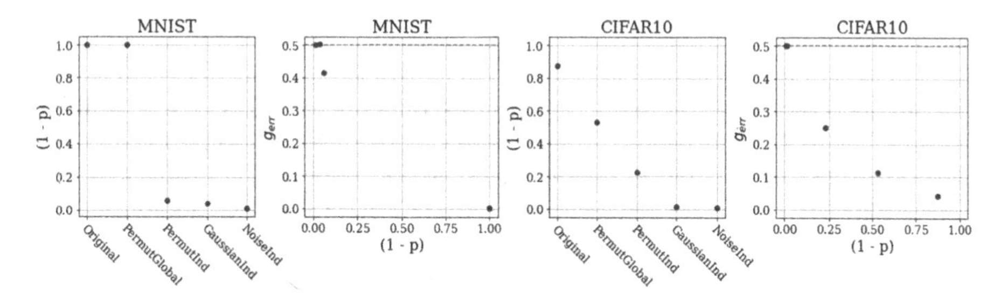

Fig. 4. $(1-p)$ over four variants of randomization (*PermutGlobal*, *PermutInd*, *GaussianInd* and *NoiseInd*) and generalization error g_{err} over $(1-p)$ for the binary versions of MNIST and CIFAR10.

both problem sets indicating that $(1-p)$ is a good approximation of the expected generalization error also on natural data.

To study the effects of different types of randomization of f, we now permutate the inputs $x \in X$ instead of the labels $y \in Y$. We consider four different variants: 1) a random permutation $\pi : x \to x$ of all input variables x_{ij} of the training data (denoted as *PermutGlobal*), 2) a random permutation of all variables for all training instances (*PermutInd*), 3) replacing a variable value by a random input value from a Gaussian distribution matching the original distribution of input values (*GaussianInd*), and 4) replacing a variable value by a value randomly drawn from a white noise distribution (*NoiseInd*). As before, we study whether the randomness of y^N is related to the generalization error.

Figure 4 plots $(1-p)$ over the four different variants of randomization and the resulting generalization error g_{err} over $(1-p)$. Again, the dashed line indicates the performance of random guessing. All results are averaged over 30 runs.

Again, we find a strong correlation (Pearson coefficient < -0.99) between generalization error g_{err} and $(1-p)$. For *PermutGlobal*, we observe a lower effect of randomization for MNIST in comparison to CIFAR10 as the neighborhood of the input data space is more relevant for CIFAR10 than MNIST. For MNIST, the value of a pixel x_{ij} also has a meaning independently of its neighboring pixels (e.g. some pixels are always activated for a specific label). In contrast for CIFAR10, destroying the neighborhood of a pixel x_{ij} by placing it next to other,

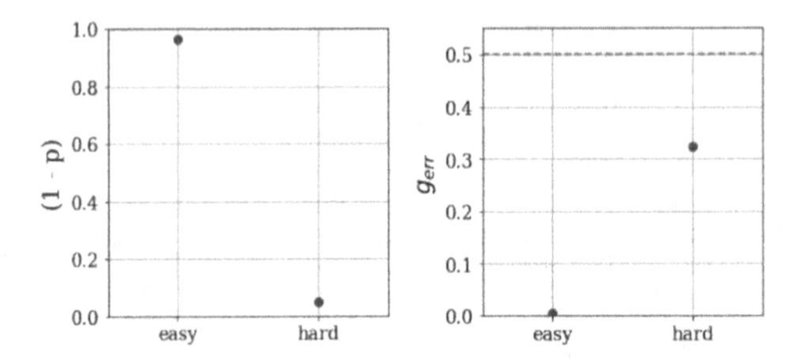

Fig. 5. $(1-p)$ for the easy and hard data samples with the corresponding generalization error g_{err}.

randomly selected pixels makes it much more difficult for the DNN to build a meaningful model. As a result, $(1-p)$ is lower for CIFAR10. For *PermutInd*, results are different as the only signal that is left after randomization is the difference in mean and standard deviation of input variables. The differences are higher in CIFAR10 training instances which makes the problem more structured (leading to a lower generalization error) in comparison to MNIST. Both cases are properly captured by $(1-p)$.

3.3 Studying Randomness of Input Data Spaces Without Randomization

While our previous experiments studied the relationship between the randomness of input data spaces measured by $(1-p)$ and generalization error for different degrees and variants of randomization, we now investigate differences in the randomness of input data spaces for easy versus hard data samples. Thus, we do not randomize neither f (Sect. 3.1) nor X (Sect. 3.2), but create data samples with different properties from $\mathcal{D}$ following an approach suggested by [5]. Consequently, we first train 100 CNNs for 1 epoch on a large data set (cats vs. dogs). Then, we select two subsets (easy versus hard) from $\mathcal{D}$ by selecting the on average 10,000 best and 10,000 worst classified examples for the easy and hard subset, respectively. We expect that Maurer's universal is a good indicator for the differences in randomness of these samples and the resulting generalization error g_{err}.

Figure 5 plots $(1-p)$ for the easy and hard data samples as well as the corresponding generalization error g_{err}. The dashed line indicates the performance of random guessing. Results are averaged over 30 runs. We find that a high value of $(1-p)$ (indicating a high randomness in y^N) correspond to a low generalization error g_{err} on the easy sample and vice versa on the hard sample confirming the prediction quality of Maurer's universal. For the easy sample, the generalization error is almost zero which corresponds to a high value of $(1-p) \approx 1$ indicating a low randomness of y^N and a high structure of the classification problem. Thus the easy data set can be learned by a DNN model with low generalization error. For the hard sample, the randomness of y^N is high indicating a low correlation between the labels of neighboring training points.

4 Related Work

Bounding the best and worst case for generalization error is a key challenge in machine learning. Traditional learning theory provides such bounds either from a complexity point of view [7,43] or using a stability based approach [10]. However, studies suggest that these generalization bounds might not be sufficient to capture the generalization problem, especially in an over-parameterized setting [8,34,50]. This leads to work on extending and sharpening the traditional bounds for neural networks by introducing norms [6,17,25,31,35,37,38] or using PAC-Bayes approaches [4,14,36,51]. A different direction of research studies the implicit regularization from gradient descent methods to explain generalization [2,19,40,41].

However, most of these approaches depend on posterior properties of a trained neural network. In contrast, [5] find that the data itself plays an important role in generalization. Therefore, other work focuses on the properties of data in context of generalization. Ma et al. [32] provide a prior estimate using properties of the true target function and [3] derive a data-depended complexity measure using the Gram matrix of the data and [16] analyze the properties of classification problems using Fourier analysis. The method suggested in this paper differs as we take a label landscape perspective to derive a generalization estimate.

5 Conclusion

This paper introduced a landscape perspective on data distributions in order to explain generalization performance of DNNs. We argued that the input data defines a label landscape and the correlation between labels of neighboring (similar) input values influences generalization. We measure the correlation of the labels of neighboring input values by performing a random walk through the input data space and use Maurer's universal to measure the randomness of the resulting label sequence y^N. A more random sequence indicates a less learnable structure in the data leading to poor generalization. At the extreme, if there is no correlation between the labels of neighboring inputs, generalization error is maximal. We performed experiments for a variety of problems to validate our hypothesis and found that the randomness (measured by Maurer's universal) of the label sequence y^N indeed can serve as an a priori indicator of the expected generalization error for a given data set. We presented results for both synthetic problems as well as real world data sets and found a high correlation between the randomness of the label sequence y^N and the generalization error. We conclude that a label landscape view on the data provides valuable insight into the generalization capability of DNN.

6 Limitations and Future Work

Our approach provides insights and an a priori indicator for generalization in a binary classification case. However, there are a few limitations due to the use of Maurer's universal test. As the test is only designed for a binary source processes, it is not applicable to multi-class problems. Therefore, in future work we will study randomness measures for integer sequences.

If the decision variables are not known, our method depends on the approximation of the underlying manifold, for which we assume an Euclidean space. Approximating such a manifold can be challenging for more difficult data sets. Studying the impact of this approximation and different distance measures for the neighborhood could lead to a better understanding of our findings.

References

1. Abadi, M., et al.: Tensorflow: a system for large-scale machine learning. In: 12th {USENIX} Symposium on Operating Systems Design and Implementation ({OSDI} 16), pp. 265–283 (2016)
2. Arora, S., Cohen, N., Hu, W., Luo, Y.: Implicit regularization in deep matrix factorization. In: Wallach, H., Larochelle, H., Beygelzimer, A., d'Alché-Buc, F., Fox, E., Garnett, R. (eds.) Advances in Neural Information Processing Systems, vol. 32. Curran Associates, Inc. (2019)
3. Arora, S., Du, S., Hu, W., Li, Z., Wang, R.: Fine-grained analysis of optimization and generalization for overparameterized two-layer neural networks. In: International Conference on Machine Learning, pp. 322–332. PMLR (2019)
4. Arora, S., Ge, R., Neyshabur, B., Zhang, Y.: Stronger generalization bounds for deep nets via a compression approach. In: International Conference on Machine Learning, pp. 254–263. PMLR (2018)
5. Arpit, D., et al.: A closer look at memorization in deep networks. In: International Conference on Machine Learning, pp. 233–242. PMLR (2017)
6. Bartlett, P.L., Foster, D.J., Telgarsky, M.J.: Spectrally-normalized margin bounds for neural networks. In: Guyon, I. (eds.) Advances in Neural Information Processing Systems, vol. 30. Curran Associates, Inc. (2017)
7. Bartlett, P.L., Mendelson, S.: Rademacher and gaussian complexities: risk bounds and structural results. J. Mach. Learn. Res. **3**, 463–482 (2002)
8. Belkin, M., Hsu, D., Ma, S., Mandal, S.: Reconciling modern machine-learning practice and the classical bias-variance trade-off. Proc. Nat. Acad. Sci. **116**(32), 15849–15854 (2019)
9. Bengio, Y., Delalleau, O., Le Roux, N.: The curse of highly variable functions for local kernel machines. In: Advances in Neural Information Processing systems, vol. 18, p. 107 (2006)
10. Bousquet, O., Elisseeff, A.: Stability and generalization. J. Mach. Learn. Res. **2**, 499–526 (2002)
11. Box, G.E., Jenkins, G.M.: Time Series Analysis: Forecasting and Control. Holden-Day, San Francisco (1976)
12. Coron, Jean -Sébastien., Naccache, David: An accurate evaluation of maurer's universal test. In: Tavares, Stafford, Meijer, Henk (eds.) SAC 1998. LNCS, vol. 1556, pp. 57–71. Springer, Heidelberg (1999). https://doi.org/10.1007/3-540-48892-8_5
13. Cybenko, G.: Approximation by superpositions of a sigmoidal function. Math. Control Signals Systems **2**(4), 303–314 (1989). https://doi.org/10.1007/BF02551274
14. Dziugaite, G.K., Roy, D.M.: Computing nonvacuous generalization bounds for deep (stochastic) neural networks with many more parameters than training data. arXiv preprint arXiv:1703.11008 (2017)

15. Elson, J., Douceur, J.R., Howell, J., Saul, J.: Asirra: a captcha that exploits interest-aligned manual image categorization. In: ACM Conference on Computer and Communications Security, vol. 7, pp. 366–374 (2007)
16. Farnia, F., Zhang, J.M., David, N.T.: A fourier-based approach to generalization and optimization in deep learning. IEEE J. Sel. Areas Inf. Theory **1**(1), 145–156 (2020)
17. Golowich, N., Rakhlin, A., Shamir, O.: Size-independent sample complexity of neural networks. In: Conference on Learning Theory, pp. 297–299. PMLR (2018)
18. Goodfellow, I., Bengio, Y., Courville, A.: Deep Learning, vol. 1. MIT press, Cambridge (2016)
19. Hardt, M., Recht, B., Singer, Y.: Train faster, generalize better: Stability of stochastic gradient descent. In: International Conference on Machine Learning, pp. 1225–1234. PMLR (2016)
20. Herrmann, S., Ochoa, G., Rothlauf, F.: Communities of local optima as funnels in fitness landscapes. In: Proceedings of the Genetic and Evolutionary Computation Conference 2016, pp. 325–331, GECCO '16, Association for Computing Machinery, New York, NY, USA (2016). https://doi.org/10.1145/2908812.2908818
21. Hinton, G.E., Salakhutdinov, R.R.: Reducing the dimensionality of data with neural networks. Science **313**(5786), 504–507 (2006)
22. Hornik, K.: Approximation capabilities of multilayer feedforward networks. Neural Netw. **4**(2), 251–257 (1991)
23. Imaizumi, M., Fukumizu, K.: Deep neural networks learn non-smooth functions effectively. In: The 22nd International Conference on Artificial Intelligence and Statistics, pp. 869–878. PMLR (2019)
24. Jones, T., Forrest, S.: Fitness distance correlation as a measure of problem difficulty for genetic algorithms. In: Proceedings of the 6th International Conference on Genetic Algorithms, pp. 184–192. Morgan Kaufmann Publishers Inc., San Francisco (1995)
25. Kawaguchi, K., Kaelbling, L.P., Bengio, Y.: Generalization in deep learning. arXiv preprint arXiv:1710.05468 (2017)
26. Kingma, D.P., Ba, J.: Adam: a method for stochastic optimization. arXiv preprint arXiv:1412.6980 (2014)
27. Kingma, D.P., Welling, M.: Auto-encoding variational bayes. arXiv preprint arXiv:1312.6114 (2013)
28. Krizhevsky, A., Hinton, G., et al.: Learning multiple layers of features from tiny images (2009)
29. LeCun, Y., Bengio, Y., Hinton, G.: Deep learning. Nature **521**(7553), 436–444 (2015)
30. LeCun, Y., Bottou, L., Bengio, Y., Haffner, P.: Gradient-based learning applied to document recognition. Proc. IEEE **86**(11), 2278–2324 (1998)
31. Liang, T., Poggio, T., Rakhlin, A., Stokes, J.: Fisher-rao metric, geometry, and complexity of neural networks. In: The 22nd International Conference on Artificial Intelligence and Statistics, pp. 888–896. PMLR (2019)
32. Ma, C., Wu, L., et al.: A priori estimates of the population risk for two-layer neural networks. arXiv preprint arXiv:1810.06397 (2018)
33. Maurer, Ueli M..: A universal statistical test for random bit generators. J. Cryptology **5**(2), 89–105 (1992). https://doi.org/10.1007/BF00193563
34. Nagarajan, V., Kolter, J.Z.: Uniform convergence may be unable to explain generalization in deep learning. In: Wallach, H., Larochelle, H., Beygelzimer, A., d'Alché-Buc, F., Fox, E., Garnett, R. (eds.) Advances in Neural Information Processing Systems, vol. 32. Curran Associates, Inc. (2019)

35. Neyshabur, B., Bhojanapalli, S., Mcallester, D., Srebro, N.: Exploring generalization in deep learning. In: Guyon, I. (eds.) Advances in Neural Information Processing Systems, vol. 30. Curran Associates, Inc. (2017)
36. Neyshabur, B., Bhojanapalli, S., Srebro, N.: A pac-bayesian approach to spectrally-normalized margin bounds for neural networks. In: International Conference on Learning Representations (2018)
37. Neyshabur, B., Li, Z., Bhojanapalli, S., LeCun, Y., Srebro, N.: The role of over-parametrization in generalization of neural networks. In: International Conference on Learning Representations (2019)
38. Neyshabur, B., Tomioka, R., Srebro, N.: Norm-based capacity control in neural networks. In: Conference on Learning Theory, pp. 1376–1401. PMLR (2015)
39. Rothlauf, F.: Design of Modern Heuristics: Principles and Application. Springer Science Business Media, Heidelberg (2011). https://doi.org/10.1007/978-3-540-72962-4
40. Smith, S.L., Le, Q.V.: A bayesian perspective on generalization and stochastic gradient descent. In: International Conference on Learning Representations (2018)
41. Soudry, D., Hoffer, E., Nacson, M.S., Gunasekar, S., Srebro, N.: The implicit bias of gradient descent on separable data. J. Mach. Learn. Res. **19**(1), 2822–2878 (2018)
42. Vapnik, V.: Principles of risk minimization for learning theory. In: Advances in Neural Information Processing Systems, pp. 831–838 (1992)
43. Vapnik, V.: The Nature of Statistical Learning Theory. Springer Science & business media, Heidelberg (2013)
44. Wald, A., Wolfowitz, J.: On a test whether two samples are from the same population. Ann. Math. Stat. **11**(2), 147–162 (1940)
45. Wolpert, D.H.: The existence of a priori distinctions between learning algorithms. Neural Comput. **8**(7), 1391–1420 (1996)
46. Wolpert, D.H.: The lack of a priori distinctions between learning algorithms. Neural Comput. **8**(7), 1341–1390 (1996)
47. Wolpert, D.H., Macready, W.G.: No free lunch theorems for search. Technical report, SFI-TR-95-02-010, Santa Fe Institute (1995)
48. Wolpert, D.H., Macready, W.G.: No free lunch theorems for optimization. IEEE Trans. Evol. Comput. **1**(1), 67–82 (1997)
49. Wright, S.: The roles of mutation, inbreeding, crossbreeding, and selection in evolution (1932)
50. Zhang, C., Bengio, S., Hardt, M., Recht, B., Vinyals, O.: Understanding deep learning requires rethinking generalization. In: International Conference on Learning Representations (2017)
51. Zhou, W., Veitch, V., Austern, M., Adams, R.P., Orbanz, P.: Non-vacuous generalization bounds at the imagenet scale: a PAC-bayesian compression approach. In: International Conference on Learning Representations (2019)

Communicating Safety of Planned Paths via Optimally-Simple Explanations

Noel Brindise(✉) and Cedric Langbort

University of Illinois Urbana-Champaign, 104 S Wright St., Urbana, IL 61801, USA
{nbrindi2,langbort}@illinois.edu

Abstract. Artificial intelligence is often used in path-planning contexts. Towards improved methods of explainable AI for planned paths, we seek optimally simple explanations to guarantee path safety for a planned route over roads. We present a two-dimensional discrete domain, analogous to a road map, which contains a set of obstacles to be avoided. Given a safe path and constraints on the obstacle locations, we propose a family of specially-defined constraint sets, named explanatory hulls, into which all obstacles may be grouped. We then show that an optimal grouping of the obstacles into such hulls will achieve the absolute minimum number of constraints necessary to guarantee no obstacle-path intersection. From an approximation of this minimal set, we generate a natural-language explanation which communicates path safety in a minimum number of explanatory statements.

Keywords: Explainable AI · Constraint optimization · Path planning · Mental model reconciliation · Human-robot interaction

1 Introduction

As autonomous systems make their way from the laboratory to the real world, automated navigation and path planning have taken on a prominent role in everyday human-AI interaction. Mutual understanding between human and robot is key in such motion- and route-planning scenarios, particularly when a human must interact directly with an autonomous system. In this paper, we consider a scenario in which a human is shown a proposed route from a start point to a goal point, supposing that the path must avoid a number of obstacles. This could represent, for example, an aircraft following a flight plan between military threats [1]. In this case, a pilot is taking a path through a region populated with potentially dangerous obstacles which they aim to avoid.

In our approach, we assume the path planner knows some bounds on the location of each obstacle, perhaps through inference or observation, and that the

This work was supported in part by the United States Department of Defense (DoD) through the National Defense Science and Engineering Graduate (NDSEG) Fellowship Program.

R. Bergmann et al. (Eds.): KI 2022, LNAI 13404, pp. 31–44, 2022.
https://doi.org/10.1007/978-3-031-15791-2_4

chosen path successfully avoids all of them. Implicit in this assumption is that some such obstacle-avoidant path exists. Meanwhile, the human in our scenario knows a priori (1) the layout of roads on the map, represented in this problem by horizontal-vertical gridlines, and (2) the locations of various landmarks, or reference points.

Given this scenario, our task is to reassure the human that the planned path does not encounter any obstacles. We do this by producing a post-hoc explanation, which we construct as a set of constraints prohibiting intersection of the obstacles with the path. Trivially, the explanation could provide all known constraints on the obstacle locations. However, to reduce the mental load for a human, we aim to produce an explanation which is as simple as possible, searching for the smallest set of constraints which, taken together, guarantee that the obstacles do not intersect with the path.

The two-dimensional grid scenario considered here constitutes a base case for more complicated problems, such as domains with probabilistic bounds on obstacle locations, where bounds are inferred from a model with uncertain dynamics. Another natural extension of this case is multi-objective optimization, in which a path is planned with explanation simplicity as an objective alongside typical optimization objectives like path speed and length.

2 Previous Work

Explanation for human-robot interaction is a growing pursuit under the umbrella of explainable artificial intelligence (xAI). xAI as a field recognizes the importance of AI system transparency in a world where human lives are shaped ever more by the actions, decisions, and predictions of artificially intelligent systems [10]. Appropriately, the field has accumulated a substantial body of literature and continues to grow [9,12].

Explanation is often formulated as a model reconciliation task, whereby humans incorporate new information into their personal mental models of a system [4]. Simpler explanations are more easily incorporated, given human susceptibility to information overload [11]; indeed, complicated explanations tend to have less of an impact on human decision making [11,13].

Explanation specifically for autonomous path planners has been approached from several angles. Post-hoc inference of linear temporal logic (LTL) constraints on system trajectories can generate a list of time- and order-dependent specifications satisfied by a planner, which may be provided to a human as explanation [3,5,7]. For Markov decision process based path planning, a tunable model has been proposed which adjusts emphasis on the mission objectives, the path segment(s) of current interest, the level of detail, and language abstractness to select an individualized list from a bank of explanatory statements [2]. The former approach focuses on the extraction of constraints, while the latter ranks an existing bank of constraints based on their saliency. In both cases, the approach supposes that the bank of explanatory constraints is itself sufficient or optimal, and explanations are made simple by de-prioritization or outright omission of complicated constraints, regardless of potential relevance [8]. In this paper, we present an algorithm which aims to address these issues, beginning with the restricted setting of path planning in a known two-dimensional environment. Our method reprocesses a large set of constraints into a smaller set without loss of necessary information, providing an optimally-simple bank of candidate constraints to be drawn from for explanation. Thus, our approach is not directly comparable to current approaches, but rather serves as an intermediate step between constraint inference and prioritization via saliency.

3 Notation and Preliminaries

3.1 Road Map Domain

Our problem considers a two-dimensional finite grid. Planned paths must follow grid lines and therefore consist of north-south (vertical) and east-west (horizontal) segments only. When grid lines are interpreted as streets, the obstacles on the map may represent a road blockage or a military threat; more abstractly, obstacles may be forbidden regions in any state space through which we wish to follow a given trajectory. An obstacle i is contained by a set Ω_i of four linear-inequality constraints, denoted ω_{i1} through ω_{i4}, whose conjunction specify a closed region. An example domain is shown in Fig. 1.

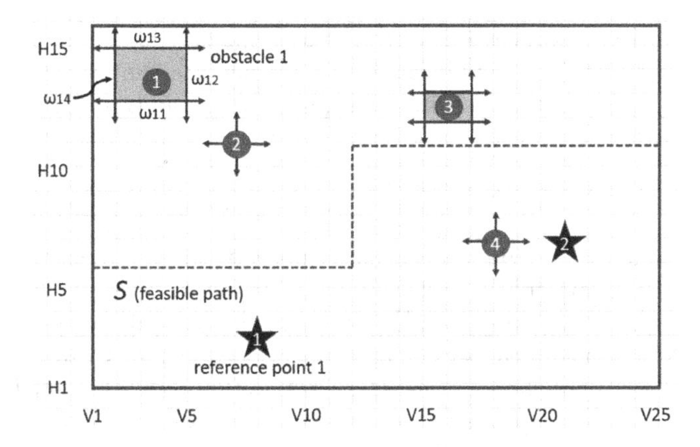

Fig. 1. The domain, a road map with a suggested route S. Obstacles are shown as shaded circles and reference points are marked with stars.

On the discrete grid, every set of linear inequality constraints prescribes a set of grid points. In this paper, we use the bracket notation [ ] to represent the set of all points p satisfying the conjunction of a set of constraints, for instance $[\Omega_i] := \{p \mid p \models \omega_{i1} \wedge \cdots \wedge \omega_{i4}\}$. We extend this notation to the path S as well, denoting the set of points in the path by $[S]$.

3.2 Form of Explanation

We aim to provide human-interpretable explanation as a set of constraints which satisfies the following general criteria:

1. *Safety:* the explanation must communicate that the path will safely avoid all of the obstacles. This relies on the assumption that a feasible (i.e., safe) path exists and has been selected by the path planner.
2. *Simplicity:* the explanation should have as few features as possible, so that it is easier for a human to understand.

To communicate safety, we can provide a human with a list of constraints on the obstacles which guarantee that no obstacle will cross the path. Obstacles which remain on one side of any grid lines may be easily bounded using linear inequalities, for instance, *Obstacle* o_1 *stays north of H13 Street* represented by $y_{o_1} \geq 13$. Meanwhile, obstacles which stay within some radius of a reference point may be explained using a nearness constraint, as in *Obstacle* o_4 *stays near Reference Point* r_2 and $|x_{o_4} - x_{r_2}| + |y_{o_4} - y_{r_2}| \leq 3$. Now, towards simplicity, we observe from Fig. 1 that not all such valid constraints contribute to proving safety. For example, *Obstacle 1 remains south of H15* is true, but it does not provide any information as to whether Obstacle 1 can intersect with S. An optimally simple explanation would need to omit such irrelevant constraints.

Linear Inequality Constraints. Towards simplicity, it is useful to create alternative safety-enforcing sets of constraints which are simpler than the true sets Ω_i. Define F_i be a set of four linear inequality constraints $f_{i1}, \ldots, f_{i4}$, where f_{ij} are expressed as

$$f_{ij} := z \leq z_0 \quad \text{or} \quad f_{ij} := z \geq z_0 \tag{1}$$

for some constant $z_0 \in \mathbb{Z}$. Here, horizontal and vertical f_{ij} take z to be the y- and x-coordinate, respectively. Additionally, $f_{i1} \wedge \cdots \wedge f_{i4}$ forms a closed region such that $[\Omega_i] \subseteq [F_i]$. As long as $[F_i] \cap [S] = \emptyset$, the obstacle remains properly constrained. Now consider the special case where one or more f_{ij} in F_i coincides with one of the domain boundaries $M_1, \ldots, M_4$. Since the domain boundary constraints are known a priori to any human user, we are able to omit any $f_{ij} = M_m$ from the explanation. This occurs in Fig. 2, supposing that $[\Omega_i] \subseteq [F_i]$ for Obstacles 1, 2, and 3. This is a useful simplification which we formalize in Sect. 4.

Nearness Constraints. Some obstacles may be constrainable by their proximity to a fixed location on the map. We denote each notable location as a reference point r, and we define the set $[P_r]$ to be the set of points within a chosen radius of r. In Fig. 2, Reference Points 1 and 2 are shown with regions of fixed radius 3, prescribed respectively by constraints P_1 and P_2. Since Obstacle 4 satisfies $[\Omega_4] \subseteq [P_2]$, it is considered *near* Reference Point 2.

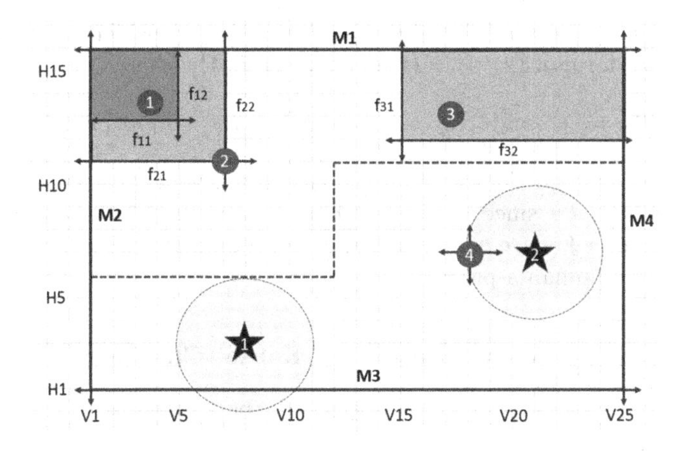

Fig. 2. A map showing F_i-constrained regions and relevant f_{ij} for Obstacles 1, 2, and 3. Also shown are the nearness neighborhoods of fixed radii around the reference points.

Obstacle Grouping. If obstacles must be individually bounded, an explanation for L obstacles will necessarily contain L lists of constraints. However, suppose a set of constraints F_i or P_r is found such that $[\Omega_l] \subseteq [F_i]$ or $[\Omega_l] \subseteq [P_r]$ for multiple obstacles o_l. For instance, by Fig. 2, it is true that the constraints shown for Obstacle 2, f_{21} and f_{22}, also contain the full region prescribed by f_{11} and f_{12}. This means that Obstacles 1 and 2 may be grouped together under the constraints $F_2 = \{f_{21}, f_{22}, f_{23}, f_{24}\}$, where $f_{23} = M_1, f_{24} = M_2$, and the total number of explanatory constraints for the two obstacles reduces from 4 to 2. This operation is called obstacle grouping.

Hull-Based Explanation. We now present the structure of the explanation E, which we define as a set of pairs of sets (O_k, B_k). Here, O_k is a subset of the obstacles from the domain; B_k is either a P_r or an F_i chosen such that B_k is satisfied by all $o_l \in O_k$. In other words, all obstacles in O_k are contained by B_k. We will refer to B_k as an *explanatory hull*, and call this form of explanation hull-based. Any hull-based E can be expressed as in (2):

$$E = \{\varepsilon_1, \ldots, \varepsilon_K\} \quad \text{where} \quad \varepsilon_k = (O_k, B_k),\ \forall k \in \{1, \ldots, K\} \tag{2}$$

Note that, for any safe path S, it is always possible to find fitting B_k for at least one arrangement of the o_i. To see this, consider the arrangement of the obstacles into singleton sets $O_1 = \{o_1\}, \ldots, O_L = \{o_L\}$; then a choice of satisfying sets B_k is simply $B_1 = F_1 := \Omega_1, \ldots, B_K = F_K := \Omega_L$. Thus, we may always construct an explanation of the form in (2) for any domain with a safe path. Such explanations are easily converted into natural language, as in Example 1.

Example 1 (Natural language explanation from E). Consider an explanation expressed in the form of (2), $E = \{(O_1, B_1), (O_2, B_2)\}$, where

$$O_1 = \{o_1\}, \quad B_1 = P_1 \quad \text{and} \quad O_2 = \{o_2, o_3, o_4\}, \quad B_2 = F_1 = \{f_{11}, f_{12}, M_1, M_2\}$$

where $f_{11} := y \geq 2$ and $f_{12} := x \leq 4$. Then, for ε_1, we return *One obstacle is near Reference Point 1* since o_1 is in P_1. For ε_2, we return *3 obstacles are north of H2 and west of V4*, since o_2, o_3, and o_4 are contained within F_1, with M_1 and M_2 known to the human a priori.

3.3 Constraint Parameterization and Subset Checking

An inequality $f_{ij} := z \leq z_0$ or $f_{ij} := z \geq z_0$ may be written as

$$f_{ij} = (d, z_0, p) \quad \text{with } d \in \{-1, 1\},\ z \in \mathbb{N},\ p \in \{-1, 1\} \tag{3}$$

where

$$d = \begin{cases} -1 & f_{ij} \text{ horiz.} \\ 1 & f_{ij} \text{ vert.} \end{cases} \quad z_0 = \begin{cases} y_0 & f_{ij} \text{ horiz.} \\ x_0 & f_{ij} \text{ vert.} \end{cases} \quad p = \begin{cases} -1 & f_{ij} := z \leq z_0 \\ 1 & f_{ij} := z \geq z_0 \end{cases}$$

and we call d the orientation and p the sign of f_{ij}. The parameterization in (3) enables us to introduce the function in (4), which we name the *subset function*:

$$\mu_s(f_1, f_2) := d_1(d_1 + d_2)\,(p_1 + p_2)\,(z_{01} - z_{02}) \tag{4}$$

We note that $d_1 = d_2, p_1 = p_2$ for orientation and sign match, respectively, and $d_1 = -d_2, p_1 = -p_2$ for mismatch. In all, (4) satisfies

$$\mu_s(f_1, f_2) > 0 \iff [f_1] \subset [f_2] \quad \text{and} \quad \mu_s(f_1, f_2) < 0 \iff [f_2] \subset [f_1]$$

where $[f_i]$ denotes the set of points satisfying f_i. Meanwhile, $\mu_s(f_1, f_2) = 0$ occurs if and only if neither constrained region constitutes a strict subset of the other. Then, for our $F = \{f_1, \ldots, f_4\}$, we see that

$$[F'] \subset [F''] \iff ([f_1'] \subset [f_1'']) \wedge \cdots \wedge ([f_4'] \subset [f_4'']), \tag{5}$$

and thus $[F'] \subset [F'']$ is true if and only if

$$\forall f' \in F', \forall f'' \in F'' \quad \mu_s(f', f'') > 0 \tag{6}$$

Because F prescribes a rectangular region, we also have the consequence that

$$\mu_s(f', f'') \geq 0 \quad \forall f' \in F', \forall f'' \in F'' \iff [F'] \subseteq [F''] \tag{7}$$

since this condition means that all sign- and orientation-matching pairs f', f'' either coincide or that $[f'] \subset [f'']$.

Path Parameterization. We represent each segment s of S^* as four linear inequalities with parameterizations as in (3). The constraints associated with a single horizontal segment with endpoints $(y, x_1), (y, x_2)$ where $x_2 > x_1$ becomes

$$s = \{s_1, s_2, s_3, s_4\} = \{(-1, y, 1), (-1, y, -1), (1, x_1, -1), (1, x_2, 1)\}. \quad (8)$$

with a similar parameterization for vertical segments. With this expression for s, we may consider Proposition 1:

Proposition 1 (Path intersection condition). *$[F] \cap [S] = \emptyset$ is true if and only if*

$$\forall f_i \in F, \forall s \in S,\ \exists s_i \in s \text{ such that } [F] \subset [s_i \wedge f_i] \quad (9)$$

Proof. Considering (8), let s_1, s_2 be the $s-$parallel linear inequalities and s_3, s_4 be the endpoint constraints. Take also some $F = \{f_1, \dots, f_4\}$ and let each s_i, f_i pair be such that $p_{f_i} = p_{s_i}, d_{f_i} = d_{s_i}$.

First, suppose there is some segment s which traverses $[F]$. This would require that $[F]$ lies between z_{s_3}, z_{s_4}, i.e.,

$$[s_3] \subseteq [f_3] \text{ and } [s_4] \subseteq [f_4] \quad (10)$$

since (10), by inspection, means that s cannot enter $[F]$. Now consider $f_1, f_2 \in F$ with $p_{f_1} = 1, p_{f_2} = -1$. If (10) holds, s will enter $[F]$ only if

$$z_{f_1} \leq z_{s_1} \leq z_{f_2} \implies [s_1] \subseteq [f_1], [s_2] \subseteq [f_2]. \quad (11)$$

Altogether, we see that $[s_i \wedge f_i] = [s_i]$ for $i = \{1, 2, 3, 4\}$, and thus, $[F] = [f_1 \wedge \cdots \wedge f_4] \supseteq [s_i]$ where $[s_i] \not\subset [s_i \wedge f_i]$ for all $f_i \in F$.

In the other direction, we have that either (10) or (11) not holding is sufficient to avoid intersection of s with F, meaning $[s_i] \supset [f_i]$ for any $s_i \in s$, $f_i \in F$ prevents intersection. In turn, $[s_i] \supset [f_i] \implies [s_i \wedge f_i] = [f_i]$, and clearly $[F] \subset [f_i]$ for any $f_i \in F$. □

3.4 Largest Possible Hulls

A useful tool for hull-based explanation is the set of *largest possible hulls* subject to the domain and the path:

Definition 1 (Largest Possible Hulls H). *A largest possible hull is a set of four linear inequality constraints prescribing a rectangular region, denoted $H = \{h_1, \dots, h_4\}$, where, for $[S] \neq \emptyset$,*

$$\begin{aligned} &\forall s \in S,\ \forall h_i \in H,\ \exists s_i \in s \text{ such that } [H] \subseteq [s_i \wedge h_i] \\ &[H] \subseteq [M] \end{aligned} \quad (12)$$

and for all H' satisfying (12), $[H] \subseteq [H'] \implies H = H'$.

We use $\mathcal{H}$ to denote the set of all largest possible hulls subject to the path and domain. We now examine the properties of these largest hulls.

Property 1. Let $b = (d_b, z_b, p_b)$ denote any constraint such that $b = s_i$ or $b = M_m$ for some $s_i \in S$ or $M_m \in M$. Then, for all $h \in H$, there exists a $b \in S \cup M$ such that $h = b$.

Proof. Suppose we have a largest possible hull $H = \{h_1, h_2, h_3, h_4\}$ where $h_1, h_2, h_3 = b_1, b_2, b_3$, but that $h_4 \neq b$ for any b. Since $[H] \subseteq [M]$, we have $\mu_s(h_4, M_4) > 0$. Therefore, we may always choose a b' where $\mu_s(h_4, b') > 0$. For $b' = M_4$,

$$[S] \subseteq [M] \implies [s_i \wedge M_4] = [s_i] \implies [h_1 \wedge h_2 \wedge h_3 \wedge b'] \subseteq [s_i]. \quad (13)$$

Meanwhile, for any $b' = s_i'$ where s_i' is the $s_i \in s$ satisfying $d_{h_4}, p_{h_4} = d_{s_i'}, p_{s_i'}$,

$$s_i = s_i' \implies [s_i \wedge s_i'] = [s_i'] \implies [h_1 \wedge h_2 \wedge h_3 \wedge b'] \subseteq [s_i'] \quad (14)$$

and for $s_i \neq s_i'$, either

$$[s_i \wedge s_i'] = [s_i'] \implies [h_1 \wedge h_2 \wedge h_3 \wedge b'] \subseteq [s_i'], \text{ or} \quad (15)$$
$$[s_i \wedge s_i'] = [s_i] \implies [h_1 \wedge h_2 \wedge h_3 \wedge b'] \subseteq [s_i] \quad (16)$$

Therefore, we may always take a $b' = s_i'$ or s_i such that the alternative hull $H' = \{h_1, h_2, h_3, b'\}$ satisfies (12). However, $\mu_s(h_4, b') > 0 \implies [H] \subseteq [H']$, but $H = H'$ only if $h_4 = b'$. Thus H is not a largest possible hull. □

Property 2. For all $H \in \mathcal{H}$, there exists an $h \in H$ such that $h = s_i$ for some $s_i \in s \in S$.

Proof. Suppose H is a largest possible hull with sides $h_1 = M_1$, $h_2 = M_2$, $h_3 = M_3$, $h_4 \neq s_i$, where there is no $s_i \in s \in S$ such that $s_i = M_1, M_2, M_3$. By Property 1, we must take $h_4 = M_4$. Clearly, since $[S] \subseteq [M] \implies [s_i \wedge M_4] = [s_i]$, we see that $[H] \subseteq [s_i \wedge M_4]$ if and only if $s_i = M_4$. This implies that either $M_4 = s_i \in H$ or H cannot be a largest hull. □

Property 3. Consider $F = \{f_1, \dots, f_4\}$ such that $[F] \subseteq [M]$ and for all $s \in S, f_i \in F$, there exists an $s_i \in s$ where $[F] \subseteq [s_i \wedge f_i]$. There must exist a largest possible hull H such that $[F] \subseteq [H]$.

Proof. Suppose there is some such F which does not satisfy $F \subseteq H$. Then by (5), F must contain at least one f_i for each H where $[f_i] \supset [h]$ for some $h \in H$. However, by the same logic in the proof of Property 1, this would mean that F either satisfies $[F] \subseteq [H]$ or F violates the $[F] \subseteq [s_i \wedge f_i]$ condition. □

We now claim that there is a surjective relationship from the set of all s_i onto $\mathcal{H}$:

Proposition 2 (Surjectivity of s_i onto H). *To each constraint s_i associated with a segment $s \in S$, there is a unique $H \in \mathcal{H}$ such that $s_i \in H$.*

Proof. Select some s_i and choose two hulls $H' = \{s_i, h'_1, h'_2, h'_3\}$, $H'' = \{s_i, h''_1, h''_2, h''_3\}$. Then, for any pair $h' = (d, z_{h'}, p)$, $h'' = (d, z_{h''}, p)$, we have either $h' = h''$ or $z_{h'} \neq z_{h''}$. By (1), both h' and h'' satisfy either $h = s_i$ or $h = M_m$. First suppose $h' = M_m$. Then $h'' \neq M_m$ allows for two scenarios:

$$\mu_s(h', h'') < 0 \implies [H''] \subset [H']$$
$$\mu_s(h', h'') > 0 \implies [H''] \not\subseteq [M]$$

both of which mean H'' violates (12). Suppose alternatively that $h' = s'_i$ for some other path segment component s'_i; then $h'' \neq s'_i$ allows for

$$\mu_s(h'', s'_i) < 0 \implies [s'_i \wedge h''] = [s'_i] \subset [h''] \implies [H''] \not\subseteq [s'_i \wedge h'']$$
$$\mu_s(h'', s'_i) > 0 \implies [H''] \subset [H']$$

Both of these scenarios also produce H'' which violate (12). Thus, any pair H', H'' of largest possible hulls with $s_i \in H', H''$ must satisfy $H' = H''$. □

Finally, we define the *explanatory hull of H*, a particular variety of the explanatory hulls B_k introduced in Sect. 3.2.

Definition 2 (Explanatory Hull of H). *The explanatory hull of H, denoted F_H, is the set of four constraints f defined by*

1. $f = h$ if $h = M_m$
2. $f = (p_h, z_h + p_h, d_h)$ otherwise

for all parameterized constraint pairs $f \in F_H$, $h \in H$ where $p_f = p_h, d_f = d_h$.

One important property of F_H is as follows:

Property 4. F_H is the largest possible $[F] \subset [H]$ such that $[F] \cap [S] = \emptyset$.

Proof. For H and corresponding F_H, $f = h$ for all $h = M_m$ and $f = (p_h, z_h + p_h, d_h)$ otherwise. Suppose that F_H may be enlarged; then, by (5), we must replace some $f \in F_H$ with an f' satisfying $\mu_s(f, f') > 0$. This is clearly not possible for any $f = M_m$. For the latter f, the smallest possible increment outward, $-p_f$, results in $f' = (p_h, z_h + p_h - p_h, d_h) = (p_h, z_h, d_h) = h$ for some $h \in H$. However, by (1), we have $h \neq M_m \implies h = s_i$, meaning that $f = s_i$. By (9), this means that the larger F must intersect S. □

4 Optimal Explanation

4.1 Formal Requirements

Now that we have defined the explanatory form and largest hulls, we may formally express safety and simplicity, the latter of which requires the introduction of a complexity cost.

Definition 3 (Safety). *Let all B_k satisfy $[B_k] \cap [S] = \emptyset$. Then an explanation E is safe if and only if, for all obstacles o_i in the domain, there exists some $(O_k, B_k) \in E$ where $o_i \in O_k$ and $[\Omega_i] \subseteq [B_k]$.*

Definition 4 (Complexity cost). *The complexity of E is given by*

$$C(E) := \sum_{k=1}^{K} c(\varepsilon_k) \text{ where } c(\varepsilon_k) := |B_k \backslash \{M1, \dots, M4\}| \quad \text{and} \quad K = |E| \tag{17}$$

By (17), each $c(\varepsilon_k)$ will equal either $|F_k \backslash \{M_1, \dots, M_4\}|$ or 1 when $B_k = P_r$.

Definition 5 (Simplicity). *The explanation E^* satisfies simplicity if it is chosen such that*

$$E^* = \arg\min_E (C(E)) \tag{18}$$

where all E satisfy the safety requirement in Definition 3.

Observe that an E^* which satisfies (18) will consist of an optimal pairing of obstacle groups $O^* = \{O_1^*, \dots, O_K^*\}$ and hulls $B^* = \{B_1^*, \dots, B_K^*\}$ to minimize $C(E)$. We will denote the full set of valid E as Σ, the search space for E^*.

4.2 Reduced Search Space Σ_{sub}

A priori, the search space for E^* contains all safety-enforcing pairs (O_k, F_i) and (O_k, P_r). To make the optimization task more tractable, we propose a subset of Σ, which we call Σ_{sub}. We first observe that all $E \in \Sigma$ may be organized into F_i- and P_r-pairs. Thus, we may rewrite (18):

$$C(E^*) = \min_{E \in \Sigma} \sum_{k=1}^{K} c(\varepsilon_k) = \min_{E \in \Sigma} \left(\sum_{\varepsilon_k = (O_k, F_i)} c(\varepsilon_k) + \sum_{\varepsilon_k = (O_k, P_r)} c(\varepsilon_k) \right) \tag{19}$$

Now, taking $\varepsilon_i = (O_{k'}, F_i)$ and $\varepsilon_s = (O_{k'}, F_s)$, if we can guarantee that

$$\varepsilon_i \in \Sigma \implies \exists \varepsilon_s \in \Sigma_{sub} \quad \text{s.t.} \quad c(\varepsilon_s) \leq c(\varepsilon_i) \tag{20}$$

for all $\varepsilon_i = (O_{k'}, F_i) \in \Sigma$, we see by (19) that $C(\varepsilon_1, \dots, \varepsilon_K, \varepsilon_s) \leq C(\varepsilon_1, \dots, \varepsilon_K, \varepsilon_i)$, and thus

$$C(E^*) = \min_{E \in \Sigma} C(E) = \min_{E \in \Sigma_{sub}} C(E). \tag{21}$$

We now present our candidate for Σ_{sub} in Theorem 1.

Theorem 1 (Hull-based Σ_{sub}). *Consider the set of all largest-possible hulls H on the domain with boundaries M and path S. Taking $\Sigma_P = \{\varepsilon_k \in \Sigma | \varepsilon_k = (O_k, P_r)\}$ and $\Sigma_{F_H} = \{\varepsilon_k \in \Sigma | \varepsilon_k = (O_k, F_H)\}$, we have that (21) is satisfied by the reduced search space*

$$\Sigma_{sub} = \Sigma_P \cup \Sigma_{F_H}. \tag{22}$$

Proof. Consider $(O_{k'}, F)$ where $[\Omega_{k'}] \subseteq [F]$, $[F] \cap [S] = \emptyset$. Then, by (9) and Property 3, $[F] \subset [H]$ for some H. Additionally, by Property 4, F_H is the largest F in H such that $[F] \cap [S] = \emptyset$. Then, $[F] \subset [H] \implies [F] \subseteq [F_H]$, and

$$[\Omega_i] \subseteq [F] \quad \forall o_i \in O_{k'} \implies [\Omega_i] \subseteq [F_H] \quad \forall o_i \in O_{k'} \tag{23}$$

From (23), we see that any valid pair $(O_{k'}, F)$ in a hull H will always correspond to a valid pair $(O_{k'}, F_H)$. To meet the conditions of (21), we need now only check that $c((O_{k'}, F_H)) \leq c((O_{k'}, F))$. Since

$$c((O_k, F)) = 4 - |\{f \in F | f = M_m\}|, \tag{24}$$

we observe that $c((O_{k'}, F)) = c((O_{k'}, F_H))$ will only occur when F and F_H share all $f \in F_H$ for which $f = M_m$; otherwise, $c((O_{k'}, F)) > c((O_{k'}, F_H))$. This means that, for any given hull H with associated F_H as described in Theorem 1, F_H will always satisfy

$$(O_{k'}, F_H) = \underset{(O_{k'}, F)}{\arg\min}\, c((O_{k'}, F)) \tag{25}$$

for valid $(O_{k'}, F_H)$ in hull H. Therefore, if we construct Σ_{sub} as in (22), we find that Σ_{sub} may be used without loss of optimality. □

4.3 Solution from Σ_{sub}

To construct Σ_{sub}, we must first build $\mathcal{H}$, the list of all H. By Proposition 2, this is possible by iterating through each $s_i \in s$ for all $s \in S$ to find every H satisfying (12); from here, every F_H for Σ_{sub} is easily calculated from $\mathcal{H}$. Recalling that the explanatory hull B_k is any F_H or P_r, we now describe the steps for optimization:

1. Find all valid $\varepsilon_k = (O_k, B_k)$ pairs, where O_k contains all obstacles o_i satisfying $[\Omega_i] \subseteq [B_k]$.
2. Identify any $\varepsilon'_{k'} = (O_{k'}, B_{k'})$ which, for some o_i such that $[\Omega_i] \subseteq [B_{k'}]$ and $o_i \in O_{k'}$, satisfy $o_i \notin O_k \quad \forall k \neq k'$. Form a list $\varepsilon'_1, \ldots, \varepsilon'_N$ of all such ε'.
3. Include ε'_1 in E^*; for each subsequent $\varepsilon'_{k'}$, take $O_{red} = O_{k'} \backslash \{O_1 \cup \cdots \cup O_{k'-1}\}$ and include $(O_{red}, B_{k'})$ in E^*.
4. Choose additional ε until all o_i are contained in some (O_k, B_k) pair in E.
5. Optionally, for every (O_k, F_k)-pair in E, we may find a *tightened hull* B_k^t. Such B_k^t are found by taking all $f \in F_k, f \neq M_m$ and replacing f by an $\omega_{ij} \in \Omega_i, o_i \in O_k$ such that all $o' \in O_k$ satisfy $[\Omega'] \subset [\omega_{ij}]$.

Step 1 identifies only those pairs which contain the largest possible O_k for which (O_k, B_k) is valid. This is acceptable because, by (17), we are able to bound the final complexity of E^* by

$$C((O_k, B_k) \in E^*) \leq C(E^*) \leq C((O_k, B_k)) + 4|O_{rem}| \tag{26}$$

where O_{rem} is the set of all obstacles $O \backslash O_k$. Since $C((O', F_H))$ is constant for all O', our lower bound is unaffected by choice of O'; thus, we always optimally select (O', F_H) where O' contains all $o \in O_{rem}$ satisfying $[\Omega] \subseteq [F_H]$.

Steps 2 and 3 ensure inclusion of all ε which contain the sole valid constraint set for a given obstacle, since omission of any such unique ε means that E does not satisfy safety.

Step 4 guarantees safety of the final E. However, solving for E^* which minimizes $C(E)$ can be shown to be equivalent to a weighted set coverage problem, which is NP-hard [6]. For the purposes of this paper, we select a greedy search heuristic over Σ_{sub} to approximate a final E^*. The greedy search iteratively selects those $\varepsilon_k \in \Sigma_{sub}$ satisfying

$$\arg\min_{\varepsilon_k}(C(\arg\max_{\varepsilon_k} |O_k|)). \tag{27}$$

The heuristic is given in more detail in the appendix, under Algorithm 1.

Algorithm 1 Greedy Optimization

Input: $\{M_1, \ldots, M_4\}$, reduced (O_k, F_H) pairs
Output: Valid E^*

```
 1: toExplain = list of reduced (O_k, F_H) pairs
 2: oInEachHull = list of obstacles o in each F_H
 3: hullsWithObstO = list of hulls occupied by each o
 4: while max(oInEachHull) > 0 do
 5:    bestComplexity = 100
 6:    mostObsts = max(oInEachHull)
 7:    for all F_H ∈hullsWithObstO do
 8:       if oInEachHull(F_H)=mostObsts and C(F_H) < bestComplexity then
 9:          bestComplexity = C(F_H)
10:          bestHull = F_H
11:       end if
12:    end for
13:    Append (oInEachHull(bestHull), bestHull) to E*
14:    Remove all o in oInEachHull(bestHull) from hullsWithObstO
15:    Set oInEachHull(bestHull) = [ ]
16: end while
17: return E*
```

Step 5 is an optional step which may increase the proximity of the constraints in ε to the true location of the obstacles O in ε. This is done to improve location accuracy in the explanation. Crucially, this does not influence optimality, since M_m-constraints are unaltered and thus $C(\varepsilon)$ remains constant.

5 Simulation

Our implementation of the method described in Sect. 4.3 is available on GitHub[1]. We present a single example case here on the domain with path, obstacles,

[1] https://github.com/n-brindise/plan_expl.

and reference points arranged as in Fig. 3. Taking a nearness radius of 2 for reference points, we return an approximation for E^* in natural language in four statements:

- 4 obstacles are north of H6
- 1 obstacle is west of V6, south of H4, and east of V5
- 1 obstacle is west of V3 and north of H3
- 2 obstacles are near Reference Point 2

Notably, for this 8-obstacle domain, the full constraint set has been reduced from $8 \times 4 = 32$ to 7. We see that the four statements indeed guarantee safety: all 8 obstacles are contained in regions through which the path does not pass.

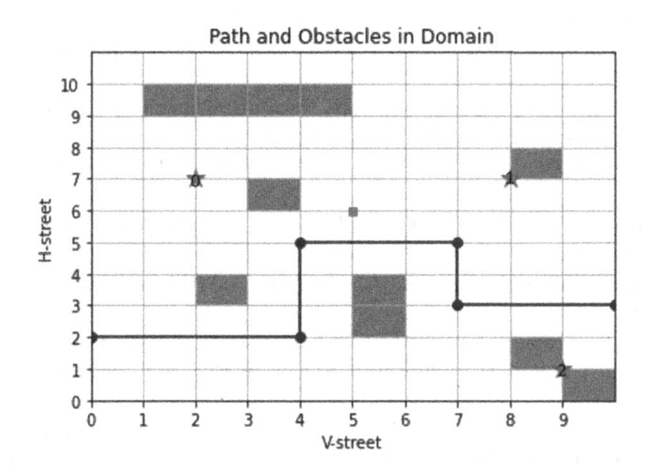

Fig. 3. Simulated domain for explanation with S shown in blue, reference points depicted as yellow stars and obstacles shown in orange. (Color figure online)

6 Conclusion and Future Work

Given a two-dimensional, finite domain with path, obstacles, and reference points as described in Sect. 3, we are able to generate explanations which guarantee no path-obstacle intersection while significantly reducing the number of explanatory statements from the nominal number of obstacle constraints. While the approach shows promise, a more extensive survey of domain types and path shapes will be necessary to characterize the best use cases for the algorithm.

The expansion of this method to a more general domain also holds promise for explanation over more realistic scenarios. Interesting cases for consideration include those with probabilistically-bounded obstacles, dynamic obstacles, or higher-dimensional, non-grid domains. Additionally, the approach may lend itself well to multi-objective path planning, which would incorporate explanation simplicity into the optimization process.

References

1. Avoidance rerouter ARR 7000 (2021). https://www.collinsaerospace.com/en/what-we-do/Military-And-Defense/Avionics/Software-Applications/Avoidance-Re-Router-Arr-7000
2. Boggess, K., Chen, S., Feng, L.: Towards personalized explanation of robot path planning via user feedback. arXiv preprint arXiv:2011.00524 (2020)
3. Brindise, N.C.: Towards explainable AI: directed inference of linear temporal logic constraints (2021). https://hdl.handle.net/2142/110849
4. Chakraborti, T., Sreedharan, S., Zhang, Y., Kambhampati, S.: Plan explanations as model reconciliation: moving beyond explanation as soliloquy. In: Proceedings of the Twenty-Sixth International Joint Conference on Artificial Intelligence, IJCAI, pp. 156–163 (2017). https://doi.org/10.24963/ijcai.2017/23
5. Gaglione, J.R., Neider, D., Roy, R., Topcu, U., Xu, Z.: Learning linear temporal properties from noisy data: a maxsat approach. arXiv preprint arXiv:2104.15083 (2021)
6. Hooker, J.N., et al.: Integrated Methods for Optimization, vol. 170. Springer, NY (2012). https://doi.org/10.1007/978-1-4614-1900-6
7. Kim, J., Muise, C., Agarwal, S., Agarwal, M.: Bayesltl (2019). https://github.com/IBM/BayesLTLcommit379924d
8. Kim, J., Muise, C., Shah, A., Agarwal, S., Shah, J.: Bayesian inference of linear temporal logic specifications for contrastive explanations. In: Proceedings of the Twenty-Eighth International Joint Conference on Artificial Intelligence, pp. 5591–5598 (2019)
9. Mittelstadt, B., Russell, C., Wachter, S.: Explaining explanations in AI. In: Proceedings of the Conference on Fairness, Accountability, and Transparency, pp. 279–288 (2019)
10. Shin, D.: The effects of explainability and causability on perception, trust, and acceptance: implications for explainable AI. Int. J. Hum Comput Stud. **146**, 102551 (2021)
11. Sultana, T., Nemati, H.R.: Impact of explainable AI and task complexity on human-machine symbiosis (2021)
12. Wells, L., Bednarz, T.: Explainable AI and reinforcement learning–a systematic review of current approaches and trends. Front. Artif. Intell. **4**, 48 (2021)
13. Zhang, W., Lim, B.Y.: Towards relatable explainable AI with the perceptual process. In: CHI Conference on Human Factors in Computing Systems. CHI 2022, Association for Computing Machinery, NY (2022). https://doi.org/10.1145/3491102.3501826

Assessing the Performance Gain on Retail Article Categorization at the Expense of Explainability and Resource Efficiency

Eduardo Brito[1,2(✉)], Vishwani Gupta[1,2], Eric Hahn[1,2], and Sven Giesselbach[1,2]

[1] Fraunhofer IAIS, Schloss Birlinghoven, 1, 53757 Sankt Augustin, Germany
[2] Competence Center for Machine Learning Rhine-Ruhr, Sankt Augustin, Germany
eduardo.alfredo.brito.chacon@iais.fraunhofer.de

Abstract. Current state-of-the-art methods for text classification rely on large deep neural networks. For use cases such as product cataloging, their required computational resources and lack of explainability may be problematic: not every online shop can afford a vast IT infrastructure to guarantee low latency in their web applications and some of them require sensitive categories not to be confused. This motivates alternative methods that can perform close to the mentioned methods while being explainable and less resource-demanding. In this work, we evaluate an explainable framework consisting of a representation learning model for article descriptions and a similarity-based classifier. We contrast its results with those obtained by DistilBERT, a solid low-resource baseline for deep learning-based models, on two different retail article categorization datasets; and we finally discuss the suitability of the different presented models when they need to be deployed considering not only their classification performance but also their implied resource costs and explainability aspects.

Keywords: Multi-label text classification · Representation learning · Resource awareness · Explainability

1 Introduction

The online shopping expansion from the last years has led to a more varying product offer so that an automated product cataloguing process is necessary when the manual maintenance becomes unsustainable. This generally involves solving a multi-class (eventually multi-label) text classification task based on the product descriptions, which is challenging due to the often very skewed label distribution: few categories are dominant, appearing on a majority of products, while many categories hardly appear on few products. This power-law distribution that appears on many natural language processing tasks can be extreme on some datasets. While the current state-of-the-art (SotA) transformer-based approaches can achieve astonishing results even in such challenging settings,

R. Bergmann et al. (Eds.): KI 2022, LNAI 13404, pp. 45–52, 2022.
https://doi.org/10.1007/978-3-031-15791-2_5

these mainly consist of very large models, whose complexity may be problematic for real-world implementations. Not only are the implementation-related costs and unexpected high energy consumption generally ignored by machine learning practitioners [6], but also the current best performing models normally run on specialized hardware (mostly GPUs or TPUs) to obtain results in a feasible time. This requirement limits the access to these models and may also be responsible for a significant carbon footprint [16]. Even ignoring training time, the computational complexity during the inference phase may be an issue when it introduces extra latency on web applications, eventually deteriorating user satisfaction. Furthermore, black-box models are difficult to inspect when issues arise ("why this unexpected label for this product?"). We present exactly such an use case where our client required a low-resource solution where some categories must not be confused (e.g. "sex toys" must not be misclassified as "toys").

We perceive a "trade-off triangle" of three aspects that we cannot fully achieve at the same time when choosing a classification model architecture: performance, explainability, and low resource requirements. In this work, we aim to quantify the performance gap between black-box transformer-based methods and more lightweight explainable models. For the former, we deliberately omit to experiment with current SotA deep networks for extreme multi-label classification. Instead, DistilBERT [15], usually performing slightly worse than its transformer-based competitors while being significantly more compact, can act as a "ceiling model" for the presented alternatives in this work. In parallel, we evaluate a k-nearest-neighbors-based classification pipeline trained on a set of various text representations of different degrees of computational complexity and explainability, including topic models and neural language models. We find this framework explainable since we can trivially show the neighbors (the most similar product descriptions) as explanations for every classified article. We test our pipeline on two retail article categorization datasets of different complexity and compare it with DistilBERT. Although we observe the expected performance gap, the difference is hardly relevant on the simpler dataset while we measure much larger CPU times for DistilBERT. We also consider the explainability aspect, questioning the suitability of huge neural networks as standard option when they need to be run in a production environment.

2 Related Work

Recent proposals for text classification are generally transformer-based models [18]. In the particular case of multi-label classification, various deep learning models are SotA [1]. All these models share a lack of explainability. In the broader context of (text) sequence classification, some approaches rely on similarity measures to a set of prototypes [4,7,12]. Although there is no consensus to measure interpretability in machine learning [11], we find these approaches as inherently explainable since the computed similarities can serve as explanations for the model decisions. Our presented similarity-based classifier fits to this class of models but we rather focus on evaluating existing representation learning methods within our framework. We also evaluate the runtime-related cost, which only partially correlates to model complexity i.e. to less explainability.

3 Experiments

3.1 Data

German Product Migration (GPM) is an internal dataset used to automate the product cataloging process for a retail online store, where certain run time and explainability requirements were requested. Each article contains a title and a description in German and is assigned a label from a three-level category hierarchy. For our experiments, we focus on the 599 categories belonging to the deepest level, ranging from clothing and accessories to health and personal care. Figure 1 shows the very skewed label distribution.

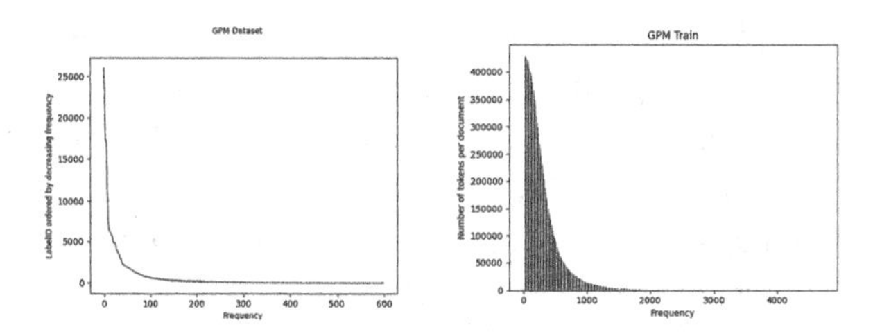

Fig. 1. Label frequency distribution and tokens per document from GPM.

AmazonCat-13K is an extreme multi-label classification dataset provided by Amazon. It consists of product reviews tagged with $\approx 13K$ product categories. The train split has 1.186.239 instances and the test split has 306.782 instances. In Fig. 2 we provide visualizations of the distribution of the number of labels associated to input texts, the distribution of the number of positive training instances per label and the distribution of the document lengths measured using a DistilBERT-tokenizer.

3.2 Methods

Nearest Neighbor. We vectorize each article title and description with a common dimensionality to make them comparable (512 for GPM, 768 for AmazonCat-13K, both determined by the pretrained models we use). Each represented article from the test set obtains the label(s) from the training set article whose vector representation is closest according to cosine distance.

Latent Dirichlet allocation (LDA) [2]. We train a topic model from the word tokens marked as (proper) nouns by the spaCy POS tagger [8] using the `de_core_web_sm` model for GPM and `en_core_web_sm` for AmazonCat-13K.

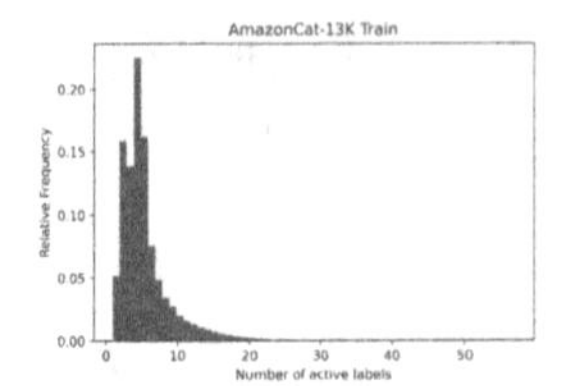

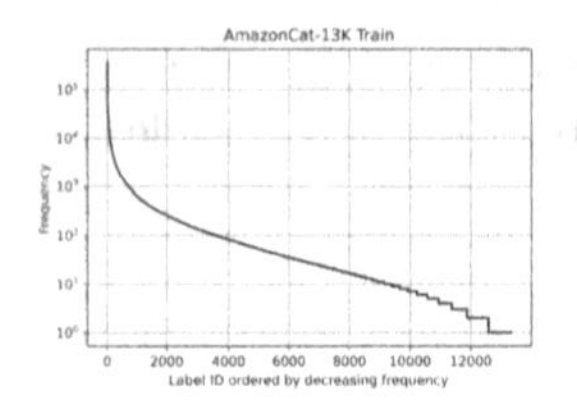

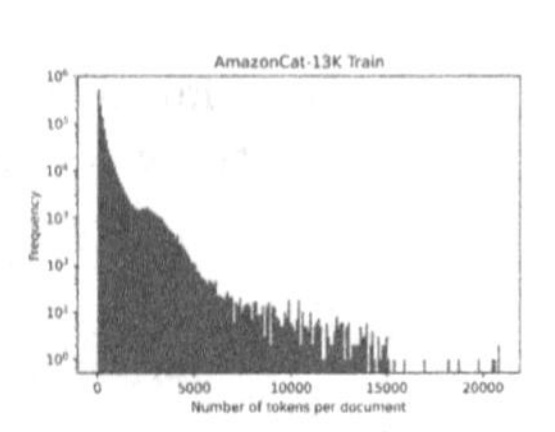

Fig. 2. Distribution of active labels, label frequency, and number of tokens per document from AmazonCat-13K.

Anchored CorEx [5]. We train two topic models with the same preprocessing pipeline as for LDA. The first one is trained in a purely unsupervised fashion. The other by assigning 10 anchor words to start the topic model training. We automatically generate anchor words by finding the words that have the highest mutual information with each label, as proposed in [9]. In the case of the AmazonCat-13K, we restrict this procedure to the 768 most frequent labels.

FastText [3]. We train a fastText embedding model for each dataset, whose obtained article representations are based on the average of N-gram features.

SentenceBERT (SBERT) [13]. We transform each text with a pretrained model: `distiluse-base-multilingual-cased-v1` for GPM, and `sentence-transformers/all-mpnet-base-v2` for AmazonCat-13K.

ML-KNN. We train a ML-KNN model [19] on the same vector representations presented in Sect. 3.2 with the scikit-multilearn implementation [17]. For each of the produced representations, we run a 3-fold cross-validation on 100,000 random articles from the training set to determine the number of neighbors k and the smoothing parameter s.

DistilBERT. We fine-tune a DistilBERT model. The architecture is given by DistilBERT as the encoder and a linear decoder with fan-out to all labels, activated with sigmoid and loss function given by binary cross-entropy.

3.3 Evaluation

Metrics. We evaluate each method applied in the single-label classification setting (GPM dataset) on the F1-score, both micro-averaged and macro-averaged on all the classes. For the multi-label setting (AmazonCat-13K), we include not only the widespread metrics P@k (precision at k) and nDCG@k (normalized discounted cumulative gain at k) with $k \in \{1, 3, 5\}$ but also their propensity-scored variants PSP and PSnDCG to better assess the performance on the less frequent classes as proposed in [10], taking the default propensity values $A = 0.55, B = 1.5$. We also measure the training time and prediction time of each method when running on a single CPU.

Results. The performance results of the evaluated methods are displayed in Table 1 for the GPM dataset and in Table 3 for the AmazonCat-13K dataset. As expected, the neural-based methods perform better on both datasets than the topic models. In particular, DistilBERT performs best on all evaluation metrics. However, the performance gap is reduced on the GPM dataset, especially on micro-averaged precision. We also observe that ML-KNN brings some performance improvement compared to the nearest neighbor baseline, although the gain is limited in several metrics, in particular when applied to the least performing topic models. We can see the training and prediction time of the different methods in Table 2. As expected, DistilBERT takes the longest CPU time.

Table 1. Performance on the GPM dataset.

Method	F-1 score	
	Macro avg.	Micro avg.
Nearest Neighbor		
+LDA	0.65	0.90
+CorEx	0.69	0.92
+Anchored CorEx	0.69	0.91
+SBERT	0.87	0.97
+FastText	0.66	0.94
DistilBERT	0.92	0.98

Table 2. Training and prediction time with AmazonCat-13K on an Intel® Xeon® Gold 6226R CPU @ 2.90 GHz.

Method	CPU time (in min.)	
	Training	Prediction
Nearest Neighbor	0.05	156
MLkNN	5,000	930
+LDA	+1,370	+25.8
+(Anchored) CorEx	+1,880	+5.22
+SBERT	+0[a]	+29.2
+FastText	+86	+9.4
DistilBERT	72,000[a]	510

[a]We omit the needed time for the pre-trained model due the given CPU-only setting.

4 Discussion

The results seem to confirm the hypothesized triangular trade-off: the most complex model (DistilBERT) outperforms by a large margin the simpler interpretable models (LDA and CorEx), while those in between in terms of explainability (SBERT somewhat more interpretable than DistilBERT due to its similarity-based nature) or regarding computational complexity (FastText) achieve intermediate results. However, the performance gap is notably smaller in the single-label classification setting of the GPM dataset. Since many online shops do not handle datasets as complex as AmazonCat-13K but rather like GPM, we question the convenience of applying the best performing available model by default disregarding significant computational and implementation costs. In some cases, the precision gain separating these models from "cheaper" models is not relevant. The combination of SBERT with a k-nearest neighbor-based classifier may be a good trade-off: it can perform close to the best model while keeping explainable predictions (an article gets a label because similar

Table 3. Performance of the evaluated methods on AmazonCat-13K evaluated as in [1]. Please note that nDCG@1=P@1 and PSnDCG@1=PSP@1.

Method	P			nDCG		PSP			PSnDCG	
	@1	@3	@5	@3	@5	@1	@3	@5	@3	@5
Nearest Neighbor										
+LDA	58.81	52.51	42.31	57.84	56.74	31.53	42.03	46.42	39.74	43.60
+CorEx	56.10	48.47	38.76	54.45	52.94	28.19	36.37	39.96	35.21	38.33
+Anchored CorEx	55.50	48.25	38.22	53.79	52.24	27.95	36.41	39.53	34.87	37.92
+SBERT	**75.94**	**67.92**	**55.58**	**74.50**	**73.45**	**41.65**	**55.51**	**62.18**	**52.29**	**57.58**
+FastText	67.98	61.54	51.18	67.42	67.31	36.81	49.50	55.96	46.63	51.81
MLkNN										
+LDA	60.93	50.87	39.62	39.62	54.36	31.89	39.86	42.57	38.24	40.93
+CorEx	56.27	49.09	38.95	54.67	53.19	28.30	36.96	39.96	35.37	40.17
+Anchored CorEx	55.93	48.52	38.51	54.11	52.62	28.17	36.64	39.84	35.09	38.21
+SBERT	**80.14**	**71.25**	**57.10**	**78.14**	**75.97**	**42.78**	**56.88**	**62.54**	**53.60**	**58.32**
+FastText	75.56	65.53	51.73	72.49	69.93	38.94	50.35	54.30	47.95	51.71
DistilBERT	**95.91**	**82.21**	**66.60**	**90.99**	**88.93**	**55.48**	**68.05**	**74.98**	**65.59**	**71.22**

articles have that label). FastText can also be an option in setups where no GPU is available. Moreover, there are still open options to increase the accuracy of the explainable framework. For instance, the SBERT representations do not incorporate the label information from Amazon-Cat-13K as DistilBERT did during training. Hence, we may easily improve the SBERT-backed model by just fine-tuning it. This is in line with some authors claiming that there is not always necessarily an accuracy-explainability trade-off when standard processes for knowledge discovery are followed [14]. Although using anchor words within anchored CorEx improved no model on any dataset (probably because of the anchor selection method), we value the possibility of selecting anchor words to control how specific topics are built for use cases where sensitive categories must not be confused.

5 Conclusion and Future Work

We evaluated several text representations on two datasets for retail product categorization within an explainable similarity-based framework. We compared them with a pure neural-based baseline not only on classification performance but also on the required training and prediction time. We additionally discussed the trade-off between obtaining a good performance, having some degree of explainability, and keeping the required computational resources low depending on the application. For future work, we plan to fine-tune an SBERT model on Amazon-Cat-13K by assigning product pairs a similarity score. We also envision a systematic model robustness inspection for specific sensitive labels. Enforcing separations via anchor words may turn anchored CorEx more valuable than our work showed.

Acknowledgements. This work was funded by the German Federal Ministry of Education and Research, ML2R - no. 01S18038B.

References

1. Bhatia, K., et al.: The extreme classification repository: multi-label datasets and code (2016). http://manikvarma.org/downloads/XC/XMLRepository.html
2. Blei, D.M., Ng, A.Y., Jordan, M.I.: Latent dirichlet allocation. J. Mach. Learn. Res. **3**, 993–1022 (2003)
3. Bojanowski, P., Grave, E., Joulin, A., Mikolov, T.: Enriching word vectors with subword information. Trans. Assoc. Comput. Linguist. **5**, 135–146 (2017). https://doi.org/10.1162/tacl_a_00051, https://aclanthology.org/Q17-1010
4. Brito, E., Georgiev, B., Domingo-Fernández, D., Hoyt, C.T., Bauckhage, C.: Ratvec: a general approach for low-dimensional distributed vector representations via rational kernels. In: LWDA, pp. 74–78 (2019)
5. Gallagher, R.J., Reing, K., Kale, D., Ver Steeg, G.: Anchored correlation explanation: Topic modeling with minimal domain knowledge. Trans. Assoc. Comput. Linguist. **5**, 529–542 (2017). https://doi.org/10.1162/tacl_a_00078, https://aclanthology.org/Q17-1037
6. García-Martín, E., Rodrigues, C.F., Riley, G., Grahn, H.: Estimation of energy consumption in machine learning. J. Parallel Distrib. Comput. **134**, 75–88 (2019)
7. Hong, D., Baek, S.S., Wang, T.: Interpretable sequence classification via prototype trajectory (2021)
8. Honnibal, M., Montani, I., Van Landeghem, S., Boyd, A.: spaCy: industrial-strength natural language processing in python (2020). https://doi.org/10.5281/zenodo.1212303
9. Jagarlamudi, J., Daumé III, H., Udupa, R.: Incorporating lexical priors into topic models. In: Proceedings of the 13th Conference of the European Chapter of the Association for Computational Linguistics, pp. 204–213. Association for Computational Linguistics, Avignon (2012). http://aclanthology.org/E12-1021
10. Jain, H., Prabhu, Y., Varma, M.: Extreme multi-label loss functions for recommendation, tagging, ranking & other missing label applications. In: Proceedings of the 22nd ACM SIGKDD International Conference on Knowledge Discovery and Data Mining, KDD 2016, pp. 935–944. Association for Computing Machinery, New York (2016). https://doi.org/10.1145/2939672.2939756
11. Molnar, C.: Interpretable machine learning (2020). http://christophm.github.io/interpretable-ml-book/
12. Pluciński, K., Lango, M., Stefanowski, J.: Prototypical convolutional neural network for a phrase-based explanation of sentiment classification. In: Kamp, M., et al. (eds.) ECML PKDD 2021. Communications in Computer and Information Science, vol. 1524, pp. 457–472. Springer, Cham (2021). https://doi.org/10.1007/978-3-030-93736-2_35
13. Reimers, N., Gurevych, I.: Sentence-BERT: sentence embeddings using Siamese BERT-networks. In: Proceedings of the 2019 Conference on Empirical Methods in Natural Language Processing and the 9th International Joint Conference on Natural Language Processing (EMNLP-IJCNLP), pp. 3982–3992. Association for Computational Linguistics, Hong Kong (2019). https://doi.org/10.18653/v1/D19-1410, http://aclanthology.org/D19-1410

14. Rudin, C.: Stop explaining black box machine learning models for high stakes decisions and use interpretable models instead. Nat. Mach. Intell. **1**(5), 206–215 (2019)
15. Sanh, V., Debut, L., Chaumond, J., Wolf, T.: Distilbert, a distilled version of bert: smaller, faster, cheaper and lighter (2020)
16. Strubell, E., Ganesh, A., McCallum, A.: Energy and policy considerations for deep learning in NLP. In: Proceedings of the 57th Annual Meeting of the Association for Computational Linguistics, pp. 3645–3650. Association for Computational Linguistics, Florence (2019). https://doi.org/10.18653/v1/P19-1355, http://aclanthology.org/P19-1355
17. Szymański, P., Kajdanowicz, T.: A scikit-based Python environment for performing multi-label classification. ArXiv e-prints (2017)
18. Vaswani, A., et al.: Attention is all you need. Adv. Neural Inf. Process. Syst. **30** (2017)
19. Zhang, M.L., Zhou, Z.H.: ML-KNN: a lazy learning approach to multi-label learning. Pattern Recogn. **40**(7), 2038–2048 (2007)

Enabling Supervised Machine Learning Through Data Pooling: A Case Study with Small and Medium-Sized Enterprises in the Service Industry

Leonhard Czarnetzki[1](✉), Fabian Kainz[2], Fabian Lächler[1], Catherine Laflamme[1], and Daniel Bachlechner[1]

[1] Fraunhofer Austria Research GmbH, Weisstraße 9, 6112 Wattens, Austria
leonhard.czarnetzki@fraunhofer.at
[2] Poool Software & Consulting GmbH, Grabenweg 3, 6020 Innsbruck, Austria

Abstract. Insufficient amounts of historical data present a major challenge in real world supervised machine learning projects. Small and Medium-Sized Enterprises (SMEs) are particularly handicapped regarding the collection of historical data. A possible solution to this problem is data pooling, where data from different entities is combined to create larger datasets that are more suitable for supervised machine learning. In this study, we investigate the potential that data pooling has for six companies from the service industry located in Germany and Austria. We find that in the studied scenario each company can benefit from the other companies' data under certain circumstances. In addition, while most companies benefit from a model that is trained with the data of all other companies, this is not always the case. This is because of specific business characteristics that can significantly affect datasets. In such a case, the key challenge is to determine which companies' data to include in the pool, i.e., to define the pooling strategy. Therefore, we analyze all possible pooling strategies in our scenario and explain selected results with insights from data distribution and feature importance analysis. We conclude that the consideration of business and data characteristics is critical to the selection of an appropriate strategy.

Keywords: Supervised machine learning · Data pooling · Transfer learning · Small and medium-sized enterprises · Service industry · Case study

1 Introduction

The hurdle for small and medium-sized enterprises (SMEs) to invest in data science and data value creation projects is still very high. In their survey on artificial intelligence (AI), Fuchs et al. [6] examine the current challenges of AI-supported processes in Austrian companies, finding that the majority of Austrian companies and especially SMEs still face major challenges with regard to AI. Fuchs et al. describe lack of competencies (40%), high investment costs (39%), uncertain added value (36%) and deficient data

R. Bergmann et al. (Eds.): KI 2022, LNAI 13404, pp. 53–59, 2022.
https://doi.org/10.1007/978-3-031-15791-2_6

(27%) as the prior hurdles that keep SMEs from investing in AI. The challenge of deficient data, which is particularly relevant for supervised machine learning, may be solvable by transfer learning, i.e., by leveraging the knowledge from different but related source domains. One application of transfer learning is data pooling, where similar data from different entities is combined to build more suitable datasets [17].

2 Current State of Research and Application

State of the art data sharing and data pooling takes place in different fields using different frameworks ensuring the trustworthy exchange of data and data sovereignty for all participants [11]. In general, we can distinguish between data sharing as cooperation among companies along value and supply chains to support, enable or optimize on a vertical level, and data sharing as horizontal cooperation among companies to enable new business models by generating value from data [15]. Data pooling is used, for example, in healthcare applications to derive information from the huge but distributed amount of data in this sector [5, 10], for economic and macroeconomic forecasting [1, 2, 9] as well as for empirical social research [12–14].

3 Methods

Project Framework. Within the scope of a publicly funded research project, Fraunhofer Austria and the software and consulting company Poool developed an algorithm to predict project success based on historical project management data from six SMEs in the service sector of the marketing and consulting industry located in Germany and Austria that use Poool's cloud-based software services. We will denote these companies with the letters A to F. Although they collected data for roughly three years, the actual amount of historical data of each company is limited. The reason for this is primarily the rather small size of the companies (5–50 employees) and the resulting limited number of completed projects per year. It is well known that an insufficient amount of training data can reduce the performance of machine learning algorithms. Additionally, the project success forecast should also work for new Poool customers that do not yet have historical data. For such cases, the pooling of data from several companies could seem promising. However, the success of a project depends on many internal and external factors that can vary substantially from company to company. Therefore, even assuming that Poool's customers engage in similar activities and deliver comparable project results, it is not clear whether the predictive power of a model based on data from one company is transferable to another. The goal of this study is to assess the potential of data pooling for predicting project success.

Data Preparation and Description. After combining the datasets from the six companies, general data preparation and feature engineering activities were conducted. All Poool customers share the same core business processes, but the specific business characteristics of companies influence the characteristics of the data they produce. Company A focuses on print production, company B has a consulting focus, companies C, E and F engage in media production and company D provides services in multiple fields. Therefore, a set of rules was used to remove outlier projects specific to the respective business area. The overall dataset consists of 1468 data points (between 97 and 300 data points for all but one company that has 600 data points) with 10 features and the target class i.e., a project with positive or negative financial outcome. Financial outcome is described as a project's revenue reduced by all costs for conducting the project. The features used mainly describe general project characteristics such as budget, team size and duration and are not company or industry specific. Note that, since all companies have the same feature and label spaces, this is a case of homogeneous transfer learning [17].

General Setup. Our general setup for the project success prediction is as follows. A Random Forest Classifier [4] is trained on a fixed training dataset and evaluated on a fixed, disjoint test dataset. Since the target class distributions of most companies are imbalanced, we randomly subsample training data points to balance the two classes. For hyperparameter selection, we perform a grid search with five-fold cross validation in the training dataset. To account for the inherent randomness of the Random Forest Classifier and the additional variance due to random subsampling of training data points, we report the experiment results as the mean and standard deviation of five runs with different seeds.

4 Results

Pooling Capability Tests. Using the described procedure, six different models were trained, each time excluding the data of one company, the target company. The target companies' data points were used as test dataset and the performance was evaluated using the weighted average F1 score. In the cases where companies B to F were the target companies, the F1 scores ranged from 0.66 to 0.79, whereas in the case where company A was the target company, the F1 score was much lower. In fact, it is comparably close to the F1 score of a model that predicts by random guessing. In the literature this phenomenon is referred to as negative transfer [16]. These results can be seen in the right most column (pool size five) in Fig. 1. The implication of this result is that companies that have not collected own data so far, can possibly benefit from a model developed with the data of the others.

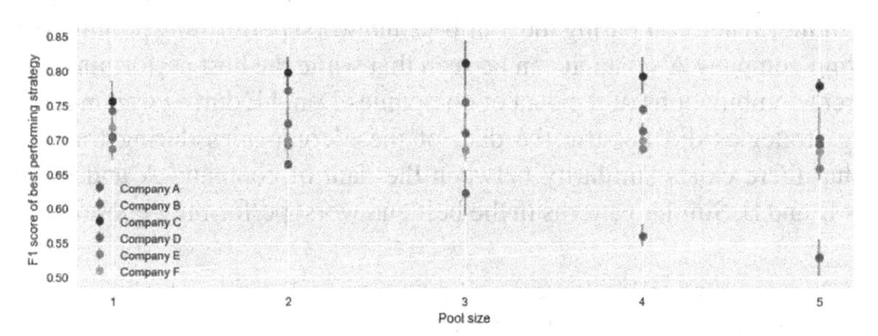

Fig. 1. The figure shows the impact of the pool size on the performance of the respective best performing pooling strategy (out of all strategies with the same pool size).

This observation raises the question whether the performance of the prediction of company A's data in terms of F1 score could be improved by choosing a different pooling strategy, i.e., a different combination of data in the pool. To test that, a separate model was trained for every possible combination of the remaining companies and tested on the data of company A. In fact, the performance of the Random Forest Classifier can be improved by choosing a different pooling strategy. For predicting data of company A, the best pooling strategy can achieve an F1 score of 0.703 ± 0.007 – an increase of roughly 0.15. Curiously, the strategy, which was used to achieve this result, uses only data points from company D. This result raises several questions: Can alternative pooling strategies improve the performance of the classifiers compared to the naive pooling strategy (a model trained on the data of all source companies) for other target companies as well? What is the impact of the pool size on the performance of the classifier with respect to F1 score? Is there an optimal pooling strategy[1] for every company? To further investigate the impact of the pooling strategy, we repeated the experiment for all other companies (as targets). Figure 1 shows the F1 scores of the best performing strategies (out of all strategies with the same pool size) for each pool size and target company. It can be seen that company A is the only company, where the performance of the best pooling strategy varies substantially with the pool size. For all other companies the variation is rather small compared to its respective F1 score. Furthermore, for all companies but C, the best performing strategy has a pool size of one or two. This suggests that the pool size of the best performing strategy and hence the optimal pool size for each company, is rather small. Note that the only exception, the pool size of the best performing strategy for predicting data of company C, has pool size three. Despite these findings, larger pool sizes still have certain advantages. As can be seen in Fig. 2, increasing the pool size increases the F1 score of the worst performing strategy. At the same time, it leads to a decrease in the standard deviation of the performances of all strategies with the same pool size. This means that as pools get larger the average performance (estimate) of the classifier stabilizes with respect to slightly varying strategies. In other words, the difference between the best and worst performing strategy with the same pool size decreases with increasing pool sizes while at the same time the F1 score of the worst performing strategy increases. While acceptable performance of pooling strategies with large pool sizes is achieved for almost all target companies, for company A the performance is still close to the F1 score of a randomly guessing model. Subsequently, we investigated the best and worst performing strategies in detail. In line with what one might expect, these strategies show patterns with respect to which company's data is used for training. When comparing the four best and worst performing pooling strategies for predicting company A's data, it can be seen that while the best performing strategies used different combinations of the data of companies D and E during training, the worst performing strategies did not use the data of these companies during training. This suggests that there exists similarity between the data of company A and the data of companies D and E. Similar patterns in the best and worst performing pooling strategies

[1] Note that since we only consider strategies where a company's data is not split, we only look at a small subset of all possible pooling strategies. Moreover, we denote a strategy as "optimal", if it achieves the best performance with respect to the subset of strategies we consider, which will most likely not be the global optimum.

can be found for the prediction of other companies' data as well. Interestingly, the companies that make up a well-performing strategy can be found by evaluating the predictive performance of a model trained exclusively on the data of a single source company. When comparing the predictive performances of the single source company strategies, we find that the best performing pooling strategies for predicting the data of company A involved only data from companies that have a good single source company strategy performance, namely companies D and E. Similar results are found for all other companies. This suggests that the performance of a single source company strategy can be used as a measure for the predictive power of one company's data for another company's data. In fact, similar methods are used for estimating domain similarity and for judging the utility of synthetic data [3, 8, 16].

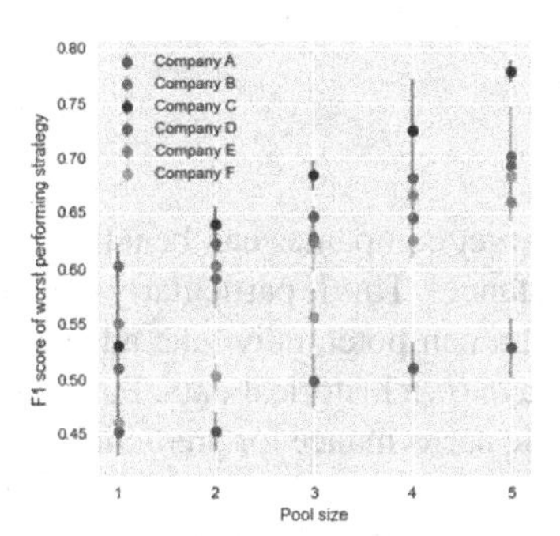

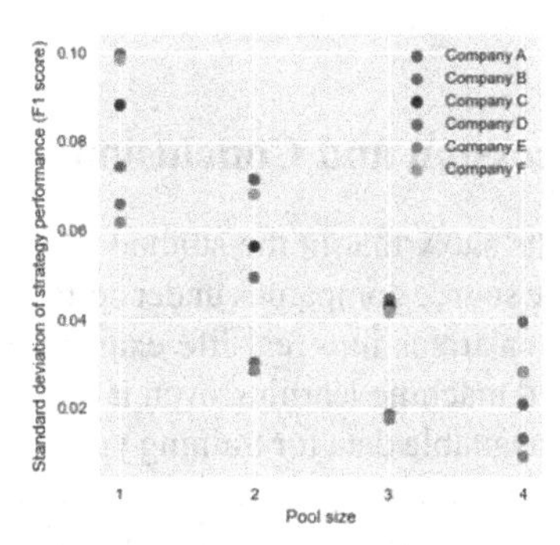

Fig. 2. The left hand side shows the performance of the worst performing strategy for each pool size. The right hand side shows the deviation of the strategy performances from the mean for each pool size.

Within the subset of pooling strategies that we considered, there is no single "one-fits-all" pooling strategy. Using all available source data for training yields acceptable performance for predicting the data of most companies, however, for company A, this approach does not work. This indicates that the data of company A is not similar enough to the data of the other (source) companies for the use of this strategy. However, a model that is trained using the data of company D predicts the data of company A sufficiently well, which motivates a further analysis of the data of these two companies.

Table 1. Overview of relative feature importance (FI) and Pearson correlation coefficients

	Company A		Company D		Pooled rest	
Feature	FI (rank)	Pearson	FI (rank)	Pearson	FI (rank)	Pearson
Budget/day	.33 (1)	.12	.23 (1)	.28	.12 (5)	.089
Budget/empl.	.13 (3)	.21	.19 (2)	.23	.15 (1)	.23
External costs	.14 (2)	−.01	.01 (9)	−.13	.12 (4)	−.21
Team size	.07 (7)	−.1	.07 (7)	−.14	.14 (2)	−.42

Data Similarity Analysis. To further investigate similarities between companies A and D and differences between them and the other companies, an in-depth analysis of their data was conducted. Table 1 presents the relative feature importance [7] of their respective models and the Pearson correlation of the respective feature with the target class. For company A, the relative importance of budget per day is significantly higher than for the other companies under investigation. Yet, budget per day also ranks highest within the model for company D supporting the evidence that companies A and D share common characteristics. The cost feature, which is used to describe the proportion of external costs in a project, a characteristic with usually high values for company A's business area (print production), has a high relative importance for company A while being mostly irrelevant (company D) or differently correlated with the target feature (pooled dataset) in other datasets. The relative feature importance scores of the model trained on the pooled dataset are similarly distributed among most of the features.

5 Discussion and Conclusion

The results show that in the studied scenario all target companies can benefit from the data of the source companies under certain circumstances. This is particularly interesting as it demonstrates in a real-life example that SMEs can potentially take advantage of supervised machine learning even if they do not have own historical data. Furthermore, using all available data for training yields acceptable performance for predicting the data of most companies. However, differences in source and target distributions can lead to failure of data pooling and need to be considered. A potential mitigation strategy is to find specific pooling strategies for groups of companies that share similar business and data characteristics. Moreover, performance estimates from evaluating single company data strategies can be helpful for guessing good pooling strategies as our analysis of the respective best and worst pooling strategies shows. However, to estimate the similarity between the source and target companies' datasets, at least a small amount of historical data is needed. In our scenario, the optimal pooling strategy was found by computing all possible pooling combinations, which is not feasible for a larger number of companies. In these cases, other transfer learning methodologies may be useful [17]. As project-based work is very common in the service sector of the marketing industry and the features used for describing projects in this study are not company or project-type-specific, we assume that the results are largely applicable to the marketing and project-based industry in general. Poool's cloud-based software services enabled us to combine the data of different companies. To benefit from industry-wide and large-scale data pooling in the future, frameworks that ensure trustworthy data exchange and data sovereignty are indispensable [10]. The importance of initiatives such as International Data Spaces and Gaia-X can therefore not be overestimated, when it comes to fully harnessing the potential of data while addressing the security and privacy concerns of participants.

Acknowledgement. This research was supported by the Tyrolean provincial government and the Austrian Research Promotion Agency under the projects my Office ML (F.22742/18-2020) and Digital Innovation Hub West (873857 and F.17913/19-2020).

References

1. Baltagi, B.H., Griffin, J.M.: Pooled estimators vs. their heterogeneous counterparts in the context of dynamic demand for gasoline. J. Econ. **77**(2), 303–327 (1997)
2. Bolhuis, M., Rayner, B.: The More the Merrier? A Machine Learning Algorithm for Optimal Pooling of Panel Data (2020). https://doi.org/10.2139/ssrn.3583406
3. Borji, A.: Pros and cons of GAN evaluation measures. Comput. Vis. Image Underst. **179**, 41–65 (2019)
4. Breiman, L.: Random forests. Mach. Learn. **45**(1), 5–32 (2001)
5. Cuggia, M., Combes, S.: The French health data hub and the German medical informatics initiatives: two national projects to promote data sharing in healthcare (1) (2019). https://doi.org/10.1055/s-0039-1677917
6. Fuchs, B., Schumacher, A., Eggeling, E., Schlund, S.: Fraunhofer Austria KI-Studie: Künstliche Intelligenz in Österreichs Unternehmen (2022)
7. Louppe, G.: Understanding random forests: from theory to practice (2014). https://doi.org/10.13140/2.1.1570.5928
8. Hittmeir, M., Ekelhart, A., Mayer, R.: On the utility of synthetic data. In: Proceedings of the 14th International Conference on Availability, Reliability and Security, pp. 1–6. ACM, New York (2019). https://doi.org/10.1145/3339252.3339281
9. Hoogstrate, A.J., Palm, F.C., Pfann, G.A., Hoogstrate, A.J.: Pooling in dynamic panel-data models: an application to forecasting GDP growth rates. J. Bus. Econ. Stat. **18**(3), 274 (2000)
10. Hulsen, T.: Sharing is caring-data sharing initiatives in healthcare. Int. J. Environ. Res. Public Health **17**, 9 (2020)
11. IDSA: Sharing Data While Keeping Data Ownership. The Potential of IDS For The Data Economy (2018)
12. Lorenz, F.O., Simons, R.L., Conger, R.D., Elder, G.H., Johnson, C., Chao, W.: Married and recently divorced mothers' stressful events and distress: tracing change across time. J. Marriage Fam. **59**(1), 219 (1997)
13. McArdle, J.J., Hamagami, F., Meredith, W., Bradway, K.P.: Modeling the dynamic hypotheses of Gf–Gc theory using longitudinal life-span data. Learn. Individ. Differ. **12**(1), 53–79 (2000)
14. McArdle, J.J., Prescott, C.A., Hamagami, F., Horn, J.L.: A contemporary method for developmental-genetic analyses of age changes in intellectual abilities. Dev. Neuropsychol. **14**(1), 69–114 (1998)
15. Otto, B., et al.: Data ecosystems. Conceptual foundations, constituents and recommendations for action (2019)
16. Zhang, W., Deng, L., Zhang, L., Wu, D.: A survey on negative transfer. IEEE Trans. Neural Netw. Learn. Syst. (2021)
17. Zhuang, F., et al.: A comprehensive survey on transfer learning. Proc. IEEE **109**(1), 43–76 (2020)

Unsupervised Alignment of Distributional Word Embeddings

Aïssatou Diallo[1](✉) and Johannes Fürnkranz[2]

[1] Department of Computer Science, University College London, London, UK
a.diallo@ucl.ac.uk
[2] Computational Data Analytics, FAW, Johannes Kepler University Linz, Linz, Austria
juffi@faw.jku.at

Abstract. Cross-domain alignment plays a key role in tasks ranging from image-text retrieval to machine translation. The main objective is to associate related entities across different domains. Recently, purely unsupervised methods operating on monolingual embeddings have successfully been used to infer a bilingual lexicon without relying on supervision. However, current state-of-the art methods only focus on point vectors although distributional embeddings have proven to embed richer semantic information when representing words. This paper investigates a novel stochastic optimization approach for aligning word distributional embeddings. Our method builds upon techniques in optimal transport to resolve the cross-domain matching problem in a principled manner. We evaluate our method on the problem of unsupervised word translation, by aligning word embeddings trained on monolingual data. We present empirical evidence to demonstrate the validity of our approach to the bilingual lexicon induction task across several language pairs.

Keywords: Unsupervised alignment · Distributional embeddings · Word translation

1 Introduction

Word embedding alignment is a fundamental Natural Language Processing task that aims at finding the correspondence between two sets of word embeddings. Word embeddings are vectorial representations of words capable of capturing the context of a word in a document, semantic and syntactic similarity as well as its relation to other words. Therefore, each embedding space exhibits different characteristics based on the semantic differences in the source of information provided as input. However, it has been first observed in [24] that continuous word embeddings exhibit similar structures across languages, even for distant ones such as English and Vietnamese. For this reason, the task of aligning two clouds of points is a crucial problem in this specific setting.

Often, the set of embeddings to align are in different languages, i.e., we face the task of cross-lingual alignment. Loosely speaking, given a source-target

R. Bergmann et al. (Eds.): KI 2022, LNAI 13404, pp. 60–74, 2022.
https://doi.org/10.1007/978-3-031-15791-2_7

language pair, the goal is to find a mapping that goes from the embedding space of one language to the embedding space of the other language. An example related to Natural Language Processing (NLP) is the task of unsupervised word translation. In this setting, the learning process can be seen as a generalization of the unsupervised cross-domain adaptation problem [5, 15, 22, 31] (Fig. 1).

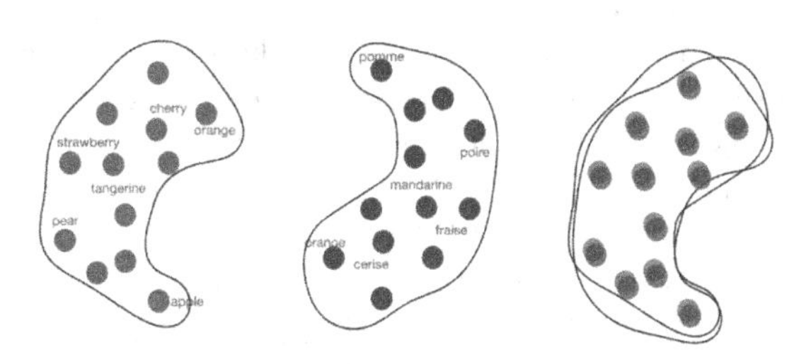

(a) Point vector embedding alignment.

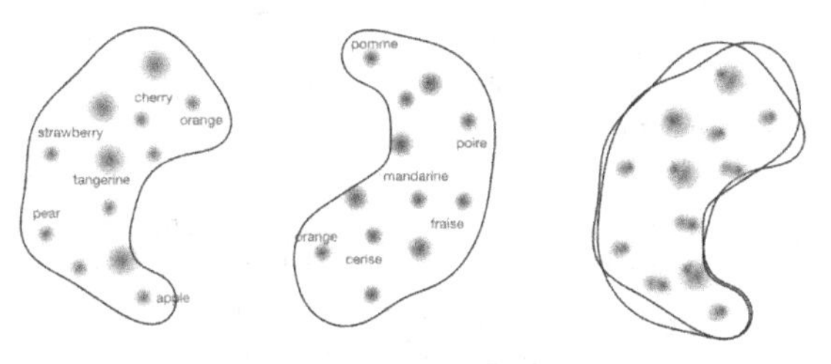

(b) Distributional vector embedding alignment.

Fig. 1. Unsupervised embedding alignment for two clouds of points in two different languages (English and French.)

Several early studies have relied on supervision from a bilingual dictionary in the form of few anchor points in order to induce the learning of the mapping [2, 17, 19]. However, recently many unsupervised approaches have been proposed and have obtained compelling results [1, 3, 16, 18, 20, 33, 34]. The unsupervised approaches frame the problem as a distance minimization between distributions using various distances, adversarial training, or domain adaptation. Generally speaking, all these methods build on the observation that mono-lingual word embeddings, or distributed representations of words, show similar geometric properties across languages. Another key point is the nature of the representation. Like other types of embeddings, word embeddings develop in two directions: point embeddings and probabilistic embeddings.

Point embeddings are powerful and compact representations that deterministically map each word into a single point in a semantic space, where the semantic similarity and other symmetric word relations are effectively captured by the relative position of points. Despite these positive properties, this projection into a single point in the embedding space brings also several limitations. Most importantly, it has been shown that a single point vector struggles to naturally model entailment among words (e.g., animal entails dog but not vice versa) or other asymmetric relations. Moreover, point vectors are typically compared by dot products, cosine-distance or Euclidean distance, which are not well suited for carrying asymmetric comparisons between objects (as is necessary to represent relations such as inclusion or entailment). Asymmetries can reveal hierarchical structures among words that can be crucial in knowledge representation and reasoning [28]. Additionally, the point vector representation fails to express the uncertainty about the concepts associated with a specific word (Fig. 2).

On the other hand, *distributional embeddings* represent each word as a probability distribution, such as a Gaussian. Such a representation is innately more expressive having the ability to additionally capture semantic uncertainties of words (as their geometric shapes) to represent words more naturally and more accurately than point vectors [32]. They allow mapping each word to soft regions in space in a manner that facilitates the modeling of uncertainty, inclusion and entailment. Nevertheless, all the approaches for unsupervised alignment of word embedding focused on point vector.

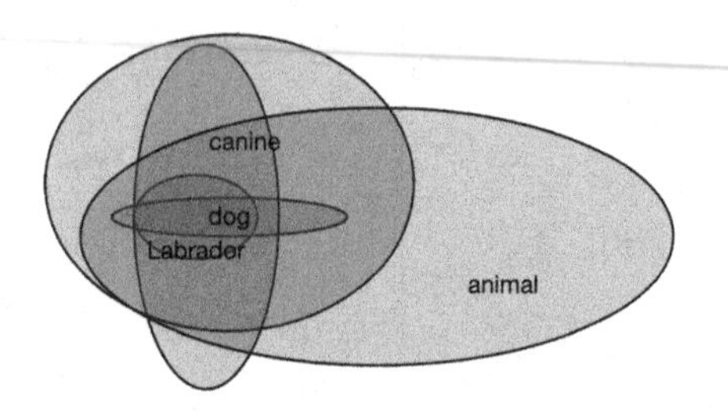

Fig. 2. Illustration of diagonal variances. Each word is defined by the position of its mean vector in the space and the dispersion is indicated by the variance. The more specific word *Labrador* has a smaller variance than the more general categories *animal* or *canine*.

In this paper, we propose an approach for aligning embedding spaces for a source and a target language in an unsupervised manner that is suited for a large set of embeddings. In particular, our algorithm shares similarities with the work of [16] where a non-linear transformation and an alignment between two point clouds are jointly learned. Experiments show the validity of the proposed approach on the bilingual lexicon induction benchmark.

The paper is organized as follows: we first discuss related works that deal with point-vector and distributional embedding models as well as alignment of word embeddings with different degrees of supervision. Then, we formulate the problem and introduce the necessary notation used throughout the paper. Finally, we present our experimental setup and discuss the results obtained.

2 Motivation and Related Work

In this section, we briefly review the relevant state-of-the-art in this area, starting first with point-based and distributional embeddings in NLP, and moving then to the problem that we study in this paper, namely the alignment of word embeddings, where we briefly recapitulate supervised and unsupervised approaches.

2.1 Point-Based Word Embeddings

One of the key problems in machine learning and natural language processing has been computing meaningful representation for high-dimensional complex data. This has been an active research area, from the traditional non-neural isometric embeddings [6,8] to the more recent and complex methods [23,25,27]. And the most widely used algorithms for learning point-based word embeddings are the continuous bag of words and skip-gram models [23], which use a series of optimization methods such as negative sampling and hierarchical softmax [26]. Another approach for learning word embeddings is through factorization of word co-occurrence matrices such as GloVe embeddings [27]. This mechanism of matrix factorization has been proved to be intrinsically linked to skip-gram and negative sampling.

2.2 Probabilistic Embedding

The work of [32] established a new trend in the representation learning field by proposing to embed words as probability distributions in $\mathbb{R}^d$. In fact, recognizing that the point-based world struggles to naturally model entailment among words (e.g., animal entails cat but not the reverse) or other asymmetric relations, probabilistic embedding emerges as a method to capture uncertainties of words, which can better capture word semantics and to express asymmetrical relationship more naturally (than dot product or cosine similarity in the point-based approach). Representing objects in the latent space as probability distributions allows more flexibility in the representation and even express multi-modality. In fact, point-vector embeddings can be considered as an extreme case of probabilistic embeddings, namely a Dirac distribution, where the uncertainty is collapsed into a single point (Fig. 3).

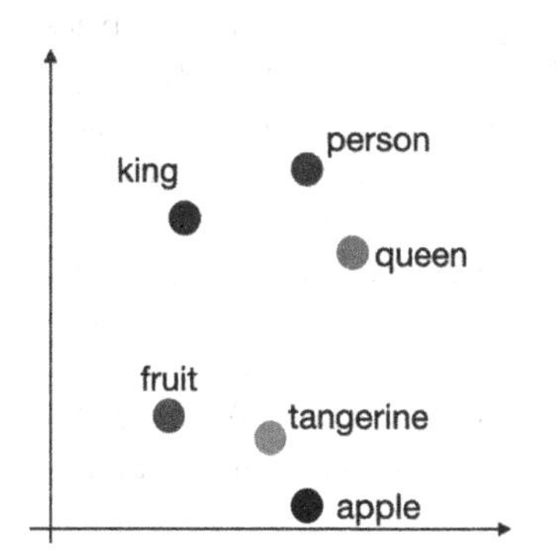

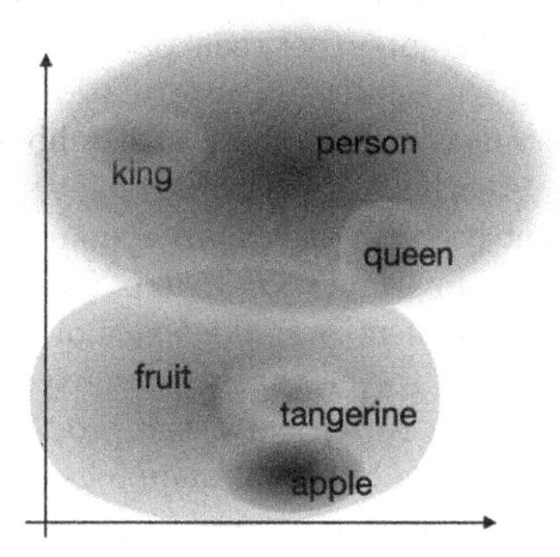

Fig. 3. Broader and more common terms have a wider dispersion than more specific ones. This characteristic is lacking in point-vector embeddings.

2.3 Minimally Supervised Alignment of Word Embeddings

As stated earlier, word embeddings allow representing word relations in a metric space. Learning the projection of a word embedding space for a given language into another embedding space is useful in many applications, in particular in aligning vocabularies for different languages. Learning these cross-lingual mappings has initially been done using seed dictionaries. In fact, most early works assumed some, albeit minimal, amount of parallel data [10,12,24]. [24] proposes a mapping from one space to the other based on the least-squares objective whereas [2,12,29] aim at finding an orthogonal transformation. Other works fall under the minimally supervised category but aim at finding a common space on which to project both sets of embeddings [11,21].

2.4 Fully Unsupervised Alignment of Word Embeddings

In recent works in the area, it has been shown that fully-unsupervised methods are able to perform on par with their supervised counterparts. The first unsupervised bilingual alignment approaches [4,20,33] were based on Generative Adversarial Networks (GANs) [14]. These methods learn a linear transformation to minimize the divergence between a target distribution (e.g. Spanish word embeddings) and a source distribution (the English word embeddings projected into the Spanish space). In the recent literature, a range of unsupervised approaches that do not rely on the use of GANs has been proposed [2,16]. Our approach relates more to those methods. [2] introduced a very simple, related initialization method that is, like our proposal, also based on Gromov-Wasserstein distances between nearest neighbors: they use these second-order statistics to build a seed dictionary directly by aligning nearest neighbors words across languages. [1] propose to learn the doubly stochastic $\boldsymbol{Y}$, the matrix that determines the mapping between the words of the languages to be aligned, as a transport mapping between the metric spaces of the words in the source and the target languages. They optimize the Gromov-Wasserstein (GW) distance, which measures how distances between pairs of words are mapped across languages. In brief, [1] learn a linear transformation to minimize Gromov-Wasserstein distances of distances between nearest neighbors, in the absence of cross-lingual supervision.

Another line of work of interest attempts to solve the unsupervised alignment problem as a domain adaptation task [31]. Their formulation searches the optimal permutation matrix for a limited number of items, specifically the 20000 most frequent over the space of doubly stochastic matrices. They rely on a Riemannian solver that allows exploiting the geometry of the doubly stochastic manifold. Empirically, the proposed algorithm outperforms the GW algorithm for learning bilingual mappings. Nevertheless, their approach is computationally more expensive.

However, all these approaches rely on point-vectors. In this paper, we argue that an unsupervised approach for aligning Gaussian embeddings can be beneficial because these types of embeddings have been proven to encode relations that normal point-vector fails to encode and could be particularly well suited for low-resource languages.

3 Approach

In this section, we describe the unsupervised alignment problem and the solution strategy for dealing with Gaussian embeddings.

3.1 Problem Formulation

In the cross-lingual alignment problem, we are given a pair of source-target languages with vocabularies V_x, with $|V_x| = n$ and V_y, with $|V_y| = m$, respectively. These vocabularies are represented by word embeddings $\boldsymbol{X} \in \mathbb{R}^{n \times d}$ and $\boldsymbol{Y} \in \mathbb{R}^{m \times d}$. The goal of the problem in its classical form is to find a mapping between the set of source embeddings and target embeddings without parallel data. In this work, we tackle the problem of finding a mapping between sets of embeddings from a pair of languages but the inputs are not point-vectors. A Gaussian embedding can be seen as a generalization of point embeddings. Concretely, Gaussian embeddings are the result of representing data points as probability distributions, namely Gaussian measures in $\mathbb{R}^d$. Each Gaussian representation $w \sim \mathcal{N}(\mu, \Sigma)$ is a tuple of a mean $\mu \in \mathbb{R}^d$ (the location vector) and a covariance matrix $\Sigma \in \mathbb{S}^d$, the set of positive semi-definite $d \times d$ matrices. The covariance matrix can be seen as the dispersion that represents the uncertainty around the position of the location vector. In this work, we focus on diagonal Gaussian embeddings, which are most used in the literature on probabilistic embeddings, but our approach can easily be extended to the general case with little effort.

The problem, then becomes, given a pair of sets of Gaussian embeddings from a source language $\mathcal{X}$ represented by $\boldsymbol{M}_x \in \mathbb{R}^{n \times d}$ and $\boldsymbol{\Sigma}_x \in \mathbb{R}_+^{n \times d}$ and from a target language $\mathcal{Y}$ $\boldsymbol{M}_y \in \mathbb{R}^{n \times d}$ and $\boldsymbol{\Sigma}_y \in \mathbb{R}_+^{n \times d}$, find a mapping $T : \mathcal{X} \rightarrow \mathcal{Y}$ such that $T(x_i \in \boldsymbol{M}_x) \approx y_j \in \boldsymbol{M}_y$.

In the next section, we begin by discussing the solution by [16] and then present our adaptation to deal with Gaussian embeddings.

3.2 Orthogonal Procrustes

The problem of finding a linear mapping between two clouds of matched vectors is known as *Procrustes*. In the classical form, it is described as:

$$\min_{\boldsymbol{W} \in \mathbb{R}^{D \times D}} \|\boldsymbol{X}\boldsymbol{W} - \boldsymbol{Y}\|_F^2$$

where $\boldsymbol{W}$ is the learned mapping and $\|\cdot\|_F$ is the Frobenius norm. This technique has been successfully applied in different fields, from analyzing sets of 2D shapes to learning a linear mapping between word vectors in two different languages with the help of a bilingual lexicon [24]. Constraints on the mapping $\boldsymbol{W}$ can be further imposed to suit the geometry of the problem. An appropriate choice of the space for the mapping T represented by $\boldsymbol{W}$ in the general case will be the space of the orthogonal matrices (rotations and reflections). Hence the problem becomes *Orthogonal Procrustes*:

$$\min_{Q \in \mathcal{O}_D} \|\boldsymbol{XQ} - \boldsymbol{Y}\|_F^2 \tag{1}$$

where $\mathcal{O}_d$ is the space of orthogonal matrices defined as :

$$\mathcal{O}_d = \{\boldsymbol{W} \in \mathbb{R}^{d \times d} | \boldsymbol{W}^T \boldsymbol{W} = \boldsymbol{I}\}. \tag{2}$$

The key advantage is that this problem has a closed-form solution. In fact, given the singular value decomposition of $\boldsymbol{XY}^T$ in $\boldsymbol{UDV}^T$, the optimal solution is

$$\boldsymbol{Q}_{opt} = \boldsymbol{UV}^T. \tag{3}$$

3.3 Wasserstein Procrustes

However, the Eq. (1) represents the supervised alignment problem, in which the learner is given a pair of sets of embedding correctly matched. If we generalize the problem, to the case in which the learner does not have access to a pair of matching embeddings, the problem at hand becomes:

$$\min_{Q \in \mathcal{O}_D, P \in \mathcal{P}_d} \|\boldsymbol{XQ} - \boldsymbol{PY}\|_F^2 \tag{4}$$

In this general case, the permutation matrix $\boldsymbol{P}$ that represents the matching is also unknown. [16] tackle this problem by jointly learning $\boldsymbol{P}$ and $\boldsymbol{W}$. While the overall problem is non-convex and computationally expensive, they propose an efficient stochastic algorithm to solve the problem and a convex relaxation which is used as an initialization for their algorithm. This convex relaxation, namely the Gold-Rangarajanng [13] relaxation is a convex approximation of the NP-hard matching problem and can be solved with the Frank Wolfe algorithm. Loosely speaking, once an initial transformation is obtained, it is used for learning the singular value decomposition. Then, the authors propose a stochastic approach in which a batch of vectors is sampled from both languages, at each step t. This is motivated by the fact that the dimension of the permutation matrix $\boldsymbol{P}$ scales quadratically with the number of points n. The approach consists in alternating the full minimization of Eq. (4) in $\boldsymbol{P}$ and a gradient-based update in $\boldsymbol{Q}$.

3.4 Wasserstein Procrustes for Gaussian Embedding

In order to adapt the learning problem for Gaussian distributions as inputs, we re-frame the problem described by Eq. (4):

$$\|\boldsymbol{M}_x \boldsymbol{R} - \boldsymbol{P}\boldsymbol{M}_y\|_F^2 + \|\boldsymbol{\Sigma}_x - \boldsymbol{P}\boldsymbol{\Sigma}_y\|_F^2 \tag{5}$$

As stated earlier, $\boldsymbol{M}_x$ and $\boldsymbol{M}_y$ represent the location (mean) vectors of the source and target Gaussian embeddings respectively, whereas $\boldsymbol{\Sigma}_x$ and $\boldsymbol{\Sigma}_y$ are

Algorithm 1: Unsupervised alignment of Gaussian embeddings

1 **for** $t = 1$ **to** T **do**
2 Draw $\boldsymbol{X}_t$ from $\boldsymbol{M}_x$ and $\boldsymbol{Y}_t$ from $\boldsymbol{M}_y$, of size b
3 Given the current $\boldsymbol{R}_t$, compute $\boldsymbol{P}_t$ between $\boldsymbol{X}_t$ and $\boldsymbol{Y}_t$
4 $\boldsymbol{P}_t = \underset{\boldsymbol{P} \in \mathcal{P}_b}{\arg\max}\ \mathrm{Tr}(\boldsymbol{R}_t \boldsymbol{X}_t \boldsymbol{P}_t \boldsymbol{Y}_t)$
5 Compute the gradient $\boldsymbol{G}_t$ w.r.t $\boldsymbol{R}_t$:
6 $\boldsymbol{G}_t = -2\boldsymbol{X}_t \boldsymbol{P}_t \boldsymbol{Y}_t$
7 Gradient step:
8 $\boldsymbol{R}_{t+1} = (\boldsymbol{R}_t - \alpha \boldsymbol{G}_t)$
9 Project on the set of orthogonal matrices:
10 $\boldsymbol{R}_{t+1} = \prod_{\mathcal{O}_d}(\boldsymbol{R}_{t+1}) = \boldsymbol{U}\boldsymbol{V}^T$
11 **for** $i = 1$ **to** L **do**
12 Draw $\boldsymbol{X}_i$ from $\boldsymbol{X}_t \boldsymbol{R}_t$ and $\boldsymbol{Y}_i$ from $\boldsymbol{P}_t \boldsymbol{Y}_t$
13 Draw $\boldsymbol{C_x}_i$ from $\boldsymbol{\Sigma}_t$ and $\boldsymbol{C_y}_i$ from $\boldsymbol{P}_t \boldsymbol{\Sigma}_t$
14 $\boldsymbol{P}_i = \underset{\boldsymbol{P} \in \mathcal{P}_b}{\arg\max}\ \mathrm{Tr}(\boldsymbol{C_x}_i^T \boldsymbol{C_y}_i)$
15 Compute the gradient $\boldsymbol{G}_i$ w.r.t $\boldsymbol{R}_i$:
16 $\boldsymbol{G}_i = -2\boldsymbol{X}_i \boldsymbol{P}_i \boldsymbol{Y}_i$
17 Gradient step:
18 $\boldsymbol{R}_{i+1} = (\boldsymbol{R}_i - \alpha \boldsymbol{G}_i)$
19 Project on the set of orthogonal matrices:
20 $\boldsymbol{R}_{i+1} = \prod_{\mathcal{O}_d}(\boldsymbol{R}_{i+1}) = \boldsymbol{U}\boldsymbol{V}^T$
21 **end for**
22 **end for**

the diagonal covariance matrices. The transformation $\boldsymbol{R}$ is derived solely from the first term of the equation. The intuition comes from the fact that the covariances represent, from a geometrical point of view, the dispersion of the embeddings. However, the permutation matrix $\boldsymbol{P}$, which identifies the matching should be based also on the covariances. This is justified by the fact that monolingual embeddings exhibit similar geometric properties across languages and taking into account the covariances of the embedding acts as a regularization of the optimization problem. Concretely, the permutation matrix $\boldsymbol{P}_t \in \mathcal{P}_d$ at step t is derived from $\boldsymbol{R}^T \boldsymbol{X}^T \boldsymbol{P} \boldsymbol{Y}$ and $\boldsymbol{\Sigma}_x^T \boldsymbol{\Sigma}_y$. The procedure is illustrated in Algorithm 1.

The second term of Eq. (5) is the contribution of the dispersion term to the Wasserstein distance of the Gaussian distributions. If we assume that the geometrical similarity of the embedding spaces is maintained across languages, then we can reasonably expect that corresponding embeddings in different languages will behave in the same way. As an example, we can consider words that describe a categorisation of elements such as the words “fruit” or “animal”. We know that the Gaussian representations of these words have a greater dispersion than their more specific counterparts, such as “pear” or “dog”. We can reasonably expect the same phenomenon to occur across languages. We propose to optimize this problem in steps:

- First, learn an optimal orthogonal matrix $\boldsymbol{R}_t$ and permutation matrix $\boldsymbol{P}_t$ only using the means of the Gaussian embeddings.
- Then, given this initial mapping and matching applications, refine the permutation matrix $\boldsymbol{P}_t$ with few iterations to match also the covariances and used this new learned $\boldsymbol{P}_i$ to derive $\boldsymbol{R}_i$.

The naive approach to optimize Eq. (5) might be to add a term to take into account the covariances at step denoted by line 3 in Algorithm 1. However, the magnitude of the cost matrix derived from the covariances is too small, and we found that the best approach will be a nested gradient descent. First, we estimate optimal $\boldsymbol{P}$ and $\boldsymbol{R}$ only from the location vectors, then we refine them with a few iterations $L << T$.

In order to quantitatively assess the quality of our approach, we consider the problem of bilingual lexicon induction for Gaussian embedding. In the next section, we describe the procedure to generate our mono-lingual probabilistic embeddings and we investigate the use of the covariance to learn the unsupervised alignment.

4 Experiments

In the following section, we present the experimental evaluation of our approach. Through this step, we seek to understand the impact of the covariance in the optimization dynamics and to evaluate the performance of our approach for the task of cross-lingual word embedding translation.

4.1 Data Generation

The first step for any unsupervised alignment algorithm is to provide the source and target embeddings. To the best of our knowledge, there aren't any trained Gaussian mono-lingual embeddings publicly available. The standard benchmark dataset for the cross-lingual is from [20] trained with *FastText* [7] on Wikipedia dumps and parallel dictionaries for 110 language pairs. The original Wikipedia dumps were not made available, which would have made it easier to retrain Gaussian embedding. We choose the following solution: we train a model using the method described in [32] with the exception that the weights for the mean component of the model are initialized with *FastText* embeddings. We fine-tune the embedding on Wikipedia dumps for each language for 3 epochs, with a learning rate $\lambda_r = 0.05$ using Adagrad for optimization. We maintain the dimensionality of the *FastText* embedding, i.e., 300 dimensions. As generally done for language modeling, we keep only the tokens appearing more than 100 times in the text (for a total average number of $210,000$ different words for all languages used).

Table 1. P@1 on five European languages: English, French, Spanish, German and Russian. Here "en-xx" refers to the average P@1 over multiple runs when English is the source language and xx is the target language. We notice, as expected that the performance is similar for closely related pairs of languages.

	en-fr	fr-en	en-es	es-en	en-de	de-en	en-ru	ru-en
μ	68.4	69.3	67.4	71.3	62.1	59.8	**33.4**	**49.7**
(μ, Σ)	**70.5**	**71.8**	**70.8**	**73.2**	**64.1**	**60.1**	29.6	41.2

4.2 Experimental Setup

After obtaining the required monolingual embeddings we proceed as follows: we first learn an alignment solely based on the means. This will be considered as the baseline that will allow us to appreciate the influence of the covariances in the computation of the alignment. We follow the same training protocol as in [16]. More precisely, we perform 5 epochs and the batch size is doubled at the beginning of each epoch while reducing the number of iterations by a factor of 4. The first epoch of our method uses a batch size of 500 and 5000 iterations. We also use the Sinkhorn solver of [9] to compute approximate solutions of optimal transport problems, with a regularization parameter of 0.05. The number of iterations in the nested step is set at 2 and the learning rate is set at 0.1 times the learning rate used in the prior step.

Since the bilingual lexicon induction problem can be seen as a retrieval problem, the standard practice is to report the precision at one (P@1). As a criterion, we compute a direct nearest-neighbor search on the mean of the Gaussian embeddings. We tried computing a distance that will take into account the covariance matrix but we noticed that the impact on the P@1 score was negligible.

Following [1], we consider the top $n = 20,000$ most frequent words in the vocabulary set for all the languages during the training stage. The inference is performed on the full vocabulary set. The obtained results are summarized in Table 1.

4.3 Discussion

In order to qualitatively assess the contribution of the covariance matrix, the results obtained considering the covariance matrices are compared to the ones without considering the covariance matrices. Overall, the performance improves when taking into account the covariances. This can also be explained by the fact that the terms containing the covariance act as a regularization. Due to the presence of a nested step, the computational time increases slightly compared to the point-vector case. However, the number of iterations in the nested loop is small, between 2 and 5, hence it is not a dramatic increase.

One explanation for the improvement of the results when taking into account the covariances might be the refinement step. In fact, it has been observed in

[2,20] that refining the alignments improves the performance by a significant margin.

A general observation is that similar pairs of languages have similar performance overall. However, some interesting points must be taken into account for the task of unsupervised bilingual dictionary induction:

- **Impact of off-the-shelf embeddings:** We rely on the *FastText* embedding for obtaining the embedding to align for different languages. However, *FastText* is trained on approximately 16M sentences in Spanish and 1M sentences in English. The Gaussian embeddings are induced from the point vector and the performance can be explained by the quality of the starting embeddings.
- **Impact of domain difference:** having a large monolingual corpus from similar domains across languages is of vital importance. In fact, it is known that when two corpora come from different topics or domains, the performance is extremely degraded. The domain dissimilarity computed by metrics such as the Jensen-Shannon divergence is significant. The term distribution is an important factor in Gaussian embeddings since the dispersion (the variance) is the direct result of the uncertainty inherent in the dataset. This is a factor that must be taken into account for aligning Gaussian embeddings in an unsupervised manner.
- **Impact of language similarity:** The morphological typology of the language is a factor that should be considered. The main considered are: *fusional*, *agglutinative*, *isolating*. A fusional language tends to form words by the fusion (rather than the agglutination) of morphemes so that the constituent elements of a word are not kept distinct. Notable examples are the Indo-European languages. An agglutinative language has words that are made up of a linear sequence of distinct morphemes and each component of meaning is represented by its own morpheme, for example, Finnish and Turkish. Finally, an isolating language is a natural language with no demonstrable genealogical relationship with other languages, examples are Vietnamese and Classical Chinese. The nature of the language pairs should be considered as it could have a bigger impact on unsupervised alignment for distributional embedding rather than the point-vector counterpart.

In general, our results are aligned with the performance shown by [16] for the same retrieval criterion. It is worth noticing that the embeddings used are the result of quick fine-tuning, their quality is far lower than the *FastText* embedding from the MUSE dataset [20]. This is valid for all pairs of languages besides the coupling "en-ru". In this specific case, we observe that the performance of the covariance approach is worse than the alignment of the only means. This can be explained by the fact that English and Russian are distant languages and the relations expressed by the dispersion in the Gaussian embeddings in English might not correspond to the same relations in Russian. In fact, as stated earlier, the covariance matrix in Gaussian embeddings from a geometrical point of view corresponds to the uncertainty in the representation. Hence broad concepts have large variance and more focused concepts have a smaller variance or dispersion.

Another explanation can be in the fact that the main assumption for unsupervised alignment approaches is that transformation is isomorphic. However, as shown in [30] this is not true for all language pairs. And since our approach is based on enforcing similarity between concepts that might not share the same dispersion this might be an explanation for the poor performance in this specific language. A way to overcome this might be to provide a small seed dictionary and turn the problem into a minimally supervised one. Few key concepts that are geometrically related even in distant languages might work as landmark points. In the following section, we present the experimental evaluation of our approach. Through this step, we seek to understand the impact of the covariance in the optimization dynamics and to evaluate the performance of our approach for the task of cross-lingual word embedding translation.

4.4 Data Generation

The first step for any unsupervised alignment algorithm is to provide the source and target embeddings. To the best of our knowledge, there aren't any trained Gaussian mono-lingual embeddings publicly available. The standard benchmark dataset for the cross-lingual is from [20] trained with *FastText* [7] on Wikipedia dumps and parallel dictionaries for 110 language pairs. The original Wikipedia dumps were not made available, which would have made it easier to retrain Gaussian embedding. We choose the following solution: we train a model using the method described in [32] with the exception that the weights for the mean component of the model are initialized with *FastText* embeddings. We fine-tune the embedding on Wikipedia dumps for each language for 3 epochs, with a learning rate $\lambda_r = 0.05$ using Adagrad for optimization. We maintain the dimensionality of the *FastText* embedding, i.e., 300 dimensions. As generally done for language modeling, we keep only the tokens appearing more than 100 times in the text (for a total average number of $210,000$ different words for all languages used).

4.5 Experimental Setup

After obtaining the required monolingual embeddings we proceed as follows: we first learn an alignment solely based on the means. This will be considered as the baseline that will allow us to appreciate the influence of the covariances in the computation of the alignment. We follow the same training protocol as in [16]. More precisely, we perform 5 epochs and the batch size is doubled at the beginning of each epoch while reducing the number of iterations by a factor of 4. The first epoch of our method uses a batch size of 500 and 5000 iterations. We also use the Sinkhorn solver of [9] to compute approximate solutions of optimal transport problems, with a regularization parameter of 0.05. The number of iterations in the nested step is set at 2 and the learning rate is set at 0.1 times the learning rate used in the prior step.

Since the bilingual lexicon induction problem can be seen as a retrieval problem, the standard practice is to report the precision at one (P@1). As a criterion, we compute a direct nearest-neighbor search on the mean of the Gaussian embeddings. We tried computing a distance that will take into account the covariance matrix but we noticed that the impact on the P@1 score was negligible.

Following [1], we consider the top $n = 20,000$ most frequent words in the vocabulary set for all the languages during the training stage. The inference is performed on the full vocabulary set. The obtained results are summarized in Table 1.

5 Conclusion

This work presents a method to align Gaussian embeddings in high-dimensional space. Our approach is motivated by the fact that Gaussian embeddings have proven to possess characteristics that are not present in normal point-based vectors. We propose to include in the optimization of the Orthogonal Procrustes method via stochastic optimization a step that takes into account the difference between matched covariances. We show that our method performs better than the solely point-vector-based approach. However, we also observed that this approach might lead to a decrease in accuracy when the pair of languages considered is too distant. In fact, in that case, the approach might force distant concepts to have similar dispersion. In future work, we would like to extend this to deal with full covariance Gaussian embeddings as well as other elliptical embeddings and find a solution to overcome the issue of distant languages.

Acknowledgments. This work has been supported by the German Research Foundation as part of the Research Training Group Adaptive Preparation of Information from Heterogeneous Sources (AIPHES) under grant No. GRK 1994/1. The work was performed while Aïssatou Diallo was at Technische Universität Darmstadt.

References

1. Alvarez-Melis, D., Jegelka, S., Jaakkola, T.S.: Towards optimal transport with global invariances. In: Proceedings of the 22nd International Conference on Artificial Intelligence and Statistics (AI-STATS), pp. 1870–1879. PMLR (2019)
2. Artetxe, M., Labaka, G., Agirre, E.: Generalizing and improving bilingual word embedding mappings with a multi-step framework of linear transformations. In: Proceedings of the AAAI Conference on Artificial Intelligence, pp. 5012–5019 (2018)
3. Artetxe, M., Labaka, G., Agirre, E.: A robust self-learning method for fully unsupervised cross-lingual mappings of word embeddings. In: Proceedings of the 56th Annual Meeting of the Association for Computational Linguistics (ACL), Volume 1: Long Papers, pp. 789–798. ACL (2018)
4. Barone, M., Valerio, A.: Towards cross-lingual distributed representations without parallel text trained with adversarial autoencoders. In: Proceedings of the 1st Workshop on Representation Learning for NLP, pp. 121–126. ACL, Berlin (2016)
5. Ben-David, S., Blitzer, J., Crammer, K., Pereira, F.: Analysis of representations for domain adaptation. In: Advances in Neural Information Processing Systems, vol. 19 (2006)

6. Blitzer, J., McDonald, R., Pereira, F.: Domain adaptation with structural correspondence learning. In: Proceedings of the 2006 Conference on Empirical Methods in Natural Language Processing (EMNLP), pp. 120–128 (2006)
7. Bojanowski, P., Grave, E., Joulin, A., Mikolov, T.: Enriching word vectors with subword information. Trans. Assoc. Comput. Linguist. **5**, 135–146 (2017)
8. Brown, P.F., Della Pietra, V.J., Desouza, P.V., Lai, J.C., Mercer, R.L.: Class-based n-gram models of natural language. Comput. Linguist. **18**(4), 467–480 (1992)
9. Cuturi, M.: Sinkhorn distances: lightspeed computation of optimal transport. In: Advances in Neural Information Processing Systems, vol. 26, pp. 2292–2300. Curran Associates, Inc. (2013)
10. Dinu, G., Lazaridou, A., Baroni, M.: Improving zero-shot learning by mitigating the hubness problem. In: Workshop Track Proceedings of the 3rd International Conference on Learning Representations (ICLR), San Diego, CA, USA (2014)
11. Faruqui, M., Dyer, C.: Improving vector space word representations using multilingual correlation. In: Proceedings of the 14th Conference of the European Chapter of the Association for Computational Linguistics, pp. 462–471 (2014)
12. Gaddy, D.M., Zhang, Y., Barzilay, R., Jaakkola, T.S.: Ten pairs to tag-multilingual POS tagging via coarse mapping between embeddings. In: Proceedings of the 2016 Conference of the North American Chapter of the Association for Computational Linguistics: Human Language Technologies (NAACL-HLT), pp. 1307–1317. ACL, San Diego (2016)
13. Gold, S., Rangarajan, A.: A graduated assignment algorithm for graph matching. IEEE Trans. Pattern Anal. Mach. Intell. **18**(4), 377–388 (1996)
14. Goodfellow, I., et al.: Generative adversarial nets. In: Advances in Neural Information Processing Systems, vol. 27, pp. 672–2680. Curran Associates, Inc. (2014)
15. Gopalan, R., Li, R., Chellappa, R.: Domain adaptation for object recognition: an unsupervised approach. In: Proceedings of the International Conference on Computer Vision, pp. 999–1006. IEEE (2011)
16. Grave, E., Joulin, A., Berthet, Q.: Unsupervised alignment of embeddings with Wasserstein Procrustes. In: The 22nd International Conference on Artificial Intelligence and Statistics, pp. 1880–1890. PMLR (2019)
17. Jawanpuria, P., Balgovind, A., Kunchukuttan, A., Mishra, B.: Learning multilingual word embeddings in latent metric space: a geometric approach. Trans. Assoc. Comput. Linguis. **7**, 107–120 (2019)
18. Jawanpuria, P., Meghwanshi, M., Mishra, B.: Geometry-aware domain adaptation for unsupervised alignment of word embeddings. In: Proceedings of the 58th Annual Meeting of the Association for Computational Linguistics, pp. 3052–3058 (2020)
19. Joulin, A., Bojanowski, P., Mikolov, T., Jégou, H., Grave, E.: Loss in translation: learning bilingual word mapping with a retrieval criterion. In: Proceedings of the 2018 Conference on Empirical Methods in Natural Language Processing (EMNLP), pp. 2979–2984. ACL, Brussels (2018)
20. Lample, G., Conneau, A., Ranzato, M., Denoyer, L., Jégou, H.: Word translation without parallel data. In: Conference Track Proceedings of the 6th International Conference on Learning Representations (ICLR), Vancouver, BC, Canada (2018)
21. Lu, A., Wang, W., Bansal, M., Gimpel, K., Livescu, K.: Deep multilingual correlation for improved word embeddings. In: Proceedings of the 2015 Conference of the North American Chapter of the Association for Computational Linguistics: Human Language Technologies (NAACL-HLT), pp. 250–256 (2015)

22. Mahadevan, S., Mishra, B., Ghosh, S.: A unified framework for domain adaptation using metric learning on manifolds. In: Berlingerio, M., Bonchi, F., Gärtner, T., Hurley, N., Ifrim, G. (eds.) ECML PKDD 2018. LNCS (LNAI), vol. 11052, pp. 843–860. Springer, Cham (2019). https://doi.org/10.1007/978-3-030-10928-8_50
23. Mikolov, T., Chen, K., Corrado, G., Dean, J.: Efficient estimation of word representations in vector space. In: Workshop Track Proceedings of the 1st International Conference on Learning Representations (ICLR), Scottsdale, Arizona, USA (2013)
24. Mikolov, T., Le, Q.V., Sutskever, I.: Exploiting similarities among languages for machine translation. arXiv preprint arXiv:1309.4168 (2013)
25. Mikolov, T., Sutskever, I., Chen, K., Corrado, G., Dean, J.: Distributed representations of words and phrases and their compositionality. In: Advances in Neural Information Processing Systems, vol. 26, pp. 3111–3119 (2013)
26. Mnih, A., Hinton, G.E.: A scalable hierarchical distributed language model. In: Advances in Neural Information Processing Systems, vol. 21, pp. 1081–1088. Citeseer (2008)
27. Pennington, J., Socher, R., Manning, C.D.: Glove: global vectors for word representation. In: Proceedings of the 2014 Conference on Empirical Methods in Natural Language Processing (EMNLP), pp. 1532–1543 (2014)
28. Roller, S., Erk, K., Boleda, G.: Inclusive yet selective: supervised distributional hypernymy detection. In: Proceedings of the 25th International Conference on Computational Linguistics (COLING): Technical Papers, pp. 1025–1036 (2014)
29. Smith, S.L., Turban, D.H.P., Hamblin, S., Hammerla, N.Y.: Offline bilingual word vectors, orthogonal transformations and the inverted softmax. In: Conference Track Proceedings of the 5th International Conference on Learning Representations (ICLR). OpenReview.net, Toulon (2017)
30. Søgaard, A., Ruder, S., Vulić, I.: On the limitations of unsupervised bilingual dictionary induction. In: Proceedings of the 56th Annual Meeting of the Association for Computational Linguistics (ACL), Volume 1: Long Papers, pp. 778–788. ACL, Melbourne (2018)
31. Sun, B., Feng, J., Saenko, K.: Return of frustratingly easy domain adaptation. In: Proceedings of the AAAI Conference on Artificial Intelligence, pp. 2058–2065 (2016)
32. Vilnis, L., McCallum, A.: Word representations via Gaussian embedding. In: Conference Track Proceedings of the 3rd International Conference on Learning Representations (ICLR), San Diego, CA, USA (2015)
33. Zhang, M., Liu, Y., Luan, H., Sun, M.: Adversarial training for unsupervised bilingual lexicon induction. In: Proceedings of the 55th Annual Meeting of the Association for Computational Linguistics (Volume 1: Long Papers), pp. 1959–1970 (2017)
34. Zhang, M., Liu, Y., Luan, H., Sun, M.: Earth mover's distance minimization for unsupervised bilingual lexicon induction. In: Proceedings of the Conference on Empirical Methods in Natural Language Processing (EMNLP), pp. 1934–1945 (2017)

NeuralPDE: Modelling Dynamical Systems from Data

Andrzej Dulny(✉), Andreas Hotho, and Anna Krause

Chair for Computer Science X Data Science, University of Würzburg, Würzburg, Germany
{andrzej.dulny,andreas.hotho,anna.krause}@informatik.uni-wuerzburg.de

Abstract. Many physical processes such as weather phenomena or fluid mechanics are governed by partial differential equations (PDEs). Modelling such dynamical systems using Neural Networks is an active research field. However, current methods are still very limited, as they do not exploit the knowledge about the dynamical nature of the system, require extensive prior knowledge about the governing equations or are limited to linear or first-order equations. In this work we make the observation that the Method of Lines used to solve PDEs can be represented using convolutions which makes convolutional neural networks (CNNs) the natural choice to parametrize arbitrary PDE dynamics. We combine this parametrization with differentiable ODE solvers to form the NeuralPDE Model, which explicitly takes into account the fact that the data is governed by differential equations. We show in several experiments on toy and real-world data that our model consistently outperforms state-of-the-art models used to learn dynamical systems.

Keywords: NeuralPDE · Dynamical systems · Spatio-temporal · PDE

1 Introduction

Deep learning methods have brought revolutionary advances in computer vision, time series prediction and machine learning in recent years. Handcrafted feature selection has been replaced by modern end-to-end systems, allowing efficient and accurate modelling of a variety of data. In particular, convolutional neural networks (CNNs) automatically learn features on gridded data, such as images or geospatial information, which are invariant to spatial translation [7]. Recurrent neural networks (RNNs) such as long short-term memory networks (LSTMs) or gated recurrent units (GRUs) are specialised for modelling sequential data, such as time series or sentences (albeit now replaced by transformers) [16].

Recently, modelling dynamical systems from data has gained attention as a novel and challenging task [12,15,18,28]. These systems describe a variety

Supplementary Information The online version contains supplementary material available at https://doi.org/10.1007/978-3-031-15791-2_8.

R. Bergmann et al. (Eds.): KI 2022, LNAI 13404, pp. 75–89, 2022.
https://doi.org/10.1007/978-3-031-15791-2_8

of physical processes such as weather phenomena [22], wave propagation [12], chemical reactions [24], and computational fluid dynamics [2]. All dynamical systems are governed by either ordinary differential equations (ODEs) involving time derivatives or partial differential equations (PDEs) involving time and spacial derivatives. Due to their chaotic nature, learning such systems from data remains challenging for current models [4].

In recent years, several approaches to model dynamical data incorporating prior knowledge about the physical system have been proposed [3,17,18,20]. However, most of the models make specific assumptions about the type or structure of the underlying differential equations: they have been designed for specific problem types such as advection-diffusion problems, require prior knowledge about the equation such as the general form or the exact equation, or are limited to linear equations. In current literature only a handful of flexible approaches exist [1,10,12].

In this work we propose NeuralPDE, a novel approach for modelling spatio-temporal data. NeuralPDE learns the dynamics of partial differential equations using convolutional neural networks as summarized in Fig. 1. The derivative of the system is used to solve the underlying equations using the Method of Lines [26] in combination with differentiable ODE solvers [5]. Our approach works on an end-to-end basis, without assuming any prior constraints on the underlying equations, while taking advantage of the dynamical nature of the data by explicitly solving the governing differential equations.

The main contributions of our work are[1]:

1. We combine NeuralODEs and the Method of Lines through usage of CNNs to account for the spatial component in PDEs.
2. We propose using general CNNs that do not require prior knowledge about the underlying equations.
3. NeuralPDEs can inherently learn continuous dynamics which can be used with arbitrary time discretizations.
4. We demonstrate that our model is applicable to a wide range of dynamical systems, including non-linear and higher-order equations.

2 Related Work

NeuralODEs [5] introduces continuous depth neural networks for parametrizing an ODE. The networks are combined with a standard ODE solver for solving the ODE. NeuralODE forms the basis for our method in the same way that numerical ODE solvers are the basis for one family of numerical PDE solvers.

Many approaches for learning dynamical systems from data operate under strong assumptions about the underlying data: Universal Differential Equations (UDE) [19], Physics Informed Neural Networks (PINN) [21], and PDE Net 2.0 [17] require prior knowledge about the generating equations. UDEs use

[1] Our code will be made publicly available upon publication.

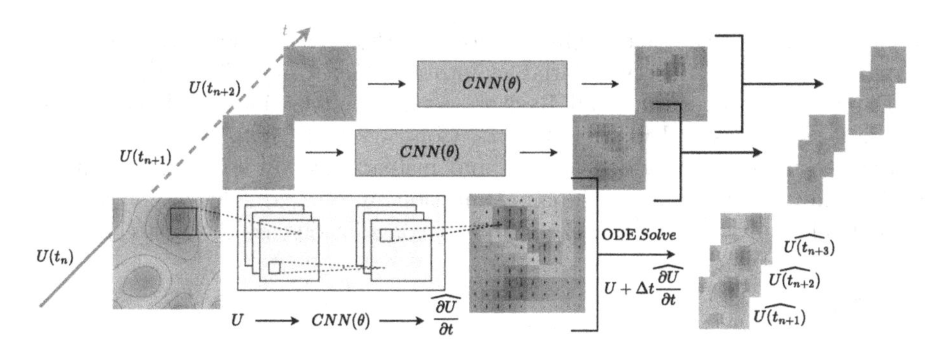

Fig. 1. NeuralPDE: combining the method of lines and NeuralODE. Our model employs a CNN to parametrize the dynamics of the system $\frac{\partial \mathcal{U}}{\partial t}$. This allows the representation of the PDE by a system of ODEs (Method of Lines) which is solved using any differentiable ODE Solver predicting multiple future states (three in the figure above). The CNN is trained using adjoint backpropagation.

separate neural networks to model each component of a PDE and have to be redesigned manually for every new PDE. PINNs are a machine learning technique for neural networks which design the loss function such, that it satisfies the initial value problem of the PDE. PDE-Net 2.0 assumes a library of available components and learns the parameters of the linear combination of these components using a ResNet-like model. Finite Volume Networks (FINN) [18] integrate the finite volume method with neural networks, but are strictly limited to advection-diffusion type equations. Our results show that none of the restrictions apply to NeuralPDE: we do not need to know the exact PDE that governed the data and make no assumption about the structure of the governing PDE.

Flexible approaches include Distana [12], hidden state models [1], and the approaches proposed by Berg [3] and Iakovlev [10]. Distana [12] describes a neural network architecture that combines two types of LSTM-based kernels: predictive kernels make predictions at given spatial positions, transitional kernels model transitions between adjacent predictive kernels. Distana proved successful in modeling wave equations and is applicable for further problems. Iakovlev et al. [10] propose using message passing graph neural networks in conjunction with Neural ODEs for modelling non-equidistant spatial grids and non-constant time intervals and evaluate their method on generated data. In contrast to our approach, they use message passing graph neural networks which are inherently computationally less efficient than our method. We provide a theoretical justification for using convolutional filters and use real-world as well as generated data for our experiments.

Berg [3] introduce a two step procedure: in the first step, the data is approximated by an arbitrary model. In the second step a differentiation operator is approximated by training a neural network on the data approximator and its derivatives up to a given order.

Ayed et al. [1] introduce the hidden state method with a learnable projection matrix to transform observed variables into a hidden state. The authors apply their method to training small ResNets as parametrizations of dynamics on toy data as well as real world data sets. Contrary to their method, we do not assume an underlying hidden process and instead directly learn the dynamic. Additionally we do not use residual connections in our parametrization, as our theoretical results show (Sect. 4) that direct convolutions are the best choice.

3 Task

Dynamical systems can be defined as a deterministic rule of evolution of a state in time [13]. At any point in time $t \in T$ the entirety of the system is assumed to be completely described by a set of space variables x from the state space X. The evolution of the system is given by the evolution function:

$$\Phi\colon T \times X \longrightarrow X \tag{1}$$

which describes the how an initial state $x_0 \in X$ is transformed into the state $x_1 \in X$ after time $t_1 \in T$ as $\Phi(t_1, x_0) = x_1$. An important property of dynamical systems is their time homogeneity, meaning the evolution of the state only depends on the current state:

$$\Phi(t_1, \Phi(t_2, x)) = \Phi(t_1 + t_2, x) \tag{2}$$

The main concern of this work is dynamical systems governed by a set of partial differential equations. These are continuous spatio-temporal systems where the state at each point in time is described by a field of k quantities on a given spatial domain $\Omega \subseteq \mathbb{R}^n$. Examples of dynamical systems that can be described by PDE include many physical systems such as weather phenomena [22] or wave propagation [12]. These systems often exhibit chaotic behaviour which makes them difficult to model with classical machine learning models [11].

We define the task of *modeling dynamical systems from data* as a spatio-temporal time series prediction task, where from one or more states used as input the model should predict the evolution of the state for the next H timesteps. As opposed to physical simulations (usually used to model such systems) where the governing equation is known, in this task the equation is assumed to be unknown. Additionally, retrieving the exact form of the equation is also not part of the task, which is the task of *learning differential operators from data* [17].

4 Neural PDE

In this section we describe our method, which combines NeuralODEs and the Method of Lines through the use of a multi-layer convolutional neural network to model arbitrarily complex PDEs. Our primary focus lies on modelling spatio-temporal data describing a dynamical system and not on recovering the exact parameters of the differential equation(s).

4.1 Method of Lines

The Method of Lines describes a numerical method of solving PDEs, where all of the spatial dimensions are discretized and the PDE is represented as a system of ordinary differential equations of one variable, for which common ODE solvers can be applied [26]. Given a partial differential equation of the form

$$\frac{\partial u}{\partial t} = f(t, u, \frac{\partial u}{\partial x}, \frac{\partial u}{\partial y}, \dots) \qquad (3)$$

where $u = u(t, x, y), x \in X, y \in Y$ is the unknown function, the spatial domain $X \times Y$ is discretized on a regular grid $X \sim \{x_1, x_2, \dots, x_N\}$ and $Y \sim \{y_1, y_2, \dots, y_M\}$. The function u can then be represented as $N \cdot M$ functions of one variable (i.e. time):

$$u(t) \simeq \begin{bmatrix} u(t, x_1, y_1) & \cdots & u(t, x_N, y_1) \\ \vdots & \ddots & \vdots \\ u(t, x_1, y_M) & \cdots & u(t, x_N, y_M) \end{bmatrix} =: \mathcal{U} \qquad (4)$$

From this representation one can derive the discretization of the spatial derivatives:

$$\frac{\partial u}{\partial x}(t, x_i, y_i) = \frac{u(t, x_{i+1}, y_i) - u(t, x_{i-1}, y_i)}{x_{i+1} - x_{i-1}} \qquad (5)$$

and

$$\frac{\partial u}{\partial y}(t, x_i, y_i) = \frac{u(t, x_i, y_{i+1}) - u(t, x_i, y_{i-1})}{y_{i+1} - y_{i-1}} \qquad (6)$$

When a fixed grid size is used for the discretization, the spatial derivatives can thus be represented as a convolutional operation [7]:

$$\mathcal{U}_x = conv(\frac{1}{2\Delta x} \begin{bmatrix} 0\,0\,0 \\ -1\,0\,1 \\ 0\,0\,0 \end{bmatrix}, \mathcal{U}) \qquad \mathcal{U}_y = conv(\frac{1}{2\Delta y} \begin{bmatrix} 0 & -1\,0 \\ 0 & 0\,0 \\ 0 & 1\,0 \end{bmatrix}, \mathcal{U}) \qquad (7)$$

where Δx and Δy are the constant grid sizes for both spatial dimensions:

$$\begin{aligned} \Delta x &= x_{i+1} - x_i, i = 1, \dots, N \\ \Delta y &= y_{i+1} - y_i, i = 1, \dots, M \end{aligned} \qquad (8)$$

Higher-order spatial derivatives can be represented in a similar fashion by a convolutional operation on the lower-order derivatives. This can be easily seen from the representation

$$\frac{\partial^{p+q} u}{\partial x^p \partial y^q} = \frac{\partial}{\partial x} \frac{\partial^{p+q-1} u}{\partial x^{p-1} \partial y^q} = \frac{\partial}{\partial y} \frac{\partial^{p+q-1} u}{\partial x^p \partial y^{q-1}} \qquad (9)$$

as higher-order derivatives are defined as derivatives of lower-order derivatives.

The original PDE can now be represented as a system of ordinary differential equations, each representing the trajectory of a single point in the spatial domain (thus the name Method of Lines):

$$\frac{d\mathcal{U}}{dt} \simeq f(t,\mathcal{U},\mathcal{U}_x,\mathcal{U}_y,\ldots) = f^*(t,\mathcal{U}) \tag{10}$$

for which any numerical ODE solver can be used.

4.2 NeuralPDEs

Our method makes the assumption that the spatio-temporal data to be modelled is governed by a partial differential equation of the form Eq. (3), but by physical constraints of the measuring process, the data has been sampled on a discrete spatial grid as in Eq. (4) and depicted in Fig. 1 on the bottom left. We also assume that the dynamics of the system only depends on the state of the system itself

$$f^*(t,\mathcal{U}) = f^*(\mathcal{U}) \tag{11}$$

As can be seen from Eq. (7), the spatial derivatives of the discretized PDE can be represented by a convolutional filter on the values of $\mathcal{U}$ and thus the whole dynamics of the system (which depends on the spatial derivatives) can be recovered from $\mathcal{U}$.

Figure 1 shows an overview of our model. Given the state of the system $\mathcal{U}_0$ at $t = t_0$, our method uses the Method of Lines representation of the underlying PDE (given by Equation (10)) and employs a multi-layer convolutional network to parametrize the unknown function f^* describing the dynamics of the system

$$\frac{d\mathcal{U}}{dt} \simeq f^*(\mathcal{U}) \simeq \mathrm{CNN}_\theta(\mathcal{U}) \tag{12}$$

Similar to NeuralODEs [5], the parametrization of the dynamics is used in combination with differentiable ODE solvers. Predictions are made by numerically solving the ODE Initial Value Problem given by

$$\begin{aligned} \frac{d\mathcal{U}}{dt} &= \mathrm{CNN}_\theta(\mathcal{U}) \\ \mathcal{U}(t_0) &= \mathcal{U}_0 \end{aligned} \tag{13}$$

for time points $t_1,\ldots,t_K$. The weights θ of the parametrization CNN_θ are updated using adjoint backpropagation as described in [5].

For higher-order equations our model is augmented with additional channels corresponding to higher order derivatives. Given the ordinary differential equation system

$$\frac{d^p\mathcal{U}}{dt^p} = f^*(t,\mathcal{U}) \tag{14}$$

we parametrize the lower-order derivatives as separate variables

$$\frac{d\mathcal{V}_1}{dt} := \frac{d\mathcal{U}}{dt} \quad \frac{d\mathcal{V}_2}{dt} := \frac{d^2\mathcal{U}}{dt^2} \quad \cdots \quad \frac{d\mathcal{V}_{p-1}}{dt} := \frac{d^{p-1}\mathcal{U}}{dt^{p-1}} \tag{15}$$

Using these auxiliary variables $\mathcal{V}_1,\ldots\mathcal{V}_{p-1}$, the original equation Equation (16) can be rewritten as a system of p first-order ODEs:

$$\frac{d\mathcal{U}}{dt} = \mathcal{V}_1 \qquad \frac{d\mathcal{V}_1}{dt} = \mathcal{V}_2 \qquad \cdots \qquad \frac{d\mathcal{V}_{p-1}}{dt} = f^*(t, \mathcal{U}) \tag{16}$$

We implement this augmentation method within NeuralPDE to represent higher-order dynamics.

5 Data

Our aim for NeuralPDE is to be applicable to the largest possible variety of dynamical data. For this, we curated a list of PDEs from related work as toy data, one simulated climate data set (PlaSim), and two reanalysis data sets (Weatherbench and Ocean Wave).

Toy Data Sets. We use several equation systems that are available from other publications as toy data sets: the advection-diffusion equation (AD), Burger's equation (B), the gas dynamics equation (GD), and the wave propagation equation (W). The equation systems and the parameters used for data generation are available from Appendix A. We use 50 simulations for different initial conditions for training, and 10 for validation and testing each.

Weatherbench [22]. Weatherbench is a curated benchmark data set for learning medium-range weather forecasting model from data. The data is derived from ERA5 archives and is accompanied by evaluation metrics, and several baseline models. Instead of the very large raw data set, we use the data set with a spacial resolution of 5.625° or 32×64. Following the recommendation of Rasp et al. [22], we use *geopotential at 500 hPa pressure* and *temperature at 850 hPa pressure* as target variables. Data from years 1979 to 2014 is used for training, 2015 and 2016 for validation and 2017 and 2018 for testing.

Ocean Wave [2]. The Ocean Wave data set contains aggregated global data on ocean sea surface waves from 1993 to 2020. The data is on an equirectangular grid with a resolution of 1/5° or approximately 20 km and with a temporal resolution of 3 h. We regrid the data to a spatial resolution of 32×64 to match Plasim and Weatherbench. We use *spectral significant wave height (Hm0)*, *mean wave from direction (VDMR)* and *wave principal direction at spectral peak (VPED)* as target variables. Data from years 1993 to 2016 is used for training, 2017 and 2018 for validation and 2019 and 2020 for testing.

PlaSim [3] The Planet Simulator (PlaSim) is a climate simulator using a medium complexity general circulation model for education and research into climate modelling and simulation. For simulation, we used the setup *plasimt21* as presented in [25], Sect. 2.1. Our simulation data contains one data point per day

[2] https://resources.marine.copernicus.eu/product-detail/GLOBAL_MULTIYEAR_WAV_001_032/INFORMATION.

[3] https://www.mi.uni-hamburg.de/en/arbeitsgruppen/theoretische-meteorologie/modelle/plasim.html.

for 200 years. We use temperature, geopotential, wind speed in x direction and wind speed in y direction at the lowest level of the simulation as our target variables. Data from the first 180 years of the simulation is used for training, the 10 following years for validation and the years 191 through 200 for testing.

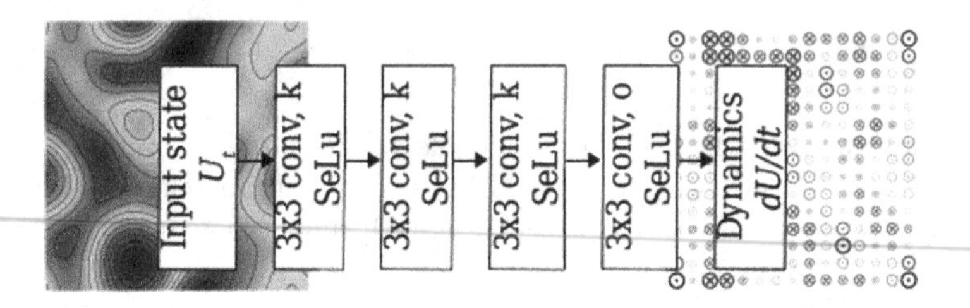

Fig. 2. Architecture of our NeuralPDE models. We use four convolutional layers with $k = 16$ channels, 3×3 kernels, SeLu activation functions, and o as the number of outputs.

6 Experiments

We train and evaluate NeuralPDE and all selected comparison methods on the seven datasets as described below.

6.1 NeuralPDE Architecture

Figure 2 shows the NeuralPDE architecture: a four layer CNN. The first convolutional layer increases the number of channels to k, the last convolutional layer reduces the number of channels down to the number of inputs. Then any number of intermediate layers each with k channels can be used to perform the main computations. After some primary experimentation we set the number of intermediate layers to 4 and the number of channels k to 16. The number of outputs o depends on the choice of equation. We train and evaluate two versions of our model using a first-order and second order dynamic as described in Eq. (16). We denote these models as *NeuralPDE-1* and *NeuralPDE-2* respectively

6.2 Comparison Models

We evaluate our model against several models from related work and simple baselines. We follow [12] in the selection of our comparison models which we shortly describe in this section. We omit models discussed in Sect. 2 which require prior knowledge about the equations.

Baseline. Persistence refers to a model that directly returns it input as output. It always takes the state at $t - 1$ as the current prediction.

CNN. Similar to [12] we use a CNN [14] consisting of multiple convolutional layers as a comparison model. We use the same architecture as for our NeuralPDE model.

ResNet. Motivated by the recent success of ResNet type architectures for modelling weather data [23], we include a simple ResNet model using identity mappings as proposed in [8]. He et al. [8] use an residual unit consisting of BatchNorm layer followed by a ReLu activation, linear layer, BatchNorm, ReLu and another linear layer, which is then connected to the input by an additive skip-connection. We stack 4 residual blocks preceded and followed by a linear CNN layer during our experiments.

Distana. [12] propose the distributed spatio-temporal artificial neural network architecture (DISTANA) to model spatio-temporal data. Their model uses a graph network with learnable prediction kernels (LSTMs) at each node to learn spatio-temporal data. We adopt the implementation of Distana from [18].

ConvLSTM. The convolutional LSTM as proposed in [27] replaces the fully connected layers within the standard LSTM model [9] with convolutional layers. It is well suited for modelling sequential grid data such as sequences of images [29], or precipitation nowcasting [27]. We thus reason it might provide a strong comparison for modelling dynamical data. We stack 4 ConvLSTM layers with 16 channels preceded and followed by a linear CNN layer for our experiments.

PDE-Net. PDE-Net 2.0 [17] is a model explicitly designed to extract governing PDEs from data. Contrary to our approach it focuses on retrieving the equation in interpretable, closed form and not on modelling the data accurately. It uses a collection of learnable convolutional filters, connected together within a symbolic polynomial network to parametrize the dynamic. Our implementation is adapted from Long et al. [17] and we use their parameters for our experiments.

Hidden State. Ayed et al. [1] propose a hidden state model, using a learnable projection to transform the input data into a higher-dimensional hidden state, where similarily to our approach, a differentiable solver is used to predict the next states. The predictions are projected again into the observed space by taking the first o dimensions, where o is the number of observed variables. We adopt the original parameters from [1] to perform our experiments. We project the observed data into a hidden state of 8 channels.

6.3 Training

Each model is trained in a closed-loop setting, where only the state of the system at t_0 is used as input for each of the models and the output $\hat{\mathcal{U}}_t$ at step t is fed again into the model to make the prediction at step $t+1$. For the higher-order models we initialize the higher-order derivatives as zeroes.

We train all our models using a horizon of 4 time steps with batch size 8, 5000 steps per epoch, and 5 epochs in total for both our *NeuralPDE* and the *Hidden State* model and 20 epochs for all other models. We use the Adam optimizer with the learning rate of 0.001.

All experiments are performed on a machine with a Nvidia RTX GPU, 16 CPUs and 32 GB RAM.

Table 1. Average RMSE 16-step predictions for all models. For our NeuralPDE model we only show the better score between the first and second order model. Bold print denotes the best model for each dataset, [1] denotes NeuralPDE-1, [2] denotes NeuralPDE-2. *AD - Advection-Diffusion, B - Burgers, GD - Gas Dynamics, OW - Ocean Wave, P - PlaSim, W - Wave Propagation, WB - Weatherbench.*

Model	AD	B	GD	W	OW	WB	P
Persistence	0.932	0.080	0.220	1.481	0.558	0.114	0.708
CNN	0.113	0.437	0.348	1.016	0.440	0.107	0.573
Distana	0.174	0.102	0.144	0.958	0.440	0.108	0.559
ConvLSTM	0.497	0.079	0.167	1.102	0.463	0.107	0.546
ResNet	0.086	0.314	0.200	1.043	**0.427**	0.103	**0.537**
PDE-Net	**0.007**	0.078	0.112	1.046	0.488	0.100	1.802
Hidden State	0.639	0.066	0.097	1.115	0.482	**0.096**	0.572
NeuralPDE-* (Ours)	0.057[1]	**0.063**[1]	**0.092**[1]	**0.846**[2]	0.435[1]	0.097[1]	0.563[1]

7 Results

All models are evaluated using a prediction horizon of 16 time steps, using a hold-out test set as described in Sect. 5. Table 1 compares the RMSE averaged over 16 prediction steps and all target variables. Bold entries denote the best model for any given dataset.

The first four datasets (AD, B, GD, W) represent generated toy datasets of four different partial differential equations. Our model achieves state-of-the-art performance on all of these datasets except on the advection-diffusion equation, where the PDE-Net 2.0 model [17] outperforms all other models by a large margin. We hypothesize that the very simple dynamic governing this equation (given by just one linear convolutional filter) makes it very easy for the explicit approach used by the PDE-Net model to learn the dynamic. On the other hand, our approach, which parametrizes the dynamic by a multilayer convolutional network is better at learning more complex systems of equations.

On the real-world datasets (OW, WB, P) NeuralPDE-1 closely matches the best state-of-the-art models on the Oceanwave and Weatherbench datasets coming in second best. The Plasim dataset shows to be particularily difficult to learn for methods which directly parametrize the underlying dynamics (Hidden State, PDE-Net 2.0, NeuralPDE). Our results show that the ResNet model achieves best performance on this dataset. We hypothesize that the large time steps of 1 day in the simulated data makes it difficult for a continuous dynamic to be learned by our model.

Figure 3 shows the comparison of all tested models over increasing prediction horizons. We only show a selection of different datasets and target variables, the full overview is available from Appendix B. For all models the prediction accuracy decreases with increasing prediction horizon.

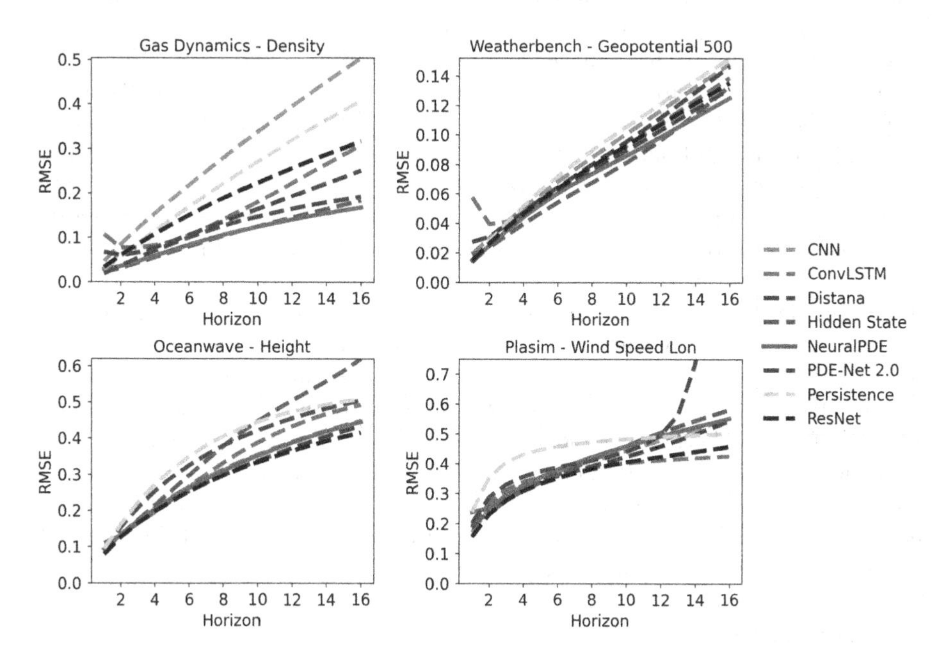

Fig. 3. Predictions over different horizons. The figure shows the RMSE for four different datasets and target variables for all tested models as a function of the prediction horizon.

8 Discussion

Our method uses a multi-layer convolutional network as a generalized approach to represent differential equations. Our experiments demonstrate that the same architecture can be applied successfully to learn a wide variety of PDE types, including linear and non-linear equations, equations in one and two dimensions, second-order equations, and coupled PDE systems of up to four equations. In our current setting, NeuralPDE achieves state-of-the-art perfomance on generated data except for very simple equations, where we hypothesize a much simpler and less overparametrized network might perform better. On the real-world datasets models that do not approximate the dynamic directly (ResNet) outperform our model and other models of this type, albeit not by a large margin.

One advantage of NeuralPDE over other flexible approaches is its inherent ability to directly capture the continous dynamics of the system. While Distana or ResNet [23] can only make discrete predictions at the next point in time, NeuralPDE can make predictions for any future point in time. This also enables the modelling of data sampled at non-equidistant points in time. In our experiment we used a fixed-step Euler solver, but in principle our method can be applied with any black-box numerical solver, including adaptive solvers like the Dormant-Prince (dopri) family of solvers [6].

Currently, NeuralPDEs only encompass periodic boundary conditions. We hypothesize that NeuralPDEs can be extended to other boundary conditions by adapting the parameterization of the convolutional layer, e.g. different padding types. Moreover, the boundary conditions need to be specified beforehand and cannot yet be learned directly from data.

The Method of Lines comes with its own set of limitations: most prominently, it cannot be used to solve elliptical second-order PDEs. These limitations apply directly to NeuralPDEs as well.

Our model is a black box model that comes with limited interpretability. While we do not directly learn the parametrization of a PDE, we could in theory extract the trained filters from the network for simple linear equations similarily to the PDE-Net [17]. However, as the system of equations grows more complex, the exact form of the PDE cannot be recovered from the learned weights.

If the order of the underlying system of equations is known, the appropriate order of our model can be chosen. This is unfortunately not the case for many real-world applications. However, as our experiments show, the first order model is a good first choice for a wide range of datasets.

9 Conclusion

In this work we proposed a novel approach to modelling dynamical data. It is based on the Method of Lines used as a numerical heuristic for solving Partial Differential Equations, by approximating the spatial derivatives using convolutional filters. In contrast to other methods, NeuralPDE does not make any assumptions about the structure of the underlying equations. Instead they rely on a deep convolutional neural network to parametrize the dynamics of the system. We evaluated our method on a wide selection of dynamical systems, including non-linear and higher-order equations and showed that it is competetive compared to other approaches.

In our future work, we will address the remaining limitations: First, we are planning to adapt NeuralPDE to learn boundary conditions from data. Second, we are going to investigate combining other methods to model spatial dynamics with neural networks. This includes other arbitrary mesh discretization methods as well as methods for continuous convolutions which could replace discretization completely.

References

1. Ayed, I., de Bézenac, E., Pajot, A., Brajard, J., Gallinari, P.: Learning dynamical systems from partial observations. CoRR abs/1902.11136 (2019). http://arxiv.org/abs/1902.11136
2. Belbute-Peres, F.D.A., Economon, T., Kolter, Z.: Combining differentiable PDE solvers and graph neural networks for fluid flow prediction. In: International Conference on Machine Learning, pp. 2402–2411. PMLR, November 2020. https://proceedings.mlr.press/v119/de-avila-belbute-peres20a.html
3. Berg, J., Nyström, K.: Data-driven discovery of PDEs in complex datasets. J. Comput. Phys. **384**, 239–252 (2019). https://doi.org/10.1016/j.jcp.2019.01.036, http://arxiv.org/abs/1808.10788, arXiv: 1808.10788
4. Bronstein, M.M., Bruna, J., LeCun, Y., Szlam, A., Vandergheynst, P.: Geometric deep learning: going beyond Euclidean data. IEEE Sig. Process. Mag. **34**(4), 18–42 (2017). https://doi.org/10.1109/MSP.2017.2693418, http://arxiv.org/abs/1611.08097, arXiv: 1611.08097 version: 1
5. Chen, R.T.Q., Rubanova, Y., Bettencourt, J., Duvenaud, D.K.: Neural ordinary differential equations. In: Advances in Neural Information Processing Systems, vol. 31. Curran Associates, Inc. (2018). https://papers.nips.cc/paper/2018/hash/69386f6bb1dfed68692a24c8686939b9-Abstract.html
6. Dormand, J., Prince, P.: A family of embedded runge-kutta formulae. J. Comput. Appl. Math. **6**(1), 19–26 (1980). https://doi.org/10.1016/0771-050X(80)90013-3, https://www.sciencedirect.com/science/article/pii/0771050X80900133
7. Goodfellow, I., Bengio, Y., Courville, A.: Deep Learning. MIT Press, Cambridge (2016). http://www.deeplearningbook.org
8. He, K., Zhang, X., Ren, S., Sun, J.: Identity mappings in deep residual networks. arXiv:1603.05027 [cs], July 2016. http://arxiv.org/abs/1603.05027, arXiv: 1603.05027
9. Hochreiter, S., Schmidhuber, J.: Long short-term memory. Neural Comput. **9**(8), 1735–1780 (1997). https://doi.org/10.1162/neco.1997.9.8.1735
10. Iakovlev, V., Heinonen, M., Lähdesmäki, H.: Learning continuous-time pdes from sparse data with graph neural networks. In: International Conference on Learning Representations (2021). https://openreview.net/forum?id=aUX5Plaq7Oy
11. Iooss, G., Helleman, R.H.G.: Chaotic behaviour of deterministic systems. North-Holland, Netherlands (1983). http://inis.iaea.org/search/search.aspx?orig_q=RN:16062000
12. Karlbauer, M., Otte, S., Lensch, H.P.A., Scholten, T., Wulfmeyer, V., Butz, M.V.: A distributed neural network architecture for robust non-linear spatio-temporal prediction. arXiv:1912.11141 [cs], December 2019. http://arxiv.org/abs/1912.11141, arXiv: 1912.11141
13. Kuznetsov, Y.A.: Introduction to Dynamical Systems, pp. 1–35. Springer, New York (1995). https://doi.org/10.1007/978-1-4757-2421-9_1
14. LeCun, Y., et al.: Handwritten digit recognition with a back-propagation network. In: Advances in Neural Information Processing Systems, vol. 2. Morgan-Kaufmann (1990). https://proceedings.neurips.cc/paper/1989/hash/53c3bce66e43be4f209556518c2fcb54-Abstract.html
15. Li, J., Sun, G., Zhao, G., Lehman, L.W.H.: Robust low-rank discovery of data-driven partial differential equations. In: Proceedings of the AAAI Conference on Artificial Intelligence, vol. 34, no. 01, pp. 767–774, April 2020. https://doi.org/10.1609/aaai.v34i01.5420, https://ojs.aaai.org/index.php/AAAI/article/view/5420

16. Lipton, Z.C., Berkowitz, J., Elkan, C.: A critical review of recurrent neural networks for sequence learning. arXiv:1506.00019 [cs], October 2015. http://arxiv.org/abs/1506.00019, arXiv: 1506.00019
17. Long, Z., Lu, Y., Dong, B.: PDE-Net 2.0: learning PDEs from data with a numeric-symbolic hybrid deep network. J. Comput. Phys. **399**, 108925 (2019). https://doi.org/10.1016/j.jcp.2019.108925, http://arxiv.org/abs/1812.04426, arXiv: 1812.04426
18. Praditia, T., Karlbauer, M., Otte, S., Oladyshkin, S., Butz, M.V., Nowak, W.: Finite volume neural network: modeling subsurface contaminant transport. arXiv:2104.06010 [cs], April 2021. http://arxiv.org/abs/2104.06010, arXiv: 2104.06010
19. Rackauckas, C., et al.: Universal differential equations for scientific machine learning. arXiv:2001.04385 [cs, math, q-bio, stat], August 2020. http://arxiv.org/abs/2001.04385, arXiv: 2001.04385
20. Raissi, M., Perdikaris, P., Karniadakis, G.E.: Physics informed deep learning (Part I): data-driven solutions of nonlinear partial differential equations. arXiv:1711.10561 [cs, math, stat], November 2017. http://arxiv.org/abs/1711.10561, arXiv: 1711.10561
21. Raissi, M., Perdikaris, P., Karniadakis, G.E.: Physics informed deep learning (Part II): data-driven discovery of nonlinear partial differential equations. arXiv:1711.10566 [cs, math, stat], November 2017. http://arxiv.org/abs/1711.10566, arXiv: 1711.10566 version: 1
22. Rasp, S., Dueben, P.D., Scher, S., Weyn, J.A., Mouatadid, S., Thuerey, N.: WeatherBench: a benchmark data set for data-driven weather forecasting. J. Adv. Model. Earth Syst. **12**(11), e2020MS002203 (2020). https://doi.org/10.1029/2020MS002203, https://agupubs.onlinelibrary.wiley.com/doi/abs/10.1029/2020MS002203
23. Rasp, S., Thuerey, N.: Data-driven medium-range weather prediction with a resnet pretrained on climate simulations: a new model for weatherbench. J. Adv. Model. Earth Syst. **13**(2), e2020MS002405 (2021). https://doi.org/10.1029/2020MS002405, https://agupubs.onlinelibrary.wiley.com/doi/abs/10.1029/2020MS002405
24. Rudolph, M.: Attaining exponential convergence for the flux error with second- and fourth-order accurate finite-difference equations. II. Application to systems comprising first-order chemical reactions. J. Comput. Chem. **26**(6), 633–641 (2005). https://doi.org/10.1002/jcc.20201, https://onlinelibrary.wiley.com/doi/abs/10.1002/jcc.20201
25. Scher, S., Messori, G.: Weather and climate forecasting with neural networks: using general circulation models (GCMS) with different complexity as a study ground. Geosci. Model Dev. **12**(7), 2797–2809 (2019). https://doi.org/10.5194/gmd-12-2797-2019, https://gmd.copernicus.org/articles/12/2797/2019/
26. Schiesser, W.E.: The numerical method of lines: integration of partial differential equations. Elsevier, July 2012. google-Books-ID: 2YDNCgAAQBAJ
27. Shi, X., Chen, Z., Wang, H., Yeung, D.Y., Wong, W.K., WOO, W.C.: Convolutional LSTM network: a machine learning approach for precipitation nowcasting. In: Advances in Neural Information Processing Systems, vol. 28. Curran Associates, Inc. (2015). https://proceedings.neurips.cc/paper/2015/hash/07563a3fe3bbe7e3ba84431ad9d055af-Abstract.html

28. So, C.C., Li, T.O., Wu, C., Yung, S.P.: Differential spectral normalization (DSN) for PDE discovery. In: Proceedings of the AAAI Conference on Artificial Intelligence, vol. 35, no. 11, pp. 9675–9684, May 2021. https://ojs.aaai.org/index.php/AAAI/article/view/17164
29. Wang, D., Yang, Y., Ning, S.: DeepSTCL: a deep spatio-temporal ConvLSTM for travel demand prediction. In: 2018 International Joint Conference on Neural Networks (IJCNN), pp. 1–8, Julu 2018. https://doi.org/10.1109/IJCNN.2018.8489530, iSSN: 2161–4407

Deep Neural Networks for Geometric Shape Deformation

Aida Farahani, Julien Vitay(✉), and Fred H. Hamker

Chemnitz University of Technology, Straße der Nationen 62, 09111 Chemnitz, Germany
{aida.farahani,julien.vitay,fred.hamker}@informatik.tu-chemnitz.de

Abstract. Geometric deep learning is a promising approach to bring the representational power of deep neural networks to 3D data. Explicit 3D representations such as point clouds or meshes can have varying and often a huge number of dimensions, what limits their use as an input to a neural network. Implicit representations such as signed distance functions (SDF) are on the contrary low-dimensional and fixed representations of the structure of a 3D shape that can be easily fed into a neural network. In this paper, we demonstrate how deep SDF neural networks can be used to precisely predict the deformation of a material after the application of a specific force. The model is trained using a set of custom finite element simulations in order to generalize to unseen forces.

Keywords: Geometric deep learning · Implicit neural representation · Geometric deformation modeling · FEM simulations · 3D data processing · Signed Distance Functions

1 Introduction

Deep neural networks have shown great potential in processing data such as images and videos. In contrast, providing the 3D structures to the neural networks is still challenging. Geometric deep learning (GDL) is a branch of machine learning (ML) that deals with 3D data for different purposes such as classification, compression, and segmentation. Although 3D models are more informative to describe the environment, common 3D representations are unfortunately not easily combined with neural networks, and methods introduced for geometric deformation processing are still under development.

The most common 3D representations are explicit, such as RGB-depth images, voxels, point clouds, and meshes. A RGB-depth map simply concatenates the depth information to the 2D image grid, which makes it Neural Network (NN)-friendly and an ideal data type for 3D pose estimation; however, the 3D modeling is partial and depends on the camera viewpoint.

Point clouds visualize the object's shape with a set of unordered 3D points sampled on the surface. The unstructured point clouds are most favored in the industry and are easily captured by 3D scanners. As a global descriptor of the

R. Bergmann et al. (Eds.): KI 2022, LNAI 13404, pp. 90–95, 2022.
https://doi.org/10.1007/978-3-031-15791-2_9

shape, they are mostly limited to segmentation and classification tasks [4,5]. A point cloud does not contain surface information; therefore, dense sampling of points is usually required. Insufficient point sampling of a shape with fine details may cause an incomplete description while increasing the number of samples becomes quickly memory inefficient. By extending the pixels of 2D images to the third dimension, voxels represent the shape structure in a 3D space. Despite the regular structure and arrangement of data units in grids that make them NN-friendly, voxels are memory inefficient and are not suitable for small localized deformations. As for point clouds, the sampling rate can highly affect memory consumption.

Polygon meshes define a shape by a set of vertices on the surface and their neighboring connections as edges. Meshes are generic data structures favored in 3D modeling for being simple, informative, and easy to use. However, bringing them into the deep learning domain either limits us to the datasets with fixed topologies [6], or the input size of the network should be equal to the largest available mesh data sample [1]. In general, methods based on mesh representations suffer from a large set of features provided for the network as the whole structure should be fed simultaneously as input. The difficulty in working with meshes stems from the multitude of possibilities to apply a mesh on a shape, as many different topologies could be defined for one geometry. Therefore for large meshes, the network size drastically increases, and training will be computationally expensive.

Recent advances in GDL and challenges of applying explicit representations to deep networks have shifted attention to implicit ones. For instance, the well-known representation "Signed Distance Functions" (SDF) refers to a regression of the 3D space based on the distance from the shape surface. Here, each point in the 3D space takes a signed value depending on whether they are inside or outside of the shape. The density of these sample points increases the resolution of the final shape and could be justified based on the needs of the problem. SDF representations are a proper feed to deep networks and are highly efficient in terms of memory consumption.

The idea of combining SDF with deep neural models was first introduced in 2019 [3] and has received significant attention since then. First, they trained a network to estimate the SDF value of a shape S for each query input position in the 3D space (Fig. 1-left). This neural network is an embedded representation of a single shape. To generalize the network to multiple shapes, they add an encoded shape S_i as a condition to the network input and estimate the distance from the shape S_i. This shape encoding is implemented using a layered autodecoder architecture. Similar to latent codes in autoencoder architecture, the "codes" of an autodecoder are embedded representations of shapes. For each query point, the network predicts the SDF value corresponding to the provided condition. This continuous interpretation of space makes it possible to reconstruct shapes in any desired resolution and preserves shape deformations without large memory requirements. This method could effectively address the problem of efficient shape embedding, and the network size is comparable to classical approaches for processing any size of meshes with arbitrary topology.

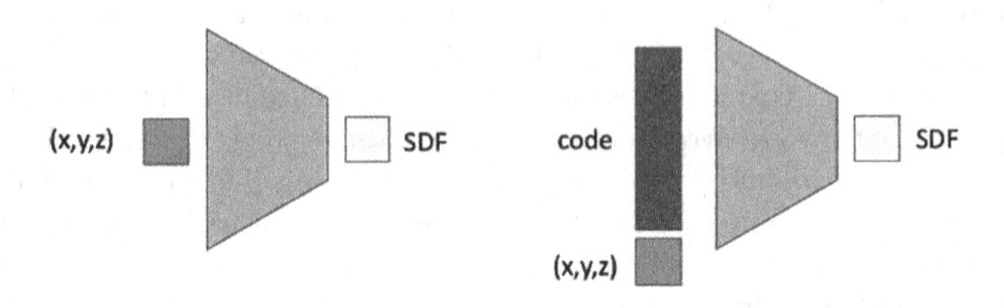

Fig. 1. Left: DeepSDF network for a single shape. Right: DeepSDF for an encoded shape and a 3D query point.

2 Shape Deformation Modeling

The term "Deformation" refers to the changes in the geometry of a structural body. Using Finite Element Analysis (FEA) has been a typical approach for many years that helps the engineers to investigate the simulation results and analyze the final product shape before production. However, the simulation process is time-consuming for large meshes with fine details and requires high computational power to solve the numerical equations. In addition, handmade tuning and re-execution of simulations are required to find the proper process parameters.

Most importantly, finding the optimized input parameters that lead to the desired results in the FEA approaches is only possible through trial and error. Neural networks can optimize input parameters that result in the desired output where cause-and-effect FE methods could not handle this functionality. Our research aims at modeling geometry deformations using neural networks. Inspired by the original DeepSDF paper [3], we designed a model combining SDF representations and deep networks to parameterize the shape deformation based on input conditions. Although the primary goal of the DeepSDF method was the efficient embedding of various shapes from different classes rather than deformations on one shape, our results showed the effectiveness of the approach for the shape deformation task.

2.1 Preparation of the Dataset

In order to train our model, we need a large dataset of deformed shapes. The publicly available datasets are mainly designed for classification or segmentation tasks and do not contain deformations or variations of shapes. Lacking appropriate data, we created our own dataset. For this purpose, we used FreeCAD, which is an open-source application for CAD design, including packages such as FEA. In FreeCAD, we defined a simple cuboid, set the material properties, and fixed the initial constraints on both ends. We then added force constraints at different positions on the surface. By also varying the force magnitude, we obtained 6228 deformed shapes corresponding to each condition. A mesh sample in the FreeCAD environment is shown in Fig. 2. The generated dataset is freely available for download[1]. The meshes then should be converted to the corresponding SDF representation. Each 3D mesh is first scaled into a unit sphere and is virtually rendered from 100 virtual cameras on the sphere surface. Then the distance from the closest mesh triangle is calculated. It is important to sample the points mostly near the surface to have an accurate sampling. We sampled 400000 points for each shape in our dataset.

Fig. 2. Left: the initial shape. Middle and right: resulting deformed meshes affected by different forces.

2.2 Implementation Results

The neural network using SDF representations with arbitrary mesh topology can handle large meshes without increasing the network size. We provide the spatial coordinate and the force vector to the network and predict the corresponding SDF value as an output. We train a fully-connected neural network with six inputs (x, y, z coordinates of a sampled point, x, y position of the force applied on the surface, and force magnitude). After a Bayesian hyperparameter search using the optuna library, the neural network is composed of 4 hidden layers (130-118-150-148) using the LeakyRelu activation function and one linear output neuron. The mean square error loss is minimized using the Adam regularizer and a learning rate of 0.0005. The network is trained for 150 epochs, with a validation set composed of 20% of the data. At the end of the training, both the training and validation losses are below 10^{-6}.

The network provides a continuous function representing the distance values for each query point in the space, so another step needs to be taken for the final shape retrieval. Marching cube (MC) [2] is the most common approach for extracting the mesh in varying resolutions. By modifying the "cube size" as an input parameter to the algorithm, the SDF values are discretized inside a unit cube to reconstruct the surface of the shape in different resolutions. In Fig. 3,

[1] https://www.tu-chemnitz.de/informatik/KI/projects/geometricdeeplearning.

two generated mesh samples and the corresponding ground truth meshes are depicted (for cube count = 95^3).

Fig. 3. The reconstructed mesh from NN prediction (in gray) and the ground-truth mesh (in color) for two different samples. (Color figure online)

We use a popular metric, the Chamfer distance (CD), to evaluate the quality of the reconstructed mesh. This metric represents the difference between two sets of points S1 and S2 sampled from both mesh surfaces. In one variation of CD, for all S1 points, the closest distance to the S2 points is averaged, and the same process is repeated in reverse. The sum of these two averages is called CD, which should be closed to zero. We randomly chose 116 samples from the test set and reconstructed the mesh from the network predictions. The mean of CDs for 30000 sample points is shown in Table 1. As expected, increasing the number of cubes in the MC algorithm (that leads to finer meshes) reduces the Chamfer distance.

The use of SDF representation has the following advantages: Any simulated mesh or CAD model could be easily converted to an SDF representation so that the existing datasets could be used for training. Contrary to explicit representations, this representation is NN-friendly and handles large-size meshes with arbitrary topology. Also, a large number of shapes could be stored as a trained neural network and save storage. After training the network, less computational power and time are needed to process the large meshes compared with numerical approaches. To our knowledge, this is the first time that this representation is combined with neural networks for processing deformable objects.

Table 1. Chamfer distance metric for different resolutions of the MC algorithm.

Cube count	Chamfer distance
85^3	0.0009595
95^3	0.0009171
105^3	0.0008937
120^3	0.0008754
130^3	0.0008747

Despite these advantages, SDF representations have two major difficulties to deal with: The 3D mesh samples have to be watertight to divide the 3D space into inside and outside regions. Unfortunately, many CAD models are not watertight, and some modifications in the algorithm are needed to be compatible with non-watertight meshes. The second issue is the additional step added at the end to discretize the space and extract an explicit representation of the shape such as a mesh or point cloud. The final step is, unfortunately, dependent on the required mesh size.

3 Conclusion and Future Work

In this paper, we showed that implicit representations could be effectively combined with neural networks to predict the shape deformations caused by an applied force. The designed network is able to be trained on very large meshes, while the size of the network is kept reasonable. The main advantage is the independence from mesh size and topology that brings the flexibility to process 3D shapes. However, the shapes provided to the network must be watertight. Future work could be suggested to find a solution for non-watertight meshes to generalize the approach. Another improvement could be proposed for the discretization phase in the end to substitute the current marching cube algorithm, which highly depends on the mesh size.

Acknowledgement. This work has been funded by the Federal Ministry of Education and Research (BMBF) - ML@Karoprod (01IS18055) and the German Research Foundation (DFG, 416228727) - SFB 1410 Hybrid Societies.

References

1. Feng, Y., Feng, Y., You, H., Zhao, X., Gao, Y.: Meshnet: mesh neural network for 3d shape representation. In: Proceedings of the AAAI Conference on Artificial Intelligence, vol. 33, pp. 8279–8286 (2019)
2. Lorensen, W.E., Cline, H.E.: Marching cubes: A high resolution 3d surface construction algorithm. ACM siggraph Comput. Graph. **21**(4), 163–169 (1987)
3. Park, J.J., Florence, P., Straub, J., Newcombe, R., Lovegrove, S.: DeepSDF: learning continuous signed distance functions for shape representation. In: Proceedings of the IEEE/CVF Conference on Computer Vision and Pattern Recognition, pp. 165–174 (2019)
4. Qi, C.R., Su, H., Mo, K., Guibas, L.J.: PointNet: deep learning on point sets for 3D classification and segmentation. arXiv:1612.00593 [cs] (2016)
5. Qi, C.R., Yi, L., Su, H., Guibas, L.J.: PointNet++: deep hierarchical feature learning on point sets in a metric space. arXiv:1706.02413 [cs] (2017)
6. Ranjan, A., Bolkart, T., Sanyal, S., Black, M.J.: Generating 3D faces using convolutional mesh autoencoders. In: Proceedings of the European Conference on Computer Vision (ECCV), pp. 704–720 (2018)

Dynamically Self-adjusting Gaussian Processes for Data Stream Modelling

Jan David Hüwel[1(✉)], Florian Haselbeck[2,3], Dominik G. Grimm[2,3,4], and Christian Beecks[1]

[1] Department of Mathematics and Computer Science, University of Hagen, Hagen, Germany
{jan.huewel,christian.beecks}@fernuni-hagen.de
[2] Technical University of Munich, Campus Straubing for Biotechnology and Sustainability, Bioinformatics, Straubing, Germany
[3] Weihenstephan-Triesdorf University of Applied Sciences, Bioinformatics, Straubing, Germany
{florian.haselbeck,dominik.grimm}@hswt.de
[4] Technical University of Munich, Department of Informatics, Garching, Germany

Abstract. One of the major challenges in time series analysis are changing data distributions, especially when processing data streams. To ensure an up-to-date model delivering useful predictions at all times, model reconfigurations are required to adapt to such evolving streams. For Gaussian processes, this might require the adaptation of the internal kernel expression. In this paper, we present dynamically self-adjusting Gaussian processes by introducing **E**vent-**T**riggered **K**ernel **A**djustments in Gaussian process modelling (ETKA), a novel data stream modelling algorithm that can handle evolving and changing data distributions. To this end, we enhance the recently introduced Adjusting Kernel Search with a novel online change point detection method. Our experiments on simulated data with varying change point patterns suggest a broad applicability of ETKA. On real-world data, ETKA outperforms comparison partners that differ regarding the model adjustment and its refitting trigger in nine respective ten out of 14 cases. These results confirm ETKA's ability to enable a more accurate and, in some settings, also more efficient data stream processing via Gaussian processes.

Keywords: Gaussian process · Time series modelling · Change point detection · Kernel search · Data stream modelling

1 Introduction

For many applications, accurate real-time analysis of data streams is essential to guarantee a constant workflow. In order to analyze data streams, incoming data

J.D. Hüwel and F. Haselbeck—Both authors contributed equally.

R. Bergmann et al. (Eds.): KI 2022, LNAI 13404, pp. 96–114, 2022.
https://doi.org/10.1007/978-3-031-15791-2_10

points need to be incorporated into an online-generated data model. However, changing data distributions at so called change points are a common challenge in data stream modelling [4]. As a consequence, an outdated prediction model might impair a downstream application. For instance, this could lead to overstocking and missed sales in demand forecasting or power supply issues in case of smart grid systems [2,11,12]. Providing an up-to-date model is therefore a major objective when modelling data streams. However, identifying the correct time point for model reconfiguration is challenging. Simply adjusting the current model periodically bypasses this challenge, but it might lead to prolonged periods with inaccurate models or increased computational costs due to unnecessary reconfigurations. Because of these drawbacks, many algorithms aim to detect change points online and consequently trigger model adjustments [3,18].

A Gaussian process (GP) is a stochastic process based on the Gaussian distribution and is commonly used as a non-parametric machine learning model [17]. GPs' probabilistic nature makes them excel at dealing with small and noisy datasets. To incorporate knowledge on the general behavior of the data, GPs use positive semi-definite covariance functions, often called kernels. If no prior knowledge about this behavior is available, an automatic kernel search can determine a fitting function for the given data [8]. However, this is usually a computationally expensive process and requires the optimization of numerous GP models. Recently, the Adjusting Kernel Search (AKS) [13] algorithm was introduced to accelerate this process on data streams. If multiple GP models are used to represent consecutive segments of a stream, it is often reasonable to assume that the models' covariance functions will be similar. AKS enables a search of similar kernels based on a given kernel expression and circumvents the construction of novel expressions from scratch.

In this paper, we enhance AKS with a novel GP-based change point detection (CPD) method in order to propose the **E**vent-**T**riggered **K**ernel **A**djustments in Gaussian process modelling (ETKA) algorithm. The major objective of this algorithm is to deliver an up-to-date GP model describing the current data behavior at all times. We evaluate ETKA based on simulated as well as real-world data and compare it to alternatives in CPD and model inference. Beyond that, we present multiple ways to further expand and improve this method.

The rest of the paper is structured as follows: In Sect. 2, we outline relevant literature about GPs, kernel search algorithms and CPD methods. Sect. 3 introduces the ETKA algorithm. In Sect. 4 we describe our experimental setup. Afterwards, in Sect. 5, we show and discuss the results, before we conclude our findings in Sect. 6.

2 Related Work

The research we present in this paper combines the fields CPD and GP-based data stream modelling. In this section, we briefly introduce relevant works from these fields. Due to a lack of space, we include a formal introduction of GPs and kernel search approaches in Appendix 1.

GPs are commonly used probabilistic machine learning models that mainly depend on their inherent kernel function. An appropriate kernel expression can be chosen by an expert based on previous knowledge about the data. Without such expert knowledge, automatic kernel search algorithms can be employed to find an optimal fit for the given data [6,8,14,15]. In 2013, Duvenaud et al. [8] introduced such an algorithm for the first time, i.e. the Compositional Kernel Search (CKS). Lloyd et al. [15] expanded the method to the Automatic Bayesian Covariance Discovery (ABCD) by including change point kernels. Since both algorithms require the exact evaluation of numerous GP models per iteration, they are restricted to small to medium-sized datasets only.

The problem of scalability was addressed in different ways: Kim et al. [14] introduced the Scalable Kernel Composition (SKC) in 2018, which performs model selection via lower and upper bounds for the GPs' marginal likelihood instead of the exact evaluations. Berns et al. developed new approaches that use a prior segmentation of the data to accelerate the search [5] or perform the segmentation themselves [6]. We aim to expand this idea by performing a dynamic segmentation via online CPD during the modelling process. Recently, Hüwel et al. [13] introduced the Adjusting Kernel Search (AKS) algorithm, which exploits prior assumptions about the data without ascertaining the kernel function. More details about AKS can be found in Appendix 1.

Haselbeck et al. [10] developed EVARS-GPR, a framework to update a GP model online at certain change points. While this approach is restricted to output scale changes, it can be seen as a predecessor of this work due to its retraining of a GP at online-detected change points.

CPD approaches can be separated into offline and online methods. The former were extensively reviewed by Truong et al. [18]. In this paper, we focus on online methods to enable real-time model adjustments. Aminikhanghahi and Cook [3] provided an elaborate overview of available approaches in this area. One widely-used method is the Bayesian Online Change Point Detection [1]. It uses Bayes' rule to determine the number of observations since the last change point and a hazard function to predict a new one. Another commonly used method is the cumulative sum (CUSUM) [16]. It tracks an accumulated deviation score over multiple data points and detects a change point when that score exceeds a custom threshold. CUSUM's high potential for adjustments allows us to use this approach for ETKA.

3 ETKA

Previous applications of AKS employed periodic adjustments of the GP model [13]. This potentially leads to extended periods of incoming data points being processed with an outdated model. Contrariwise, unnecessarily frequent model reconfigurations cause increased computational costs. For this reason, we present the Event-Triggered Kernel Adjustments in Gaussian process Modelling (ETKA) algorithm, an enhancement of AKS with a novel GP-based online CPD approach.

The combination of a kernel search algorithm with a CPD method is obvious, but the nature of GPs results in specific requirements for an optimal CPD method. Changes should not be detected in primary statistics of the incoming data, such as its mean or variance, as a GP model does not need to be adjusted to regular changes of that kind. Rather, we need to find changes in the abstract behavior and tendencies underlying the data. For example, if the periodicity changes, we want to adjust the model to find a new period length value. Aside from an accurate modelling, there are two additional requirements for CPD: the method should be simple, computationally efficient and easily comprehensible in order to maintain high explainability.

Within ETKA, we achieve these goals by employing a CUSUM approach [16] based on the current GP's performance. We hypothesize that data that differs from the current model's prediction for multiple points in a row signifies a change point. In this case, a kernel search using AKS is triggered to adjust the model to the novel data. The exact procedure of ETKA is explained below and presented in Algorithm 1.

Algorithm 1: The Event-Triggered Kernel Adjustment

Data: $\boldsymbol{D} = (x_i, y_i)_{i=1,..,N}$, base kernel set B, window size w, tolerance δ, threshold ε, CKS iterations i_{CKS}, AKS iterations i_{AKS}

Result: change points CP, kernels $\mathcal{K}$

1 $CP \leftarrow []$
2 $\mathcal{K} \leftarrow []$
3 $s \leftarrow 0$
4 $k, \sigma^2 \leftarrow CKS\,(\boldsymbol{D}_{1:w}\,, B, i_{CKS})$
5 **for** $i = w+1, .., N$ **do**
6 $\quad \hat{y}_i \leftarrow k\left(x_i, \boldsymbol{x}_{i-w,\ldots,i-1}\right)\left[k\left(\boldsymbol{x}_{i-w,\ldots,w-1}, \boldsymbol{x}_{i-w,\ldots,w-1}\right) + \sigma^2 \boldsymbol{I}\right]^{-1} \boldsymbol{y}_{i-w,\ldots,w-1}$
7 $\quad s \leftarrow max(0, s + |y_i - \hat{y}_i| - \delta \cdot 2 \cdot \sigma)$
8 $\quad$ **if** $s > \varepsilon$ **then**
9 $\quad\quad s \leftarrow 0$
10 $\quad\quad \mathcal{K} \leftarrow \mathcal{K} \cup [k]$
11 $\quad\quad CP \leftarrow CP \cup [x_i]$
12 $\quad\quad k, \sigma^2 \leftarrow AKS(\boldsymbol{D}_{i-w,..,w}\,, B, i_{AKS}, k)$

First, we construct a GP model using CKS on an initial window of w data points. By using this model with kernel $k : \mathbb{R} \times \mathbb{R} \to \mathbb{R}$, we make a prediction $\hat{y}_i$ for the next value starting from $i = w + 1$ after this initial window of length w as follows [17]:

$$\hat{y}_i = k\left(x_i, \boldsymbol{x}_{i-w,\ldots,i-1}\right)\left[k\left(\boldsymbol{x}_{i-w,\ldots,w-1}, \boldsymbol{x}_{i-w,\ldots,w-1}\right) + \sigma^2 \boldsymbol{I}\right]^{-1} \boldsymbol{y}_{i-w,\ldots,w-1} \quad (1)$$

Then, we calculate the absolute deviation $|y_i - \hat{y}_i|$ between the observed value y_i and our prediction $\hat{y}_i$. Afterwards, the window slides one position further and the consecutive data points are used for the next prediction step employing the current GP. The accumulating error is used together with the GP's noise σ^2 and a tolerance factor $\delta \in \mathbb{R}$ to compute a change point score s:

$$s \leftarrow max\left(0, s + |y_i - \hat{y}_i| - \delta \cdot 2 \cdot \sigma\right) \tag{2}$$

If this score surpasses a certain threshold $\varepsilon \in \mathbb{R}_{>0}$, a change point is detected at x_i and s is reset to zero. With this CPD approach, incoming data points need to be within the inner $100 \cdot \delta\%$ of the GP's confidence interval in order to count as accurately predicted. The further a data point is outside this interval, the more it increases s and the faster a change point is detected. When the need for a model adjustment is triggered due to a detected change point, the GP's kernel is adjusted with AKS on the current window w. Then, the procedure with the CUSUM-based CPD and a potential trigger of AKS continues with this updated model. In settings where other kernel search methods are considered more useful, this step can easily be substituted with the corresponding approach.

4 Experimental Setup

In this section, we present the experimental settings that we employed to produce the results shown in Sect. 5. All experiments were conducted on an 11th generation Intel Core i9-11900H processor with 8 cores à 2.50 GHz. For reproducibility, we published all code and data on GitHub[1].

4.1 Simulated Data

For the development and evaluation of ETKA under controlled and predefined settings, we generated artificial datasets[2]. The simulations are based on univariate time series of length n with values $\boldsymbol{b} \in \mathbb{R}^n$ that follow a periodicity of length n_{per} and have an amplitude of size a. Furthermore, we consider the multiplicative components linear trend $\boldsymbol{l} \in \mathbb{R}^n$ and random noise $\boldsymbol{\eta} \in \mathbb{R}^n$. The size of the linear trend $\boldsymbol{l}$ is defined by the coefficient m of a linear model. The random noise $\boldsymbol{\eta}$ is sampled from a normal distribution with a mean value of 1 and a standard deviation s. These components enable the simulation of time series data $\boldsymbol{y} \in \mathbb{R}^n$ with

$$y_i = b_i \cdot l_i \cdot \eta_i \tag{3}$$

where b_i is the value of the base signal $\boldsymbol{b}$ at index i, l_i the factor of the linear trend $\boldsymbol{l}$ at index i and η_i the factor of the random noise $\boldsymbol{\eta}$ at index i. A factor can be left out to simulate that a component is not present. Finally, we can generate change points by fading between time series $\boldsymbol{y}$ of different properties. The abruptness of the change is adjustable via a fading window w_{fade}. We used this framework to model for instance changes of the period length (parameter n_{per} of $\boldsymbol{b}$), the output scale (parameter a of $\boldsymbol{b}$), the size of the linear trend

[1] https://github.com/JanHuewel/ETKA.

[2] https://github.com/Nike-Inc/timeseries-generator.

(parameter m of $\boldsymbol{l}$) or the noise level (parameter s of $\boldsymbol{\eta}$). We both considered time series with single and multiple change points, all having a length of 2000 data points. An overview of all simulation settings can be found in Table 3 in Appendix 2.

4.2 Real-World Data

We further included 14 real-world datasets from various domains, see Table 4 in the appendix. Besides the different domains, the datasets show a wide range regarding the number of samples reaching from 180 (*Call centre*) up to 7718 (*Airquality*). Examining the minimum (min) and maximum (max) target values of each time series, we further observe that many datasets have a large value range. Beyond that, several datasets have a standard deviation (std) that is rather high relative to their value range, indicating a strong variation in the time series. In summary, this collection of common time series datasets with varying characteristics and domains enables a broad evaluation of ETKA.

4.3 Evaluation

As comparison partners in our experiments, we included alternatives to ETKA in both main aspects: the choices when to adjust the model and how to adjust the model. Hence, we compare the previously described CPD-based approach to data-agnostic periodic model adjustments (*PER AKS*). These model adjustments are done three times in equidistant intervals within the dataset after the initial kernel search. Furthermore, only the data in the current window is used, disregarding everything before that. By doing so, we examine the concept of CPD-driven adjustments at the cost of frequent model predictions needed to enable CPD. Furthermore, as mentioned in Sect. 2, different kernel search algorithms exist [6,8,14,15]. For this proof of concept, we included a hyperparameter optimization (no change of the GP's kernel expression) after a detected change point as a comparison partner (*CPD HPO*). This approach also performs its retraining on the current window exclusively.

The two evaluation criteria we consider in this work are runtime and the mean absolute error of the prediction. Regarding the former, we measured the total runtime for processing each dataset, excluding the initial model construction as it is identical for all comparison partners. With respect to the prediction performance, we calculated the mean absolute error $\frac{1}{N-w}\sum_{i=w+1}^{N}|y_i - \hat{y}_i|$ on every step after the initial model construction. Finally, we set the results in relation to all comparison partners. Thereby, a negative value represents a shorter runtime respective more accurate modelling of ETKA, i.e. an improvement.

For all experiments, we use a window size w equivalent to 20% of the whole dataset. We allow kernels consisting of up to three base kernels and employ one iteration of adjustments when using AKS. The base kernel set B consists of the periodic, the linear and the squared exponential kernel. Before the initial kernel search with CKS, the data is rescaled using a Z-normalization.

5 Experimental Results

In this section, we provide an overview of our experimental results, both on simulated as well as on real-world data, and discuss our findings.

5.1 Simulated Data

As outlined in Sect. 4.1, we conducted experiments based on simulated data to evaluate ETKA under controlled and predefined settings. We further considered different configurations of the online CPD integrated in ETKA. These differ with respect to the defining parameters, i.e. the tolerance factor δ and the threshold ε. With the former, one can control how strict ETKA's CPD is, i.e. a higher value increases the tolerance range for deviations between real and prediction values. A change point is declared, if the threshold ε is exceeded, so a higher value allows higher change point scores and leads to a less sensitive CPD.

We show detailed results with absolute evaluation values for all simulated datasets and CPD configurations in Figs. 2 and 3 in Appendix 3. In Table 1, we provide a summary of these results. Besides the absolute evaluation values of ETKA, all results are shown in relation to the comparison partners *PER AKS* and *CPD HPO*. The table shows averaged values over all simulated datasets and its standard deviations. As *CPD HPO* also differs based on δ and ε, the values of the relative comparison with ETKA cannot be compared across CPD settings, but give an impression which algorithm is in advantage for the specific CPD.

We observe that ETKA constantly outperforms both comparison partners with respect to the prediction error. The predictions of ETKA tend to be more accurate with a more sensitive CPD. In comparison with *PER AKS*, ETKA shows the largest improvement for $\delta = 0.5$ and $\varepsilon = 5.0$. With respect to *CPD HPO*, ETKA's top result is achieved for $\delta = 0.5$ as well, but with $\varepsilon = 7.0$. In terms of the absolute prediction error of ETKA, we see that the best overall result is achieved for $\delta = 0.5$ and $\varepsilon = 5.0$. Regarding the runtime, *CPD HPO* is in all cases more computationally efficient, while ETKA is faster or at least on a par with *PER AKS*. As expected, we observe that the runtime of ETKA decreases with a less sensitive CPD due to higher values for δ and ε. This decrease is larger for an increasing δ with a constant ε than the other way round.

Assessing Fig. 2 in Appendix 3 for the best performing CPD configuration with $\delta = 0.5$ and $\varepsilon = 5.0$, we observe that ETKA outperforms its comparison partners on the majority of the simulated datasets (*PER AKS* is best in 7 and *CPD HPO* in 4 out of 24 simulation settings). Furthermore, in cases for which ETKA does not deliver the best outcome, it is generally close to the top performer. We further see that results for lower noise levels tend to be better (scenarios with a variable noise are also less noisy than those with $s = 0.05$ as the maximum of s is 0.05 for them).

The main goal of ETKA is an up-to-date GP model at all times. For that reason, the prediction error is more important than the runtime. Consequently, we set δ to 0.5 and ε to 5.0 at the cost of longer runtimes for the experiments on real-world data.

Table 1. Summary of the results on simulated data. The table shows the results of ETKA in terms of the prediction error respective the runtime, as well as both evaluation values of ETKA in relation to *PER AKS* and *CPD HPO*. All results are given for four different configurations regarding the CPD with different tolerance factors δ and thresholds ε. Each cell shows the mean and standard deviation over all simulated datasets. For the abolsute ETKA results, a smaller value is better, both for the prediction error and the runtime. For the relative values, a negative value reflects an improvement by ETKA. All cases for which ETKA outperforms its comparison partners are highlighted in bold.

δ	ε	ETKA results		Runtime vs.		Prediction error vs.	
		Prediction error	Runtime [s]	*PER AKS*	*CPD HPO*	*PER AKS*	*CPD HPO*
0.5	5.0	0.1403 (±0.1120)	1336 (±385)	+0.96% (±28.70%)	+114.71% (±105.65%)	**−16.68% (±23.42%)**	**−16.37% (±20.73%)**
0.5	7.0	0.1413 (±0.1117)	1121 (±316)	**−10.79% (±28.88%)**	+92.25% (±80.69%)	**−13.95% (±23.33%)**	**−21.40% (±23.13%)**
0.7	5.0	0.1472 (±0.1137)	922 (±288)	**−24.86% (±29.07%)**	+56.84% (±58.70%)	**−6.40% (±32.91%)**	**−21.02% (±24.28%)**
0.7	7.0	0.1522 (±0.1110)	919 (±317)	**−28.21% (±24.83%)**	+49.07% (±47.07%)	**−1.27% (±33.12%)**	**−20.17% (±22.00%)**

5.2 Real-World Data

Besides simulated data, we evaluated ETKA on real-world data from different domains (see Sect. 4.2). An overview of the results with absolute evaluation values is shown in Fig. 4 in Appendix 3. In Table 2, we provide the comparison of ETKA with *PER AKS* and *CPD HPO*. On average, ETKA outperforms the comparison partners in terms of the prediction error by 2.73% and 6.19%, respectively. Considering the individual datasets, ETKA is more accurate than *PER AKS* and *CPD HPO* in 9 out of 14 respective 10 out of 14 cases. For two respectively three cases, ETKA is only slightly outperformed. Both comparison partners deliver better predictions than ETKA for *Unemployment*. Beyond that, *PER AKS* is the top performer for *Internet* and *Gas production*.

We observe the largest improvement of ETKA over both comparison partners in terms of the prediction error for *Airline*, which we show in Fig. 1a. ETKA detected two change points, so less refittings than for *PER AKS* were employed. Furthermore, the reconfigurations that ETKA performed are closer to the actual changes in the dataset, which leads to advantages regarding the prediction error. For *Gas production* shown in Fig. 1b, ETKA performs significantly better than *CPD HPO*, but is outperformed by *PER AKS*. By chance, the periodical refitting points of *PER AKS* are accurate, leading to better predictions.

Regarding the runtime overall, both *PER AKS* and *CPD HPO* are more efficient, with the latter requiring the lowest runtimes as expected. Furthermore, this disadvantage of ETKA is clearer for the real-world data than it was for simulated data. However, when focusing on the datasets for which the runtime of ETKA is more than 100% higher than for *PER AKS* (*Wheat, Internet, Radio* and *Airquality*), we observe lower prediction errors in three out of fourcases.

Table 2. Results overview on real-world data. The table shows the results of ETKA in terms of runtime respective prediction error in relation to *PER AKS* and *CPD HPO*. A negative value reflects an improvement by ETKA, both for the runtime and the prediction error. All cases for which ETKA outperforms its comparison partners are highlighted in bold.

Dataset	Runtime vs.		Prediction error vs.	
	PER AKS	*CPD HPO*	*PER AKS*	*CPD HPO*
Solar irradiance	+62.51%	+918.00%	+1.90%	**−1.31%**
Mauna Loa	**−86.96%**	**−0.91%**	**−1.07%**	+0.00%
Airline	+0.79%	+2059.33%	**−21.46%**	**−26.37%**
Wheat	+161.54%	+1327.28%	**−2.83%**	+2.15%
Temperature	**−31.83%**	+159.92%	**−0.57%**	**−6.66%**
Internet	+134.97%	+183.77%	+12.51%	**−19.76%**
Call centre	+48.67%	+1811.77%	**−2.39%**	**−1.96%**
Radio	+228.26%	+2458.84%	**−19.30%**	**−7.51%**
Gas production	+16.29%	+531.39%	+8.66%	**−23.77%**
Sulphuric	+23.14%	+631.26%	**−6.32%**	**−11.77%**
Unemployment	**−1.81%**	+760.48%	+7.21%	+15.44%
Births	**−38.92%**	+79.1%	**−2.47%**	**−1.46%**
Wages	**−59.78%**	+129.56%	+1.27%	+0.41%
Airquality	+130.22%	+37.77%	**−13.44%**	**−4.19%**
Summary	+41.93% (± 91.77%)	+792.03% (± 819.75%)	**−2.73% (± 9.84%)**	**−6.19% (± 11.16%)**

In comparison with *PER AKS*, *Radio* and *Airquality* are within the three datasets with the largest improvement on the prediction error. *Internet*, which already was noticeable with respect to the prediction error, is also problematic regarding the runtime.

5.3 Discussion

With respect to synthetic data, we overall observe a broad applicability of ETKA. Despite different change point patterns and noise levels, ETKA mostly outperforms its comparison partners. We further observe that ETKA does not perform worse in case of multiple changes in comparison with single changes. This indicates that the CPD within ETKA delivers the intended results and is applicable on data with multiple change points. Simulation settings B and C (see Table 3 and Fig. 2 in the Appendix) lead to the worst results for all three noise levels, while ETKA performs best for four of these six datasets. Setting B triggers a slow change of the periodicity. In contrast, settings E and H contain abrupt shifts of the periodical length, for which all three prediction models show lower prediction errors. Furthermore, the slow change of the amplitude a for setting C is problematic, whereas an abrupt shift of a for setting H is captured better

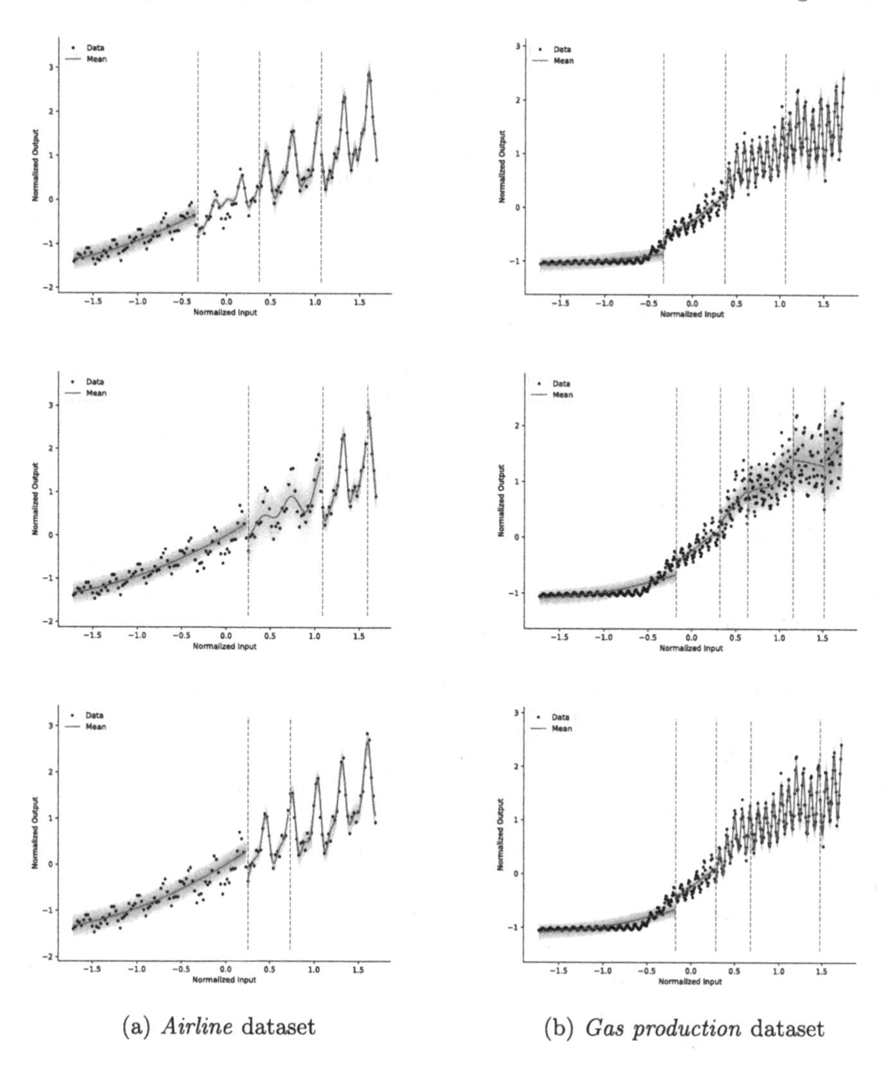

(a) *Airline* dataset (b) *Gas production* dataset

Fig. 1. Comparison of the model predictions and segmentation by the three different approaches for *Airline* and *Gas production*. The top-most plot shows the results of *PER AKS*, the second row contains *CPD HPO*'s results and the final row ETKA's outcome. Each plot shows the observed data, mean prediction of the GP model as well as the confidence intervals. Points at which the model is refitted are marked with a red vertical line. The confidence interval of the GP is shown in separate colors for each segment for visual clarity.

by all methods. This lets us assume that slow changes are problematic for all three prediction models. A potential explanation for this phenomenon for both CPD-based approaches (ETKA and *CPD HPO*) is that abrupt changes increase the change point score s faster. Consequently, this might lead to quicker model adjustments. In contrast to abrupt changes, slow shifts can lead to prolonged inaccuracies without triggering the model adjustment.

On real-world data, ETKA still delivered the best performance in terms of the mean absolute error. However, it comes at the cost of higher runtimes compared to both alternative approaches. The primary goal of ETKA is to deliver an up-to-date model at all times. As we therefore chose a rather sensitive CPD configuration based on the results on simulated data, higher runtimes were expected. A notable exception is the *Mauna Loa* dataset, for which ETKA had the lowest runtime. The prediction error comparison to *CPD HPO* let us infer that no change point was detected. Furthermore, the slightly lower prediction error of ETKA in comparison with *PER AKS* indicates that model adjustments are not beneficial for *Mauna Loa*. Hence, it might be valid that ETKA does not detect any change points. On the other datasets, the runtime difference varies greatly, while the performance improvement is, albeit not constant, more stable. For *Airline*, *Radio* and *Airquality*, we observe the highest improvement in terms of the prediction error. As outlined, ETKA detected less change points more accurately than its comparison partners for the former. In contrast to the improved prediction error, *Radio* and *Airquality* lead to higher runtimes for ETKA. Both datasets contain several change points, which were detected by ETKA and lead to multiple model adjustments. Consequently, ETKA delivered more accurate GP models, however at the cost of computational resources. Periodical refittings highly depend on coincidence regarding an appropriate timing of model reconfigurations, as for instance observed for *Gas production*. Not depending on a by chance well chosen time for refitting is a big advantage of ETKA's CPD-based approach. With respect to *Internet*, we observe both a higher runtime as well as prediction error for ETKA in comparison with *PER AKS*. This is probably caused by poor results of the initial kernel search using CKS, indicating data that would require a higher complexity of the kernel expression. For such data, AKS and consequently ETKA might intuitively be advantageous as its search is based on the current kernel expression instead of restarting from scratch.

In settings focused on fast processing, ETKA has multiple options to trade potential prediction performance for lowered computational costs, e.g. by employing a smaller window size w and a reduced set of base kernels. In contrast, more iterations of AKS per adjustment or a more sensitive CPD can increase the model quality at the cost of longer processing times. The effect of the CPD configuration can be seen in our results on simulated data, cf. Table 1. As expected, settings with higher values for δ and ϵ lead to lower runtimes at the cost of a higher prediction error.

ETKA's performance is highly dependent on the integrated CPD and consequently on its parameters. For this study, the CPD parameters were determined using simulated data. Despite having generated a broad variety of change point patterns, this might not lead to the optimal parameters for all settings. Therefore, future work enabling a parameter determination based on the processed dataset could improve ETKA and the evaluation of other state-of-the-art CPD approaches is an interesting point for future research. Beyond that, for severe changes of the data behavior, a kernel search from scratch might be more efficient than AKS. A classification of detected change points could enable ETKA to always choose the most appropriate kernel search approach. A further potential improvement could be a dynamically determined window size w in contrast to the fixed value we applied. This could be beneficial both in terms of efficiency and prediction performance.

6 Conclusion

In this paper, we enhanced AKS, a recently-introduced kernel search algorithm for GPs with a novel CUSUM-based CPD approach. The resulting algorithm, ETKA, offers the ability to automatically deliver an up-to-date GP model for data streams. In our experiments, ETKA proved to be broadly applicable to data of different behavior and noise levels. Compared to intuitive alternatives, ETKA delivered improved predictions. Especially on simulated data, the results were significantly better. On real-world datasets, the improvement was smaller. Overall, ETKA reached its main goal of an up-to-date model at all times and is therefore especially well suitable for applications for which an accurate modelling is paramount.

Funding Information. This research was supported by the research training group "Dataninja" (Trustworthy AI for Seamless Problem Solving: Next Generation Intelligence Joins Robust Data Analysis) funded by the German federal state of North Rhine-Westphalia.
The project is supported in parts by funds of the Federal Ministry of Food and Agriculture (BMEL), Germany based on a decision of the Parliament of the Federal Republic of Germany via the Federal Office for Agriculture and Food (BLE) under the innovation support program [grant number 2818504A18, DGG].

Appendix 1: Background

In the following, we formally introduce the foundations of ETKA, namely GPs and the AKS algorithm.

A GP is a non-parametric probabilistic machine learning model [9,17]. A model $GP(m, k)$ is uniquely defined by its mean function $m : \mathbb{R} \to \mathbb{R}$ and its covariance function or kernel $k : \mathbb{R} \times \mathbb{R} \to \mathbb{R}$. While the mean function can often be set to constant zero [17], the kernel contains assumptions about the GP's behavior. There are various kernels that are well understood and describe specific patterns, like the periodic $\mathcal{K}_{PER}$ and the linear kernel $\mathcal{K}_{LIN}$. These simple kernels will be referred to as base kernels throughout this paper.

Kernels often depend on parameters such as a lengthscale or a period length. These parameters are referred to as hyperparameters of the GP model and can be (locally) optimized for given data. One possible measure of performance for such optimization is the GP's log marginal likelihood $\mathcal{L}(GP(m, k), \boldsymbol{D})$ on the data $\boldsymbol{D} = (x_i, y_i)_{i=1,..,N}$ [17]. While there exist alternative measures [8,17], we will use the log marginal likelihood for model optimization in our experiments.

Individual base kernels can be combined to more complex kernel expression via addition or multiplication [8]. This way, a kernel that optimally describes the data's behavior can be constructed by experts. Alternatively, an algorithmic approach to find such an optimal kernel expression automatically can be employed. An example for this type of algorithm is CKS [8], which is depicted in Algorithm 2.

Algorithm 2: Pseudocode of the Compositional Kernel Search

Data: $\boldsymbol{D} = (x_i, y_i)_{i=1,..,N}$, base kernel set B, max iterations i_{max}
Result: Kernel expression K

1 $K \leftarrow argmax_{b \in B} \left(\mathcal{L}(GP(0, b), \boldsymbol{D})\right)$
2 **for** $i = 1, .., i_{max}$ **do**
3 $\quad C \leftarrow AddViaAddition(K, B)$
4 $\quad C \leftarrow C \cup AddViaMultiplication(K, B)$
5 $\quad C \leftarrow C \cup ReplaceKernel(K, B)$
6 $\quad K \leftarrow argmax_{c \in C} \left(\mathcal{L}(GP(0, c), \boldsymbol{D})\right)$

In each iteration, the algorithm adjusts the current best kernel expression K given a set of base kernels B by

1. adding any base kernel to any subexpression of K,
2. multiplying any base kernel to any subexpression of K or
3. replacing any base kernel in K with a different base kernel from B.

The abstract functions $AddViaAddition$, $AddViaMultiplication$ and $ReplaceKernel$ in Algorithm 2 correspond to these three options. The best performing candidate from the thus generated set is used as the basis for the next iteration. This way, the algorithm can build arbitrarily complex kernels at the cost of multiple model optimizations and evaluations per iteration. Since the candidate generation can not lower the kernel's complexity, any searches for new kernels need to start from scratch.

Our goal is to model consecutive segments of potentially infinite time series. We utilize the AKS algorithm [13] as it has inherent advantages in this specific setting. In particular, the AKS algorithm is able to adjust a given kernel to fit new data instead of starting from zero. This is accomplished by adding a fourth possibility in the candidate generation: the removal of a base kernel from the current expression. This procedure is depicted Algorithm 3.

Algorithm 3: Pseudocode of the Adjusting Kernel Search

Data: $\boldsymbol{D} = (x_i, y_i)_{i=1,..,N}$, base kernel set B, max iterations i_{max}, starting kernel K_0
Result: Kernel expression K

1 $K \leftarrow K_0$
2 **for** $i = 1, .., i_{max}$ **do**
3 $\quad C \leftarrow AddViaAddition(K, B)$
4 $\quad C \leftarrow C \cup AddViaMultiplication(K, B)$
5 $\quad C \leftarrow C \cup ReplaceKernel(K, B)$
6 $\quad C \leftarrow C \sup RemoveKernel(K)$
7 $\quad K \leftarrow argmax_{c \in C} \left(\mathcal{L}(GP(0, c), \boldsymbol{D})\right)$

It has been shown before that AKS can lead to a faster modelling process than CKS [13]. Especially in high-complexity models, the computational cost of constructing a kernel from zero are much higher than a few iterations of adjustment. However, for low-complexity models, the larger set of candidates in AKS can lead to longer processing compared to CKS, even if fewer iterations are needed.

Appendix 2: Simulated and Real-World Data Characteristics

Table 3. Overview of simulated datasets. Artificial datasets were generated at three different noise levels. For the third setting, the standard deviation s changed for each η_i. Various scenarios with single and multiple changes of the time series components as well as several configurations regarding the type of the change were simulated. We considered both abrupt as well as slower occurring changes.

		y_1			w_{fade}	y_2			w_{fade}	y_3			w_{fade}	y_4		
		b		l		b		l		b		l		b		l
		n_{per}	a	m		n_{per}	a	m		n_{per}	a	m		n_{per}	a	m
η with $s = 0.01$	$A_{0.01}$	100	0.1	–	500	100	0.1	1								
	$B_{0.01}$	100	0.1	–	[600, 650]	200	0.1	–								
	$C_{0.01}$	100	0.1	–	[1000, 1050]	100	0.2	–								
	$D_{0.01}$	100	0.1	1	[500, 550]	100	0.1	5								
	$E_{0.01}$	100	0.1	1	500	200	0.1	2	1250	200	0.1	–				
	$F_{0.01}$	100	0.1	–	500	100	0.2	–	1250	200	0.2	–				
	$G_{0.01}$	100	0.1	4	500	100	0.1	1	1250	100	0.1	−0.1	1500	100	0.1	0.5
	$H_{0.01}$	100	0.1	2	500	200	0.1	0.5	1250	200	0.2	0.5	1500	100	0.1	0.2
η with $s = 0.05$	$A_{0.05}$	100	0.1	–	500	100	0.1	1								
	$B_{0.05}$	100	0.1	–	[600, 650]	200	0.1	–								
	$C_{0.05}$	100	0.1	–	[1000, 1050]	100	0.2	–								
	$D_{0.05}$	100	0.1	1	[500, 550]	100	0.1	5								
	$E_{0.05}$	100	0.1	1	500	200	0.1	2	1250	200	0.1	–				
	$F_{0.05}$	100	0.1	–	500	100	0.2	–	1250	200	0.2	–				
	$G_{0.05}$	100	0.1	4	500	100	0.1	1	1250	100	0.1	−0.1	1500	100	0.1	0.5
	$H_{0.05}$	100	0.1	2	500	200	0.1	0.5	1250	200	0.2	0.5	1500	100	0.1	0.2
η_i with $s_{max} = 0.05$	A_{var}	100	0.1	–	500	100	0.1	1								
	B_{var}	100	0.1	–	[600, 650]	200	0.1	–								
	C_{var}	100	0.1	–	[1000, 1050]	100	0.2	–								
	D_{var}	100	0.1	1	[500, 550]	100	0.1	5								
	E_{var}	100	0.1	1	500	200	0.1	2	1250	200	0.1	–				
	F_{var}	100	0.1	–	500	100	0.2	–	1250	200	0.2	–				
	G_{var}	100	0.1	4	500	100	0.1	1	1250	100	0.1	−0.1	1500	100	0.1	0.5
	H_{var}	100	0.1	2	500	200	0.1	0.5	1250	200	0.2	0.5	1500	100	0.1	0.2

Table 4. Overview of the real-world data for our experiments. We considered 14 datasets from various domains and show some characteristics below. Most datasets were obtained from Lloyd et al. [15]. The *Airquality* data was published by De Vito et al. [7].

Dataset	Length	Mean	Std	Min	Max
Solar irradiance	391	1365.82	0.24	1365.52	1366.68
Mauna Loa	702	352.30	26.18	313.21	407.65
Airline	144	280.30	119.97	104.00	622.00
Wheat prices	370	107.88	67.21	11.00	381.00
Temperature	1000	11.21	4.14	−0.80	26.30
Internet	1000	46355.45	22058.93	13486.74	125058.79
Call centre	180	492.50	189.54	161.00	872.00
Radio	240	8.08	2.44	4.30	13.50
Gas production	476	21415.27	18678.34	1646.00	66600.00
Sulphuric production	462	131.34	41.26	42.00	228.00
Unemployment	408	520.28	261.22	122.00	1350.00
Births	1000	248.53	42.06	136.00	366.00
Wages	735	8.76	8.18	2.15	49.99
Airquality	7718	246.90	212.98	2.00	1479.00

Appendix 3: Results Overviews

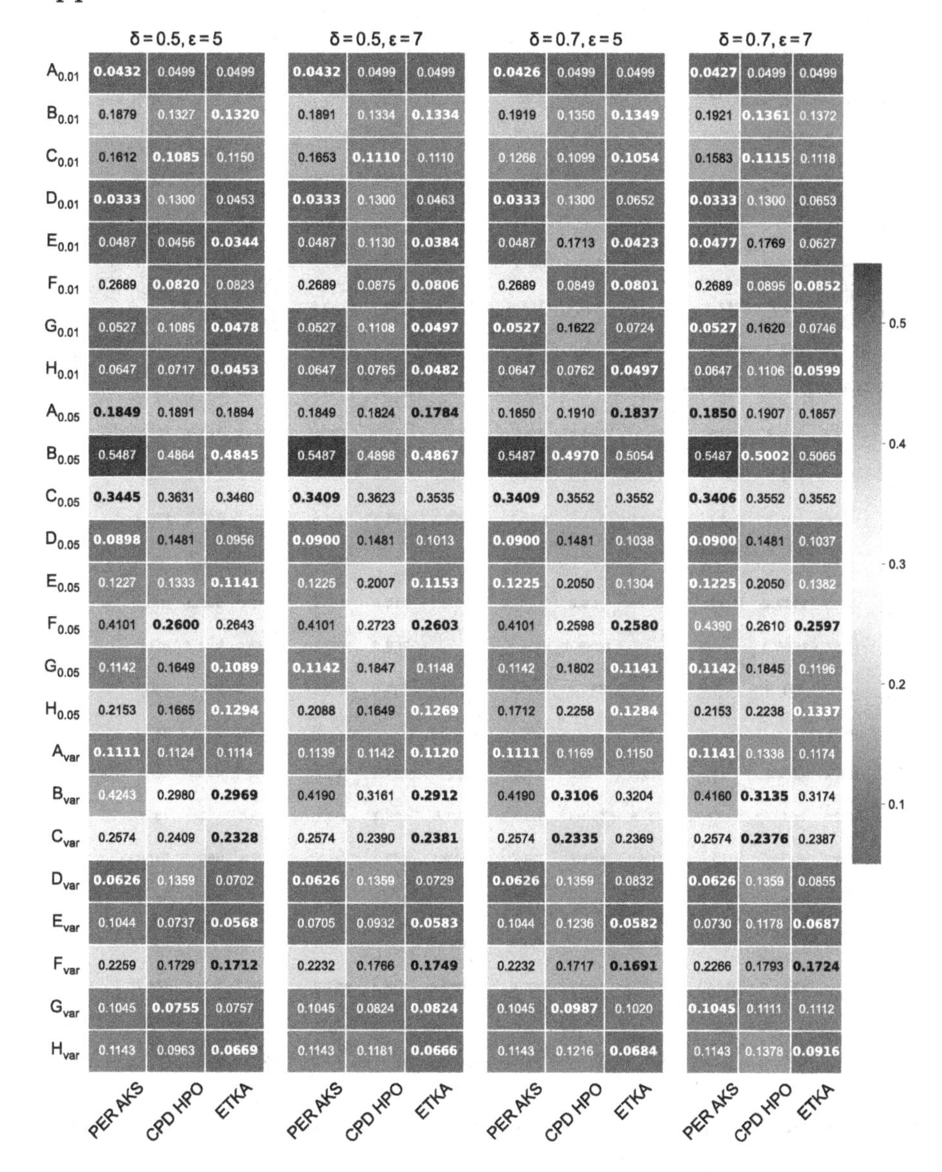

Fig. 2. Results overview in terms of the prediction error on all simulated data and CPD configurations. Each cell shows the result for the model given on the horizontal axis and the simulated dataset given on the vertical axis. Smaller values reflect a better performance. The best performing model for each CPD configuration and dataset is highlighted in bold.

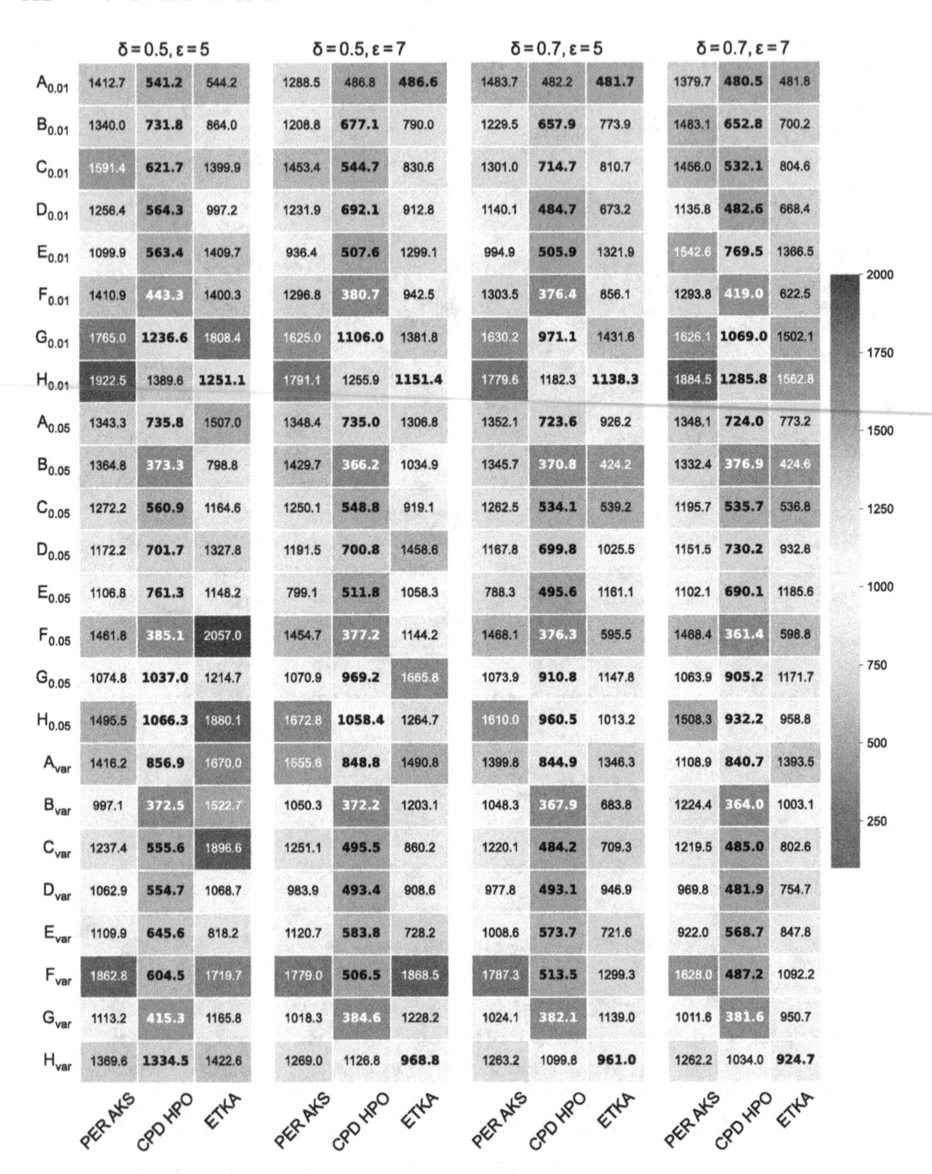

Fig. 3. Results overview in terms of runtime on all simulated data and CPD configurations. Each cell shows the result for the model given on the horizontal axis and the simulated dataset given on the vertical axis. Smaller values are considered better. The most efficient model for each CPD configuration and dataset is highlighted in bold.

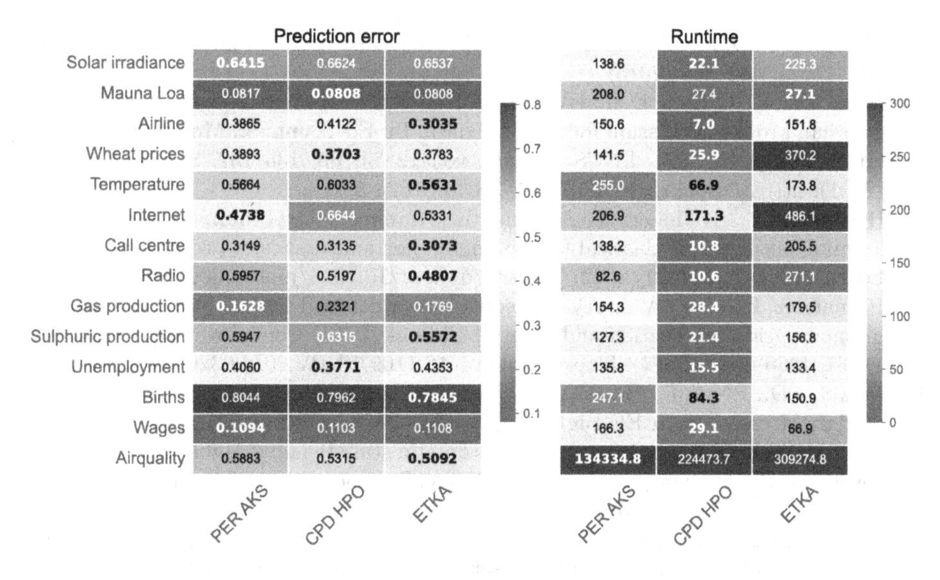

Fig. 4. Results overview for real-world data. Each cell shows the result for the model given on the horizontal axis and the dataset given on the vertical axis. Smaller values reflect a better performance. The best performing respective most efficient model for each dataset is highlighted in bold.

References

1. Adams, R.P., MacKay, D.J.C.: Bayesian online changepoint detection (2007)
2. Alagu Dharshini, M.P., Antelin Vijila, S.: Survey of machine learning and deep learning approaches on sales forecasting. In: 2021 4th International Conference on Computing and Communications Technologies (ICCCT), pp. 59–64 (2021). https://doi.org/10.1109/ICCCT53315.2021.9711878
3. Aminikhanghahi, S., Cook, D.J.: A survey of methods for time series change point detection. Knowl. Inf. Syst. **51**(2), 339–367 (2016). https://doi.org/10.1007/s10115-016-0987-z
4. Bahri, M., Bifet, A., Gama, J., Gomes, H.M., Maniu, S.: Data stream analysis: foundations, major tasks and tools. Wiley Interdiscip. Rev. Data Min. Knowl. Discov. **11**(3), e1405 (2021)
5. Berns, F., Beecks, C.: Automatic gaussian process model retrieval for big data. In: CIKM, pp. 1965–1968. ACM (2020)
6. Berns, F., Schmidt, K., Bracht, I., Beecks, C.: 3cs algorithm for efficient gaussian process model retrieval. In: ICPR, pp. 1773–1780. IEEE (2020)
7. De Vito, S., Massera, E., Piga, M., Martinotto, L., Di Francia, G.: On field calibration of an electronic nose for benzene estimation in an urban pollution monitoring scenario. Sens. Actuators B: Chem. **129**(2), 750–757 (2008). https://doi.org/10.1016/j.snb.2007.09.060
8. Duvenaud, D., Lloyd, J.R., Grosse, R.B., Tenenbaum, J.B., Ghahramani, Z.: Structure discovery in nonparametric regression through compositional kernel search. In: ICML (3). JMLR Workshop and Conference Proceedings, vol. 28, pp. 1166–1174. JMLR.org (2013)

9. Ghahramani, Z.: Probabilistic machine learning and artificial intelligence. Nature **521**(7553), 452–459 (2015)
10. Haselbeck, F., Grimm, D.G.: EVARS-GPR: EVent-triggered augmented refitting of gaussian process regression for seasonal data. In: Edelkamp, S., Möller, R., Rueckert, E. (eds.) KI 2021. LNCS (LNAI), vol. 12873, pp. 135–157. Springer, Cham (2021). https://doi.org/10.1007/978-3-030-87626-5_11
11. Haselbeck, F., Killinger, J., Menrad, K., Hannus, T., Grimm, D.G.: Machine learning outperforms classical forecasting on horticultural sales predictions. Mach. Learn. Appl. **7**, 100239 (2022). https://doi.org/10.1016/j.mlwa.2021.100239
12. Hernandez, L., et al.: A survey on electric power demand forecasting: future trends in smart grids, microgrids and smart buildings. IEEE Commun. Surv. Tutorials **16**(3), 1460–1495 (2014). https://doi.org/10.1109/SURV.2014.032014.00094
13. Hüwel, J.D., Berns, F., Beecks, C.: Automated kernel search for gaussian processes on data streams. In: IEEE BigData, pp. 3584–3588. IEEE (2021)
14. Kim, H., Teh, Y.W.: Scaling up the automatic statistician: scalable structure discovery using gaussian processes. In: AISTATS. Proceedings of Machine Learning Research, vol. 84, pp. 575–584. PMLR (2018)
15. Lloyd, J., Duvenaud, D., Grosse, R., Tenenbaum, J., Ghahramani, Z.: Automatic construction and natural-language description of nonparametric regression models. In: Proceedings of the AAAI Conference on Artificial Intelligence, vol. 28 (2014)
16. Page, E.S.: Continuous inspection schemes. Biometrika **41**(1/2), 100–115 (1954)
17. Rasmussen, C.E., Williams, C.K.I.: Gaussian Processes for Machine Learning. Adaptive Computation and Machine Learning, MIT Press, Cambridge (2006)
18. Truong, C., Oudre, L., Vayatis, N.: Selective review of offline change point detection methods. Sig. Process. **167**, 107299 (2020)

Optimal Fixed-Premise Repairs of $\mathcal{EL}$ TBoxes

Francesco Kriegel(✉)

Theoretical Computer Science, Technische Universität Dresden, Dresden, Germany
francesco.kriegel@tu-dresden.de

Abstract. Reasoners can be used to derive implicit consequences from an ontology. Sometimes unwanted consequences are revealed, indicating errors or privacy-sensitive information, and the ontology needs to be appropriately repaired. The classical approach is to remove just enough axioms such that the unwanted consequences vanish. However, this is often too rough since mere axiom deletion also erases many other consequences that might actually be desired. The goal should not be to remove a minimal number of axioms but to modify the ontology such that only a minimal number of consequences is removed, including the unwanted ones. Specifically, a repair should rather be logically entailed by the input ontology, instead of being a subset. To this end, we introduce a framework for computing fixed-premise repairs of $\mathcal{EL}$ TBoxes. In the first variant the conclusions must be generalizations of those in the input TBox, while in the second variant no such restriction is imposed. In both variants, every repair is entailed by an optimal one and, up to equivalence, the set of all optimal repairs can be computed in exponential time. A prototypical implementation is provided. In addition, we show new complexity results regarding gentle repairs.

Keywords: Description logic · Optimal repair · TBox repair · Generalized-conclusion repair · Fixed-premise repair

1 Introduction

Description Logics (DLs) [4] are logic-based languages with model-theoretic semantics that are designed for knowledge representation and reasoning. Several DLs are fragments of first-order logic, but with restricted expressivity such that reasoning problems usually remain decidable. Knowledge represented as a DL ontology consists of a terminological part (the schema, TBox) and an assertional part (the data, ABox). The TBox expresses global knowledge on the underlying domain of interest, such as implicative rules and integrity constraints, and the ABox expresses local knowledge, such as assignment of objects to classes or relations between objects. DLs differ in their expressivity and there is always a trade-off to complexity of reasoning. Many reasoning tasks in lightweight DLs such as $\mathcal{EL}$ [3] and *DL-Lite* [12] are in P and thus tractable, but are N2EXP-complete in the very expressive DL $\mathcal{SROIQ}$ [16,18], which is the logical foundation of the OWL 2 Web Ontology Language.[1] However, the latter is a worst-case complexity, and efficient reasoning techniques [34] can often avoid reaching it.

[1] https://www.w3.org/TR/owl2-primer/.

R. Bergmann et al. (Eds.): KI 2022, LNAI 13404, pp. 115–130, 2022.
https://doi.org/10.1007/978-3-031-15791-2_11

Reasoners can be used to derive implicit consequences from an ontology. Sometimes unwanted consequences are revealed, indicating errors or privacy-sensitive information, and the ontology needs to be appropriately repaired. The classical approach is to remove just enough axioms such that the unwanted consequences vanish [14,29]. In particular, optimal classical repairs can be obtained by means of axiom pinpointing [10,11,31,32]: firstly, one determines all minimal subsets of the given ontology that entail the unwanted consequences (so-called *justifications*), secondly, one constructs a minimal set that contains at least one axiom from each justification (a so-called *hitting set*) and, thirdly, one removes from the erroneous ontology all axioms in the hitting set. In a similar way, inconsistency or incoherence of ontologies can be resolved—a task also called ontology debugging [17,22,30,33]. Proof visualizations can be used to guide the process of ontology repair [1], and it can be distributed and parallelized by means of decomposition [26]. Furthermore, there are connections to belief revision [13].

The classical repair approach is often too rough since mere axiom deletion also erases too many other consequences that might actually be desired. The goal should not be to remove a minimal number of axioms but to modify the ontology such that only a minimal number of consequences is removed, including the unwanted ones. Alternative repair techniques that are less dependent on the syntax should therefore be designed. To this end, a repair need not be a subset of the input ontology anymore, but must only be logically entailed by it.

A framework for constructing *gentle repairs* based on axiom weakening was developed [8]. The main difference to the classical repair approach is that, instead of being removed completely, one axiom from each justification is replaced by a logically weaker one such that the unwanted consequences cannot be derived anymore. The framework can be applied to every monotonic logic, and one only needs to devise a suitable weakening relation on axioms.[2] In terms of belief revision, gentle repairs correspond to pseudo-contractions [27].

In the DL $\mathcal{EL}$ [3], concept descriptions are built from concept names and role names by conjunction and existential restriction, and a TBox is a finite set of concept inclusions (CIs), which are axioms of the form $C \sqsubseteq D$ where the premise C and the conclusion D are concept descriptions. For instance, the CI $\mathsf{MountainBike} \sqsubseteq \exists \mathsf{hasPart}.\mathsf{SuspensionFork} \sqcap \exists \mathsf{isSuitableFor}.\mathsf{OffRoadCycling}$ expresses that every mountain bike has a suspension fork and is suitable for off-road cycling. Such axioms can be weakened by specializing the premise or by generalizing the conclusion. Two weakening relations $\succ^{\mathsf{syn}}$ and $\succ^{\mathsf{sub}}$ for $\mathcal{EL}$ CIs were devised [8], which instantiate the gentle repair framework for $\mathcal{EL}$ TBoxes.

Repairs of $\mathcal{EL}$ TBoxes can also be obtained by axiomatizing the logical intersection of the input TBox and the theory of a countermodel to the unwanted consequences [15], e.g., by means of the framework for axiomatizing $\mathcal{EL}$ closure operators [19]. Such a countermodel can either be manually specified by the knowledge engineer or be automatically obtained by transforming a canonical model of the TBox, e.g., with the methods for repairing quantified ABoxes [9].

[2] There is always the trivial weakening relation that replaces each axiom with a tautology, for which each gentle repair is a classical repair.

The axiomatization method is very precise since it can introduce new premises in the resulting repair if necessary [15, Example 18]. From a theoretical perspective, this is a clear advantage simply because thereby a large amount of knowledge can be retained in the repair. From a practical perspective, however, this can be seen as a disadvantage as the resulting repairs might get considerably larger than the input TBox. In order to prevent such an increase in size, I have further proposed to construct a repair from a countermodel $\mathcal{J}$ in a slightly different manner [15]: namely one keeps all premises unchanged and only generalizes the conclusions by means of $\mathcal{J}$, which yields an approach very close to the gentle repairs for the weakening relation $\succ^{\mathsf{sub}}$.

The goal of this article is to elaborate the latter idea in detail. We introduce a framework for computing *generalized-conclusion repairs* of $\mathcal{EL}$ TBoxes, where the premises must not be changed and the conclusions can be generalized. We first devise a canonical construction of such repairs from polynomial-size seeds, and then show that each generalized-conclusion repair is entailed by an optimal one and that, up to equivalence, the set of all optimal generalized-conclusion repairs can be computed in exponential time.

As an example, consider the TBox consisting of the single concept inclusion $\mathsf{Bike} \sqsubseteq \exists\mathsf{hasPart}.\mathsf{SuspensionFork} \sqcap \exists\mathsf{isSuitableFor}.\mathsf{OffRoadCycling}$, which differs from the above in that the premise is replaced by Bike. It entails the false CIs $\mathsf{Bike} \sqsubseteq \exists\mathsf{hasPart}.\mathsf{SuspensionFork}$ and $\mathsf{Bike} \sqsubseteq \exists\mathsf{isSuitableFor}.\mathsf{OffRoadCycling}$. The (unique) optimal generalized-conclusion repair consists of the single CI $\mathsf{Bike} \sqsubseteq \exists\mathsf{hasPart}.\top \sqcap \exists\mathsf{isSuitableFor}.\top$. In contrast, the classical repair approach deletes the single CI completely, yielding an empty repair, which only entails tautologies but does not entail that every bike has a part and is suitable for something.

In addition to developing the framework of generalized-conclusion repairs, we introduce *fixed-premise repairs*. The difference to the generalized-conclusion repairs is that the conclusions of CIs need not be generalizations anymore; only the premises must remain the same and the input TBox must entail each CI in the repair. Thereby even more consequences can be retained. Employing the same seeds as before, we show that every fixed-premise repair is entailed by an optimal one and that the set of all optimal fixed-premise repairs can be computed in exponential time.

Clearly, the above generalized-conclusion repair is not satisfactory if additional knowledge would be expressed in the given TBox, such as $\mathsf{SuspensionFork} \sqsubseteq \mathsf{Fork}$ and $\mathsf{OffRoadCycling} \sqsubseteq \mathsf{Cycling}$. Both additional CIs are obviously true in real world and should thus be retained in an optimal repair. Taking this into account, the (unique) optimal fixed-premise repair additionally contains the CI $\mathsf{Bike} \sqsubseteq \exists\mathsf{hasPart}.\mathsf{Fork} \sqcap \exists\mathsf{isSuitableFor}.\mathsf{Cycling}$, and it preserves more consequences than the above generalized-conclusion repair, e.g., that every bike is suitable for cycling.

An experimental implementation is available.[3] In addition, we provide new complexity results regarding gentle repairs w.r.t. the weakening relation $\succ^{\mathsf{sub}}$. Due to space constraints, proofs can only be found in the extended version [20].

[3] https://github.com/francesco-kriegel/right-repairs-of-el-tboxes.

2 Preliminaries

Fix a *signature* Σ, which is a disjoint union of a set Σ_{C} of *concept names* and a set Σ_{R} of *role names*. In $\mathcal{EL}$, *concept descriptions* are inductively constructed by means of the grammar rule $C ::= \top \mid A \mid C \sqcap C \mid \exists r.C$ where A ranges over Σ_{C} and r over Σ_{R}. A *concept inclusion (CI)* is of the form $C \sqsubseteq D$ for concept descriptions C and D, where we call C the *premise* and D the *conclusion*. A *terminological box (TBox)* $\mathcal{T}$ is a finite set of concept inclusions. The set of all premises in $\mathcal{T}$ is denoted by $\mathsf{Prem}(\mathcal{T})$.

The semantics is defined via models. An *interpretation* $\mathcal{I}$ consists of a *domain* $\mathsf{Dom}(\mathcal{I})$, which is a non-empty set, and an *interpretation function* $\cdot^{\mathcal{I}}$ that maps each concept name A to a subset $A^{\mathcal{I}}$ of $\mathsf{Dom}(\mathcal{I})$ and that maps each role name r to a binary relation $r^{\mathcal{I}}$ over $\mathsf{Dom}(\mathcal{I})$. The interpretation function is extended to all concept descriptions in the following recursive manner: $\top^{\mathcal{I}} := \mathsf{Dom}(\mathcal{I})$, $(C \sqcap D)^{\mathcal{I}} := C^{\mathcal{I}} \cap D^{\mathcal{I}}$, and $(\exists r.C)^{\mathcal{I}} := \{\, x \mid (x, y) \in r^{\mathcal{I}} \text{ for some } y \in C^{\mathcal{I}} \,\}$. Furthermore, $\mathcal{I}$ *satisfies* a CI $C \sqsubseteq D$ if $C^{\mathcal{I}} \subseteq D^{\mathcal{I}}$, written $\mathcal{I} \models C \sqsubseteq D$, and $\mathcal{I}$ is a *model* of a TBox $\mathcal{T}$ if it satisfies all CIs in $\mathcal{T}$, written $\mathcal{I} \models \mathcal{T}$. We say that $\mathcal{T}$ *entails* $C \sqsubseteq D$ if $C \sqsubseteq D$ is satisfied in every model of $\mathcal{T}$, denoted as $\mathcal{T} \models C \sqsubseteq D$. We then also say that C *is subsumed by* D w.r.t. $\mathcal{T}$ and write $C \sqsubseteq^{\mathcal{T}} D$. Subsumption in $\mathcal{EL}$ can be decided in polynomial time [3]. With $C \sqsubset^{\mathcal{T}} D$ we abbreviate $C \sqsubseteq^{\mathcal{T}} D$ and $D \not\sqsubseteq^{\mathcal{T}} C$. Given sets $\mathcal{K}$ and $\mathcal{L}$ of $\mathcal{EL}$ concept descriptions, we say that $\mathcal{K}$ *is covered by* $\mathcal{L}$ w.r.t. $\mathcal{T}$ and write $\mathcal{K} \leq^{\mathcal{T}} \mathcal{L}$ if, for each $K \in \mathcal{K}$, there is some $L \in \mathcal{L}$ such that $K \sqsubseteq^{\mathcal{T}} L$.

An *atom* is either a concept name or an existential restriction $\exists r.C$. Order and repetitions of atoms in conjunctions as well as nestings of conjunctions are irrelevant. In this sense, each concept description C is a conjunction of atoms, which we call the *top-level conjuncts* of C, and the set of these is denoted by $\mathsf{Conj}(C)$. Furthermore, we sometimes write $\bigsqcap\{C_1, \ldots, C_n\}$ for $C_1 \sqcap \cdots \sqcap C_n$. The (unique) *reduced form* C^{r} of a concept description C is obtained by exhaustively removing occurrences of atoms that subsume (w.r.t. $\emptyset$) another atom in the same conjunction. C is equivalent to C^{r}, and two concept descriptions are equivalent iff they have the same reduced form [21]. The subsumption order $\sqsubseteq^{\emptyset}$ restricted to reduced concept descriptions is a partial order and not just a pre-order [9].

We denote by $\mathsf{Sub}(\alpha)$ the set of all concept descriptions that occur as subconcepts in α, and $\mathsf{Atoms}(\alpha)$ is the set of atoms occurring in α. Given a set $\mathcal{K}$ of atoms, $\mathsf{Max}(\mathcal{K})$ denotes the subset consisting of all $\sqsubseteq^{\emptyset}$-maximal atoms, i.e., $\mathsf{Max}(\mathcal{K}) := \{\, K \mid K \in \mathcal{K} \text{ and there is no } K' \in \mathcal{K} \text{ such that } K \sqsubset^{\emptyset} K' \,\}$. If all atoms in $\mathcal{K}$ are reduced, then $\mathsf{Max}(\mathcal{K})$ does not contain $\sqsubseteq^{\emptyset}$-comparable atoms.

Let $\mathcal{I}$ be an interpretation and X a subset of $\mathsf{Dom}(\mathcal{I})$. A *most specific concept description (MSC)* of X w.r.t. $\mathcal{I}$ is a concept description C that satisfies $X \subseteq C^{\mathcal{I}}$ and, for each concept description D, $X \subseteq D^{\mathcal{I}}$ implies $C \sqsubseteq^{\emptyset} D$. The MSC of X w.r.t. $\mathcal{I}$ is unique up to equivalence and is denoted as $X^{\mathcal{I}}$. Due to cycles in the interpretation, MSCs might not be expressible in $\mathcal{EL}$, but MSCs always exist in an extension of $\mathcal{EL}$ with greatest fixed-points, e.g., in $\mathcal{EL}_{\mathsf{si}}$ [23]. The latter DL extends $\mathcal{EL}$ with *simulation quantifiers* $\exists^{\mathsf{sim}}(\mathcal{I}, x)$ where the semantics of such concept descriptions is defined by: $y \in (\exists^{\mathsf{sim}}(\mathcal{I}, x))^{\mathcal{J}}$ if there is a simulation from $\mathcal{I}$ to $\mathcal{J}$ that contains (x, y). As shown in [19, Proposition 4.1.6], the MSC $X^{\mathcal{I}}$ is equivalent to $\exists^{\mathsf{sim}}(\wp(\mathcal{I}), X)$, where the *powering* $\mathcal{P}(\mathcal{I})$ has domain $\mathsf{Dom}(\wp(\mathcal{I})) := \wp(\mathsf{Dom}(\mathcal{I}))$, and $A^{\wp(\mathcal{I})}$ consists of all subsets X such that $X \subseteq A^{\mathcal{I}}$, and $r^{\wp(\mathcal{I})}$ consists of all pairs (X, Y) such that Y is a minimal hitting set of $\{\, \{\, y \mid (x, y) \in r^{\mathcal{I}} \,\} \mid x \in X \,\}$. A CI $C \sqsubseteq D$ is satisfied in $\mathcal{I}$ iff $C^{\mathcal{I}\mathcal{I}} \sqsubseteq^{\emptyset} D$, and $X \subseteq C^{\mathcal{I}}$ is equivalent to $X^{\mathcal{I}} \sqsubseteq^{\emptyset} C$ for each subset $X \subseteq \mathsf{Dom}(\mathcal{I})$ and for each $\mathcal{EL}_{\mathsf{si}}$ concept description C.

A *least common subsumer (LCS)* of concept descriptions C and D is a concept description E such that $C \sqsubseteq^{\emptyset} E$ as well as $D \sqsubseteq^{\emptyset} E$ and, for each concept description F, $C \sqsubseteq^{\emptyset} F$ and $D \sqsubseteq^{\emptyset} F$ implies $E \sqsubseteq^{\emptyset} F$. The LCS of C and D is unique up to equivalence and we denote it by $C \vee D$. It can be computed as the product of the graphs representing C and D. In particular, the LCS of an $\mathcal{EL}$ concept description C and an $\mathcal{EL}_{\mathsf{si}}$ concept description $\exists^{\mathsf{sim}}(\mathcal{I}, x)$ is always expressible in $\mathcal{EL}$ and the following recursion allows us to construct it:

$$C \vee \exists^{\mathsf{sim}}(\mathcal{I}, x) \equiv^{\emptyset} \bigsqcap\{\, A \mid A \in \mathsf{Conj}(C) \text{ and } x \in A^{\mathcal{I}} \,\} \\ \sqcap \bigsqcap\{\, \exists r.(D \vee \exists^{\mathsf{sim}}(\mathcal{I}, y)) \mid \exists r.D \in \mathsf{Conj}(C) \text{ and } (x, y) \in r^{\mathcal{I}} \,\}.$$

Furthermore, the MSC $X^{\mathcal{I}}$ is equivalent to the LCS of all $\exists^{\mathsf{sim}}(\mathcal{I}, x)$ where $x \in X$.

3 Generalized-Conclusion Repairs of $\mathcal{EL}$ TBoxes

In this section we develop the framework for computing generalized-conclusion repairs of $\mathcal{EL}$ TBoxes. We begin with defining basic notions.

Definition 1. *Let $\mathcal{T}$ and $\mathcal{U}$ be $\mathcal{EL}$ TBoxes. We say that $\mathcal{U}$ is a* generalized-conclusion weakening (GC-weakening) *of $\mathcal{T}$, written $\mathcal{T} \succeq_{\mathsf{GC}} \mathcal{U}$ if, for each CI $C \sqsubseteq D$ in $\mathcal{U}$, there is a CI $E \sqsubseteq F$ in $\mathcal{T}$ such that $C = E$ and $F \sqsubseteq^{\emptyset} D$.*

GC-weakening is strictly stronger than entailment, i.e., $\mathcal{T} \succeq_{\mathsf{GC}} \mathcal{U}$ implies $\mathcal{T} \models \mathcal{U}$ but the converse need not hold. For instance, $\{A \sqcap B \sqsubseteq \exists r.(A \sqcap B), C \sqsubseteq A \sqcap \exists r.A\}$ has the GC-weakening $\{A \sqcap B \sqsubseteq \exists r.A \sqcap \exists r.B,\ C \sqsubseteq \exists r.A\}$, and it entails $\{A \sqcap B \sqsubseteq \exists r.(A \sqcap \exists r.A)\}$, which is not a GC-weakening.

Definition 2. *A* repair request *$\mathcal{P}$ is a finite set of $\mathcal{EL}$ concept inclusions. A TBox $\mathcal{T}$* complies *with $\mathcal{P}$ if it does not entail any CI in $\mathcal{P}$, i.e., it holds that $\mathcal{T} \not\models C \sqsubseteq D$ for each $C \sqsubseteq D \in \mathcal{P}$. A* countermodel *to $\mathcal{P}$ is an interpretation in which none of the CIs in $\mathcal{P}$ is satisfied.*

Definition 3. *Given an $\mathcal{EL}$ TBox $\mathcal{T}$ and a repair request $\mathcal{P}$, a* generalized-conclusion repair (GC-repair) *of $\mathcal{T}$ for $\mathcal{P}$ is an $\mathcal{EL}$ TBox $\mathcal{U}$ that is a GC-weakening of $\mathcal{T}$ and complies with $\mathcal{P}$. We further call $\mathcal{U}$* optimal *if there is no other GC-repair $\mathcal{V}$ such that $\mathcal{V} \succeq_{\mathsf{GC}} \mathcal{U}$ but $\mathcal{U} \not\succeq_{\mathsf{GC}} \mathcal{V}$.*

Throughout the whole section we assume that $\mathcal{T}$ is an $\mathcal{EL}$ TBox and that $\mathcal{P}$ is a repair request, and the goal is to construct a generalized-conclusion repair (preferably an optimal one). Of course, if $\mathcal{P}$ contains a tautology, then no repair exists. We therefore assume that this is not the case. Without loss of generality, all concept descriptions in $\mathcal{T}$ and $\mathcal{P}$ must be reduced.

Induced Countermodels. In the first step, we transform a canonical model of the input TBox $\mathcal{T}$ into countermodels to $\mathcal{P}$, which are used in the next section to devise a canonical construction of generalized-conclusion repairs. The construction of each countermodel is guided by a repair seed.

Definition 4. *A* repair seed *is a TBox $\mathcal{S}$ that complies with $\mathcal{P}$ and consists of CIs of the form $C \sqsubseteq F$ for a premise $C \in \mathsf{Prem}(\mathcal{T})$ and an atom $F \in \mathsf{Atoms}(\mathcal{P}, \mathcal{T})$ where $C \sqsubseteq^{\mathcal{T}} F$.*

The completion algorithm for $\mathcal{EL}$ is a decision procedure for the subsumption problem (and also for the instance problem). In the correctness proof a canonical model of the TBox is constructed that involves all subconcepts occurring in the TBox [3]. While this algorithm works in a rule-based manner, thus implicitly constructing the canonical model step by step, there is also a closed-form representation [25]. Resembling the latter we define the *canonical model* $\mathcal{I}$ with domain $\mathsf{Dom}(\mathcal{I}) := \{ x_C \mid C \in \mathsf{Sub}(\mathcal{P}, \mathcal{T}) \}$ and its interpretation function is given by $A^{\mathcal{I}} := \{ x_C \mid C \sqsubseteq^{\mathcal{T}} A \}$ for each $A \in \Sigma_{\mathsf{C}}$ and $r^{\mathcal{I}} := \{ (x_C, x_D) \mid C \sqsubseteq^{\mathcal{T}} \exists r.D \}$ for each $r \in \Sigma_{\mathsf{R}}$. Then $\mathcal{I}$ is a model of $\mathcal{T}$, and $x_C \in E^{\mathcal{I}}$ iff $C \sqsubseteq^{\mathcal{T}} E$ for each subconcept $C \in \mathsf{Sub}(\mathcal{P}, \mathcal{T})$ and for each $\mathcal{EL}$ concept description E [20].

The transformation of the canonical model $\mathcal{I}$ is based on modification types. These describe how copies of objects in the domain of $\mathcal{I}$ are modified in order to create objects of a countermodel.

Definition 5. *Let $x_C \in \mathsf{Dom}(\mathcal{I})$. A* modification type *for x_C is a subset $\mathcal{K}$ of $\mathsf{Atoms}(\mathcal{P}, \mathcal{T})$ where $x_C \in K^{\mathcal{I}}$ for each $K \in \mathcal{K}$, and $K_1 \not\sqsubseteq^{\emptyset} K_2$ for each two $K_1, K_2 \in \mathcal{K}$. Given a repair seed $\mathcal{S}$, we say that $\mathcal{K}$* respects *$\mathcal{S}$ if additionally $\{D\} \leq^{\mathcal{S}} \mathcal{K}$ implies $\{D\} \leq^{\emptyset} \mathcal{K}$ for each $D \in \mathsf{Sub}(\mathcal{P}, \mathcal{T})$ where $x_C \in D^{\mathcal{I}}$.*

Each repair seed $\mathcal{S}$ induces a countermodel to $\mathcal{P}$. Its domain consists of all copies of objects in the canonical model $\mathcal{I}$ that are annotated with an $\mathcal{S}$-respecting modification type. The definition of the interpretation function guarantees that each such copy does not satisfy any atom in the modification type.

Definition 6. *Let $\mathcal{S}$ be a repair seed. The* induced countermodel *$\mathcal{J}_{\mathcal{S}}$ has the domain $\mathsf{Dom}(\mathcal{J}_{\mathcal{S}})$ consisting of all objects $x_{C,\mathcal{K}}$ where $x_C \in \mathsf{Dom}(\mathcal{I})$ and $\mathcal{K}$ is a modification type for x_C that respects $\mathcal{S}$, and its interpretation function is defined by $A^{\mathcal{J}_{\mathcal{S}}} := \{ x_{C,\mathcal{K}} \mid x_C \in A^{\mathcal{I}} \text{ and } A \notin \mathcal{K} \}$ for each concept name $A \in \Sigma_{\mathsf{C}}$ and $r^{\mathcal{J}_{\mathcal{S}}} := \{ (x_{C,\mathcal{K}}, x_{D,\mathcal{L}}) \mid (x_C, x_D) \in r^{\mathcal{I}} \text{ and } \mathsf{Succ}(\mathcal{K}, r, x_D) \leq^{\emptyset} \mathcal{L} \}$ for each role name $r \in \Sigma_{\mathsf{R}}$, where $\mathsf{Succ}(\mathcal{K}, r, x_D) := \{ E \mid \exists r.E \in \mathcal{K} \text{ and } x_D \in E^{\mathcal{I}} \}$.*

We can show that an object $x_{C,\mathcal{K}}$ satisfies an $\mathcal{EL}$ concept description E in $\mathcal{J}_{\mathcal{S}}$ iff x_C satisfies E in $\mathcal{I}$ and $\mathcal{K}$ does not contain an atom subsuming E [20]. Now consider an unwanted CI $C \sqsubseteq D$ in the repair request $\mathcal{P}$. Since $\mathcal{S}$ complies with $\mathcal{P}$, there is a top-level conjunct D' in D such that $\mathcal{S} \not\models C \sqsubseteq D'$. We can thus construct an $\mathcal{S}$-respecting modification type $\mathcal{K}$ for x_C that contains an atom subsuming D' but none subsuming C. It follows that the copy $x_{C,\mathcal{K}}$ satisfies the premise C but not the conclusion D, i.e., $\mathcal{J}_{\mathcal{S}}$ is indeed a countermodel to $C \sqsubseteq D$.

Proposition 7. *For each repair seed $\mathcal{S}$, the induced countermodel $\mathcal{J}_{\mathcal{S}}$ is a countermodel to $\mathcal{P}$.*

Canonical Generalized-Conclusion Repairs. Next, we show how each repair seed $\mathcal{S}$ induces a GC-repair. We obtain it by generalizing each conclusion according to countermodel $\mathcal{J}_\mathcal{S}$, namely we take each concept inclusion $C \sqsubseteq D$ in the given TBox $\mathcal{T}$ and replace D with the least common subsumer of D and the most specific concept description E for which the CI $C \sqsubseteq E$ is satisfied in $\mathcal{J}_\mathcal{S}$.

Definition 8. *Each repair seed $\mathcal{S}$ induces the TBox*

$$\mathsf{rep}_\mathsf{GC}(\mathcal{T},\mathcal{S}) := \{\, C \sqsubseteq D \vee C^{\mathcal{J}_\mathcal{S}\mathcal{J}_\mathcal{S}} \mid C \sqsubseteq D \in \mathcal{T} \,\}.$$

The following lemma shows that $\mathsf{rep}_\mathsf{GC}(\mathcal{T},\mathcal{S})$ has exactly those TBoxes as GC-weakenings that are GC-weakenings of $\mathcal{T}$ and of which $\mathcal{J}_\mathcal{S}$ is a model.

Lemma 9. $\mathsf{rep}_\mathsf{GC}(\mathcal{T},\mathcal{S}) \succeq_\mathsf{GC} \mathcal{U}$ *iff* $\mathcal{T} \succeq_\mathsf{GC} \mathcal{U}$ *and* $\mathcal{J}_\mathcal{S} \models \mathcal{U}$

As $\mathsf{rep}_\mathsf{GC}(\mathcal{T},\mathcal{S})$ is a GC-weakening of itself, we infer that $\mathcal{J}_\mathcal{S}$ is a model of $\mathsf{rep}_\mathsf{GC}(\mathcal{T},\mathcal{S})$. According to Proposition 7, $\mathcal{J}_\mathcal{S}$ is a countermodel to $\mathcal{P}$, and so $\mathsf{rep}_\mathsf{GC}(\mathcal{T},\mathcal{S})$ complies with $\mathcal{P}$. It is further easy to see that $\mathsf{rep}_\mathsf{GC}(\mathcal{T},\mathcal{S})$ is a GC-weakening of $\mathcal{T}$. We have thus shown that the following holds.

Proposition 10. *If $\mathcal{S}$ is a repair seed, then $\mathsf{rep}_\mathsf{GC}(\mathcal{T},\mathcal{S})$ is a GC-repair.*

If the repair request $\mathcal{P}$ does not contain a tautological CI, then the empty set is already a repair seed, i.e., $\mathsf{rep}_\mathsf{GC}(\mathcal{T},\emptyset)$ is a GC-repair of $\mathcal{T}$ for $\mathcal{P}$. Furthermore, the induced GC-repairs are complete in the sense that every GC-repair is a GC-weakening of $\mathsf{rep}_\mathsf{GC}(\mathcal{T},\mathcal{S})$ for some repair seed $\mathcal{S}$.

Proposition 11. *If $\mathcal{U}$ is a GC-repair of $\mathcal{T}$ for $\mathcal{P}$, then there is a repair seed $\mathcal{S}$ such that $\mathsf{rep}_\mathsf{GC}(\mathcal{T},\mathcal{S}) \succeq_\mathsf{GC} \mathcal{U}$.*

Proof Sketch. Given a GC-repair $\mathcal{U}$, a repair seed $\mathcal{S}^*_\mathcal{U}$ is obtained as the least fixed point of the equation $\mathcal{S} = \{C \sqsubseteq F \mid C \sqsubseteq D' \in \mathcal{U},\ F \in \mathsf{Atoms}(\mathcal{P},\mathcal{T}),$ and $D' \sqsubseteq^\mathcal{S} F\}$. It has the important property that $\{D'\} \leq^{\mathcal{S}^*_\mathcal{U}} \mathcal{K}$ implies $\{C\} \leq^{\mathcal{S}^*_\mathcal{U}} \mathcal{K}$ for each CI $C \sqsubseteq D' \in \mathcal{U}$ and for each modification type $\mathcal{K}$. With this property we can easily show that $x_{E,\mathcal{K}} \in C^{\mathcal{J}_{\mathcal{S}^*_\mathcal{U}}}$ implies $x_{E,\mathcal{K}} \in (D')^{\mathcal{J}_{\mathcal{S}^*_\mathcal{U}}}$ for each CI $C \sqsubseteq D' \in \mathcal{U}$, and thus the induced countermodel $\mathcal{J}_{\mathcal{S}^*_\mathcal{U}}$ is a model of $\mathcal{U}$. Lemma 9 yields that $\mathcal{U}$ is a GC-weakening of $\mathsf{rep}_\mathsf{GC}(\mathcal{T},\mathcal{S}^*_\mathcal{U})$. □

Each repair seed is of polynomial size, and there are at most exponentially many seeds. Even with a naïve approach, we can compute all seeds in exponential time and thus also all induced GC-repairs. Then we must filter out the non-optimal ones, e.g., by comparing each two repairs w.r.t. $\succeq_\mathsf{GC}$. Each comparison needs polynomial time [3], and we obtain the following main result.

Theorem 12. *The set of all optimal GC-repairs of an $\mathcal{EL}$ TBox $\mathcal{T}$ for a repair request $\mathcal{P}$ can be computed in exponential time, and each GC-repair is a GC-weakening of an optimal one.*

In the below example, an optimal GC-repair is not polynomial-time computable.

Example 13. For the repair request $\{\exists r.A \sqsubseteq \exists r.B\}$, the TBox $\{\exists r.A \sqsubseteq \exists r.(P_1 \sqcap Q_1 \sqcap \cdots \sqcap P_n \sqcap Q_n),\ P_1 \sqcap Q_1 \sqsubseteq B,\ \ldots,\ P_n \sqcap Q_n \sqsubseteq B\}$ has the optimal GC-repair $\{\exists r.A \sqsubseteq \bigsqcap\{\, \exists r.(X_1 \sqcap \cdots \sqcap X_n) \mid X_i \in \{P_i, Q_i\}$ for each $i \in \{1,\ldots,n\}\,\},\ P_1 \sqcap Q_1 \sqsubseteq B,\ \ldots,\ P_n \sqcap Q_n \sqsubseteq B\}$. It has exponential size.

Computing a Canonical Generalized-Conclusion Repair. In the last step, we are concerned with the question how the GC-repair induced by a seed $\mathcal{S}$ can efficiently be computed. Recall that, as explained in the preliminaries, each conclusion $D \vee C^{\mathcal{J}_{\mathcal{S}}\mathcal{J}_{\mathcal{S}}}$ can be obtained as the product of the $\mathcal{EL}$ concept description D and the $\mathcal{EL}_{\mathsf{si}}$ concept description $\exists^{\mathsf{sim}}(\wp(\mathcal{J}_{\mathcal{S}}), C^{\mathcal{J}_{\mathcal{S}}})$, or alternatively as the product of D and all $\exists^{\mathsf{sim}}(\mathcal{J}_{\mathcal{S}}, x_{E,\mathcal{K}})$ where $x_{E,\mathcal{K}} \in C^{\mathcal{J}_{\mathcal{S}}}$. However, computing the induced GC-repair $\mathsf{rep}_{\mathsf{GC}}(\mathcal{T}, \mathcal{S})$ in this way is very inefficient since $\mathcal{J}_{\mathcal{S}}$ has exponential size.

The first important observation is that the concept description $C^{\mathcal{J}_{\mathcal{S}}\mathcal{J}_{\mathcal{S}}}$ is already equivalent to $\exists^{\mathsf{sim}}(\mathcal{J}_{\mathcal{S}}, x_{C,\mathcal{S}[C]})$ where $\mathcal{S}[C]$ is the largest modification type for x_C that respects $\mathcal{S}$ and does not contain an atom subsuming C. This follows from the fact that there is a simulation on $\mathcal{J}_{\mathcal{S}}$ that contains the pair $(x_{C,\mathcal{S}[C]}, x_{E,\mathcal{K}})$ for each object $x_{E,\mathcal{K}}$ in the extension $C^{\mathcal{J}_{\mathcal{S}}}$. Secondly, in order to compute the LCS $D \vee \exists^{\mathsf{sim}}(\mathcal{J}_{\mathcal{S}}, x_{C,\mathcal{S}[C]})$ it is not necessary to start from $x_{C,\mathcal{S}[C]}$ in the product construction, but it suffices to start from $x_{D,\mathcal{S}[C \sqsubseteq D]}$ where $\mathcal{S}[C \sqsubseteq D]$ is the largest modification type for x_D that respects $\mathcal{S}$ and does not contain an atom subsuming C. Thirdly, when computing the product of D and $\exists^{\mathsf{sim}}(\mathcal{J}_{\mathcal{S}}, x_{D,\mathcal{S}[C \sqsubseteq D]})$ we do not need to consider all objects $x_{E,\mathcal{K}}$ that are reachable from $x_{D,\mathcal{S}[C \sqsubseteq D]}$ in $\mathcal{J}_{\mathcal{S}}$, but only those where E is a filler of an existential restriction that occurs in D. As main result we obtain the following proposition.

Definition 14. *Given a subconcept $E \in \mathsf{Sub}(\mathcal{P}, \mathcal{T})$ and a modification type $\mathcal{K}$ for x_E that respects $\mathcal{S}$, we define the* restriction $E{\restriction_{\mathcal{K}}}$ *by the following recursion.*

$$E{\restriction_{\mathcal{K}}} := \bigsqcap\{\, A \mid A \in \mathsf{Conj}(E) \text{ and } A \notin \mathcal{K} \,\} \sqcap \bigsqcap\left\{ \exists r.F{\restriction_{\mathcal{L}}} \;\middle|\; \begin{array}{l} \exists r.F \in \mathsf{Conj}(E), \text{ and } \mathcal{L} \text{ is a } \leq^{\emptyset}\text{-minimal mod. type} \\ \text{for } x_F \text{ that respects } \mathcal{S} \text{ and where } \mathsf{Succ}(\mathcal{K}, r, x_F) \leq^{\emptyset} \mathcal{L} \end{array} \right\}$$

Proposition 15. *Given a repair seed $\mathcal{S}$, it holds that $D \vee C^{\mathcal{J}_{\mathcal{S}}\mathcal{J}_{\mathcal{S}}} \equiv^{\emptyset} D{\restriction_{\mathcal{S}[C \sqsubseteq D]}}$ for each CI $C \sqsubseteq D$ in $\mathcal{T}$, and thus the induced GC-repair $\mathsf{rep}_{\mathsf{GC}}(\mathcal{T}, \mathcal{S})$ is equivalent to the TBox $\{\, C \sqsubseteq D{\restriction_{\mathcal{S}[C \sqsubseteq D]}} \mid C \sqsubseteq D \in \mathcal{T} \,\}$, where the modification type $\mathcal{S}[C \sqsubseteq D]$ is defined as $\mathsf{Max}\{\, K \mid K \in \mathsf{Atoms}(\mathcal{P}, \mathcal{T}),\ C \not\sqsubseteq^{\mathcal{S}} K, \text{ and } D \sqsubseteq^{\mathcal{T}} K \,\}$.*

Two Observations. The below example illustrates that entailment between repair seeds need not imply entailment between the induced GC-repairs.

Example 16. For the TBox $\mathcal{T} := \{A \sqsubseteq B,\ C \sqsubseteq \exists r.(A \sqcap B)\}$ and the repair request $\mathcal{P} := \{C \sqsubseteq \exists r.B\}$, there are two optimal GC-repairs: $\mathcal{U}_1 := \{A \sqsubseteq B, C \sqsubseteq \exists r.\top\}$, induced by the seed $\mathcal{S}_1 := \{A \sqsubseteq B\}$, and $\mathcal{U}_2 := \{A \sqsubseteq \top,\ C \sqsubseteq \exists r.A\}$, induced by $\mathcal{S}_2 := \emptyset$. Now, $\mathcal{U}_1$ does not entail $\mathcal{U}_2$, although $\mathcal{S}_1$ entails $\mathcal{S}_2$.

The next example shows that, possibly contradicting intuition, it does not suffice that a repair seed consists only of CIs $C \sqsubseteq F$ where $C \sqsubseteq D \in \mathcal{T}$ and $F \in \mathsf{Atoms}(\mathcal{P}, \mathcal{T})$ such that $D \sqsubseteq^{\emptyset} F$. We definitely sometimes need CIs $C \sqsubseteq F$ where $C \sqsubseteq^{\mathcal{T}} F$, as per Definition 4. Notably, the only optimal repair in the following example can be described by the latter CIs.

Example 17. Consider the TBox $\mathcal{T} := \{A \sqsubseteq \exists r.\exists r.(B \sqcap C),\ \exists r.B \sqsubseteq B\}$ and the repair request $\mathcal{P} := \{A \sqsubseteq \exists r.\exists r.C\}$. The unique optimal GC-repair is $\{A \sqsubseteq \exists r.\exists r.B,\ \exists r.B \sqsubseteq B\}$. It is induced only by the seeds $\{A \sqsubseteq \exists r.B,\ \exists r.B \sqsubseteq B\}$ and $\{A \sqsubseteq B,\ A \sqsubseteq \exists r.B,\ \exists r.B \sqsubseteq B\}$. Specifically the seed CI $A \sqsubseteq \exists r.B$ would not be allowed if we simplified the definition of a seed as explained above.

Another GC-repair is $\{A \sqsubseteq \exists r.\exists r.B,\ \exists r.B \sqsubseteq \top\}$, which is induced by the empty seed $\emptyset$, but also by $\{A \sqsubseteq B\}$, $\{A \sqsubseteq \exists r.B\}$, and $\{A \sqsubseteq B,\ A \sqsubseteq \exists r.B\}$.

The above example also shows that a repair need not entail its seed, and that a repair can be induced by multiple seeds. Conducted experiments support the claim that each GC-repair might be induced by a unique seed with minimal cardinality and such that every CI in the seed is also entailed by the repair.

4 Fixed-Premise Repairs of $\mathcal{EL}$ TBoxes

We have seen in the introduction that simply generalizing the conclusions of the input TBox $\mathcal{T}$ might not yield satisfactory repairs. Therefore, we will now construct repairs that can retain more consequences. It is still required that each premise in the repair is also a premise in $\mathcal{T}$, but apart from that we do not impose further conditions except that the repair must, of course, be entailed by $\mathcal{T}$.

Definition 18. *Consider TBoxes $\mathcal{T}$ and $\mathcal{U}$. We say that $\mathcal{T}$* fixed-premise entails (FP-entails) *$\mathcal{U}$, written $\mathcal{T} \models_{\mathsf{FP}} \mathcal{U}$, if $\mathsf{Prem}(\mathcal{T}) = \mathsf{Prem}(\mathcal{U})$ and $\mathcal{T} \models \mathcal{U}$.*

$\mathcal{T} \succeq_{\mathsf{GC}} \mathcal{U}$ implies $\mathcal{T} \models_{\mathsf{FP}} \mathcal{U}$ and the latter implies $\mathcal{T} \models \mathcal{U}$, but the converse implications need not hold. This means that the relation $\models_{\mathsf{FP}}$ is between $\succeq_{\mathsf{GC}}$ and $\models$. Thus, repairs based on this new relation are, usually, better than GC-repairs.

Definition 19. *Let $\mathcal{T}$ be an $\mathcal{EL}$ TBox and $\mathcal{P}$ a repair request. A* fixed-premise repair (FP-repair) *of $\mathcal{T}$ for $\mathcal{P}$ is an $\mathcal{EL}$ TBox $\mathcal{U}$ that is FP-entailed by $\mathcal{T}$ and complies with $\mathcal{P}$. We further call $\mathcal{U}$* optimal *if there is no other FP-repair $\mathcal{V}$ such that $\mathcal{V} \models_{\mathsf{FP}} \mathcal{U}$ and $\mathcal{U} \not\models_{\mathsf{FP}} \mathcal{V}$.*

Obviously, each GC-repair is an FP-repair but the converse does not hold.

By reusing the notion of a repair seed as well as the results on GC-repairs in Sect. 3, we obtain the following characterization of (optimal) FP-repairs. First of all, each repair seed $\mathcal{S}$ induces an FP-repair: we take each CI $C \sqsubseteq D$ in the input TBox $\mathcal{T}$ and replace the conclusion D with the most specific concept description E for which the CI $C \sqsubseteq E$ is satisfied in the induced countermodel $\mathcal{J}_{\mathcal{S}}$. Note that now D is not generalized anymore by computing an LCS.

Definition 20. *Each repair seed $\mathcal{S}$ induces the TBox*

$$\mathsf{rep}_{\mathsf{FP}}(\mathcal{T}, \mathcal{S}) := \{\, C \sqsubseteq C^{\mathcal{J}_{\mathcal{S}}\mathcal{J}_{\mathcal{S}}} \mid C \in \mathsf{Prem}(\mathcal{T}) \,\}.$$

Recall that each conclusion $C^{\mathcal{J}_{\mathcal{S}}\mathcal{J}_{\mathcal{S}}}$ is equivalent to the $\mathcal{EL}_{\mathsf{si}}$ concept description $\exists^{\mathsf{sim}}(\mathcal{J}_{\mathcal{S}}, x_{C,\mathcal{S}[C]})$, where $\mathcal{S}[C]$ is the largest modification type for x_C that respects $\mathcal{S}$ and does not cover $\{C\}$, i.e., $\mathcal{S}[C] := \mathsf{Max}\{ K \mid K \in \mathsf{Atoms}(\mathcal{P},\mathcal{T}),\ C \not\sqsubseteq^{\mathcal{S}} K, \text{ and } C \sqsubseteq^{\mathcal{T}} K \}$. Analogously to the GC-repairs, every TBox $\mathsf{rep}_{\mathsf{FP}}(\mathcal{T},\mathcal{S})$ is an FP-repair and each FP-repair is FP-entailed by $\mathsf{rep}_{\mathsf{FP}}(\mathcal{T},\mathcal{S})$ for some repair seed $\mathcal{S}$.

Proposition 21. *For each repair seed $\mathcal{S}$, the TBox $\mathsf{rep}_{\mathsf{FP}}(\mathcal{T},\mathcal{S})$ is an FP-repair.*

Proposition 22. *For each FP-repair $\mathcal{U}$ of $\mathcal{T}$ for $\mathcal{P}$, there is a repair seed $\mathcal{S}$ such that $\mathsf{rep}_{\mathsf{FP}}(\mathcal{T},\mathcal{S}) \models_{\mathsf{FP}} \mathcal{U}$.*

We obtain the following main result of this section. Its proof is analogous to Theorem 12, but uses the argument that entailment between $\mathcal{EL}_{\mathsf{si}}$ TBoxes can be decided in polynomial time [23].

Theorem 23. *The set of all optimal FP-repairs of an $\mathcal{EL}$ TBox $\mathcal{T}$ for a repair request $\mathcal{P}$ can be computed in exponential time, and each FP-repair is FP-entailed by an optimal one.*

We have seen in Example 17 that a repair seed might not be entailed by its induced GC-repair. This is *not* the case for its induced FP-repair.

Lemma 24. *Each repair seed $\mathcal{S}$ is entailed by its induced FP-repair $\mathsf{rep}_{\mathsf{FP}}(\mathcal{T},\mathcal{S})$.*

Contrary to the GC-repairs, not every FP-repair is an $\mathcal{EL}$ TBox but might require cyclic $\mathcal{EL}_{\mathsf{si}}$ concept descriptions [23] as conclusions to be optimal. For instance, consider the TBox $\{A \sqsubseteq \exists r.A\}$ that is also the repair request. The unique optimal FP-repair consists of the single CI

$$A \sqsubseteq \exists^{\mathsf{sim}}(\ \to\overset{A}{\bigcirc}\overset{r}{\longrightarrow}\bigcirc\overset{r}{\longrightarrow}\overset{A}{\bigcirc}\circlearrowright r\ ).$$

If a standard $\mathcal{EL}$ TBox is required as result, one might rewrite the repair by introducing fresh concept names (used as quantified monadic second-order variables). For the above optimal repair this yields the TBox $\exists\{X,Y,Z\}.\{A \sqsubseteq X,\ X \sqsubseteq A \sqcap \exists r.Y,\ Y \sqsubseteq \exists r.Z,\ Z \sqsubseteq A \sqcap \exists r.Z\}$. One could also try to compute a uniform interpolant [24,28] of the latter in order to get rid of the additional symbols and so obtain a usual $\mathcal{EL}$ TBox. Alternatively, one could unfold the cyclic conclusions into $\mathcal{EL}$ concept descriptions up to a certain role-depth bound.

If the TBox $\mathcal{T}$ is cycle-restricted [2], then the canonical model $\mathcal{I}$ is acyclic and so is the induced countermodel $\mathcal{J}_{\mathcal{S}}$ for each repair seed $\mathcal{S}$. The FP-repair $\mathsf{rep}_{\mathsf{FP}}(\mathcal{T},\mathcal{S})$ then only has acyclic $\mathcal{EL}_{\mathsf{si}}$ concept descriptions as conclusions and these can be rewritten into $\mathcal{EL}$ concept descriptions.

5 Complexity of Maximally Strong $\succ^{\mathsf{sub}}$-Weakenings

As mentioned in the introduction, a framework for computing *gentle repairs* based on axiom weakening was developed, and two weakening relations that operate on $\mathcal{EL}$ CIs were introduced [8]. We briefly recall the *modified gentle repair algorithm*. As input, fix an ontology $\mathcal{O}$ that is partitioned into a static part $\mathcal{O}_\mathsf{s}$ and a refutable part $\mathcal{O}_\mathsf{r}$ as well as an axiom α, the unwanted consequence, that follows from $\mathcal{O}$ but not already from $\mathcal{O}_\mathsf{s}$. A *repair* is an ontology $\mathcal{O}'$ such that $\mathcal{O} \models \mathcal{O}'$ but $\mathcal{O}_\mathsf{s} \cup \mathcal{O}' \not\models \alpha$. In order to obtain such a repair, we repeatedly compute a justification J for α and replace one axiom $\beta \in J$ by a weaker one.[4] Specifically, a *justification* for α is a minimal subset $J \subseteq \mathcal{O}_\mathsf{r}$ such that $\mathcal{O}_\mathsf{s} \cup J \models \alpha$. After at most exponentially many iterations a repair has been obtained.

A *weakening relation* is a pre-order $\succ$ on axioms such that $\beta \succ \gamma$ implies that γ is weaker than β. Such relations are used to guide the selection of a weaker axiom in the above iteration. Specifically, when processing a justification J for α and a selected axiom $\beta \in J$, we should replace β by a *maximally strong weakening*, which is an axiom γ such that $\beta \succ \gamma$ and $\mathcal{O}_\mathsf{s} \cup (J \setminus \{\beta\}) \cup \{\gamma\} \not\models \alpha$, but $\mathcal{O}_\mathsf{s} \cup (J \setminus \{\beta\}) \cup \{\delta\} \models \alpha$ for all δ where $\beta \succ \delta \succ \gamma$. This prevents the loss of too many other consequences (apart from α). However, maximally strong weakenings need not exist for every weakening relation.

The *syntactic weakening relation* $\succ^\mathsf{syn}$ on $\mathcal{EL}$ CIs removes subconcepts from the conclusions. Maximally strong $\succ^\mathsf{syn}$-weakenings always exist in all directions,[5] all of them can be computed in exponential time, one can be computed in polynomial time, and recognizing them is coNP-complete.

The *semantic weakening relation* $\succ^\mathsf{sub}$ replaces conclusions of $\mathcal{EL}$ CIs by more general concepts, i.e., $C \sqsubseteq D \succ^\mathsf{sub} C' \sqsubseteq D'$ if $C = C'$, $D \sqsubseteq^\emptyset D'$, and $C' \sqsubseteq D' \not\models C \sqsubseteq D$. It has only been known that maximally strong $\succ^\mathsf{sub}$-weakenings always exist in all directions (see footnote 5) all of them can effectively be computed, and recognizing them is coNP-hard. As a side result from Sect. 3, we obtain the following.

Proposition 25. *If the unwanted consequence α is a CI, then all maximally strong $\succ^\mathsf{sub}$-weakenings of an axiom β in a justification J for α can be computed in exponential time.*

The following modification of [8, Example 30] shows that a single maximally strong $\succ^\mathsf{sub}$-weakening cannot always be computed in polynomial time.

Example 26. Take the ontology $\mathcal{O}$ with $\mathcal{O}_\mathsf{s} := \{\, P_i \sqcap Q_i \sqsubseteq B \mid i \in \{1,\ldots,n\}\,\}$ and $\mathcal{O}_\mathsf{r} := \{\beta\}$ for $\beta := \exists r.A \sqsubseteq \exists r.(P_1 \sqcap Q_1 \sqcap \cdots \sqcap P_n \sqcap Q_n)$, and the unwanted consequence $\alpha := \exists r.A \sqsubseteq \exists r.B$. Then $J := \{\beta\}$ is a justification for α. There is exactly one maximally strong $\succ^\mathsf{sub}$-weakening of β in J, namely $\exists r.A \sqsubseteq \bigsqcap\{\, \exists r.(X_1 \sqcap \cdots \sqcap X_n) \mid X_i \in \{P_i, Q_i\}$ for each $i \in \{1,\ldots,n\}\,\}$. Since this weakening has exponential size, it cannot be computed in polynomial time.

Finally, recognizing maximally strong $\succ^\mathsf{sub}$-weakenings is also in coNP.

Proposition 27. *The problem of deciding whether an $\mathcal{EL}$ CI γ is a maximally strong $\succ^\mathsf{sub}$-weakening of an $\mathcal{EL}$ CI β in a justification J for α is coNP-complete.*

---

[4] We say that γ is *weaker* than β if β entails γ but γ does not entail β.

[5] That is, each weakening of an axiom β in a justification J is weaker than a maximally strong weakening of β in J—where a *weakening* of β in J is an axiom γ such that $\beta \succ \gamma$ and $\mathcal{O}_\mathsf{s} \cup (J \setminus \{\beta\}) \cup \{\gamma\} \not\models \alpha$.

6 Conclusion

We have introduced a framework for computing generalized-conclusion repairs of $\mathcal{EL}$ TBoxes, where the premises must not be changed and the conclusions can be generalized. Up to equivalence, the set of all optimal generalized-conclusion repairs can be computed in exponential time. Each generalized-conclusion repair is entailed by an optimal one and, furthermore, each optimal generalized-conclusion repair can be described by a repair seed that has polynomial size. In addition, we have extended the framework to the fixed-premise repairs, with the difference that the conclusions need not be generalizations anymore. This usually leads to better repairs, but with the disadvantage that the conclusions in an optimal repair might be cyclic and can thus only be expressed in an extension of $\mathcal{EL}$ with greatest fixed-point semantics or by introducing fresh concept names. Not affected by the latter, all optimal fixed-premise repairs can be computed in exponential time too, and each fixed-premise repair is entailed by an optimal one, which is induced by a polynomial-size repair seed. An experimental implementation is available, which interacts with the user to construct the seed from which the repair is built.

An interesting task for future research is to combine this approach to repairing TBoxes with the approach to repairing quantified ABoxes [5]. This should be possible by, firstly, adapting the notion of a repair seed such that it can additionally contain concept assertions and role assertions and, secondly, suitably adapting the transformation of the saturation/canonical model into a countermodel from which the final repair is constructed. Another interesting question is how the approach can be extended to more expressive DLs, such as $\mathcal{EL}$ with the bottom concept $\bot$, nominals $\{a\}$, inverse roles r^-, and role inclusions $R_1 \circ \cdots \circ R_n \sqsubseteq S$. Ideas from the latest extension of quantified ABox repairs to the DL $\mathcal{ELROI}(\bot)$ might be helpful [6,7]. An extension with nominals would immediately add support for ABox axioms, since each concept assertion $C(a)$ is equivalent to the CI $\{a\} \sqsubseteq C$ and each role assertion is equivalent to $\{a\} \sqsubseteq \exists r.\{b\}$. Furthermore, it should not be hard to add support for a partitioning of the TBox into a static and a refutable part, or for a set of wanted consequences that must still be entailed by the repair. Also, it would be interesting to find a suitable partial order on repair seeds such that minimality of the seed is equivalent to optimality of the induced repair, similar to the qABox repairs [9]. Last, it would be interesting to investigate whether and how the quality of the repairs can be improved if also new premises can be introduced by the repair process. Currently, this can be done by manually extending the input TBox to be repaired.

Acknowledgements. The author thanks Franz Baader for his valuable comments that helped to improve this article. He further thanks the anonymous reviewers for their helpful feedback. The author is funded by Deutsche Forschungsgemeinschaft (DFG) in Project 430150274 ("Repairing Description Logic Ontologies").

References

1. Alrabbaa, C., Baader, F., Dachselt, R., Flemisch, T., Koopmann, P.: Visualising proofs and the modular structure of ontologies to support ontology repair. In: Borgwardt, S., Meyer, T. (eds.) Proceedings of the 33rd International Workshop on Description Logics (DL 2020) co-located with the 17th International Conference on Principles of Knowledge Representation and Reasoning (KR 2020), Online Event [Rhodes, Greece], 12–14 September 2020. CEUR Workshop Proceedings, vol. 2663. CEUR-WS.org (2020). https://ceur-ws.org/Vol-2663/paper-2.pdf
2. Baader, F., Borgwardt, S., Morawska, B.: A goal-oriented algorithm for unification in $\mathcal{EL}$ w.r.t. cycle-restricted TBoxes. In: Kazakov, Y., Lembo, D., Wolter, F. (eds.) Proceedings of the 2012 International Workshop on Description Logics, DL-2012, Rome, Italy, 7–10 June 2012. CEUR Workshop Proceedings, vol. 846 (2012). https://ceur-ws.org/Vol-846/paper_1.pdf
3. Baader, F., Brandt, S., Lutz, C.: Pushing the $\mathcal{EL}$ envelope. In: Kaelbling, L.P., Saffiotti, A. (eds.) Proceedings of the Nineteenth International Joint Conference on Artificial Intelligence, Edinburgh, Scotland, 30 July–5 August 2005, pp. 364–369. Professional Book Center (2005). https://ijcai.org/Proceedings/05/Papers/0372.pdf
4. Baader, F., Horrocks, I., Lutz, C., Sattler, U.: An Introduction to Description Logic. Cambridge University Press, Cambridge (2017). https://doi.org/10.1017/9781139025355
5. Baader, F., Koopmann, P., Kriegel, F., Nuradiansyah, A.: Computing optimal repairs of quantified ABoxes w.r.t. static $\mathcal{EL}$ TBoxes. In: Platzer, A., Sutcliffe, G. (eds.) CADE 2021. LNCS (LNAI), vol. 12699, pp. 309–326. Springer, Cham (2021). https://doi.org/10.1007/978-3-030-79876-5_18
6. Baader, F., Kriegel, F.: Pushing optimal ABox repair from $\mathcal{EL}$ towards more expressive Horn-DLs. In: Kern-Isberner, G., Lakemeyer, G., Meyer, T. (eds.) Proceedings of the 19th International Conference on Principles of Knowledge Representation and Reasoning, KR 2022, Haifa, Israel, 31 July–5 August 2022, pp. 22–32 (2022). https://doi.org/10.24963/kr.2022/3
7. Baader, F., Kriegel, F.: Pushing optimal ABox repair from $\mathcal{EL}$ towards more expressive Horn-DLs (extended version). LTCS-Report 22-02, Chair of Automata Theory, Institute of Theoretical Computer Science, Technische Universität Dresden, Dresden, Germany (2022). https://doi.org/10.25368/2022.131
8. Baader, F., Kriegel, F., Nuradiansyah, A., Peñaloza, R.: Making repairs in description logics more gentle. In: Thielscher, M., Toni, F., Wolter, F. (eds.) Principles of Knowledge Representation and Reasoning: Proceedings of the Sixteenth International Conference, KR 2018, Tempe, Arizona, 30 October–2 November 2018, pp. 319–328. AAAI Press (2018). https://aaai.org/ocs/index.php/KR/KR18/paper/view/18056
9. Baader, F., Kriegel, F., Nuradiansyah, A., Peñaloza, R.: Computing compliant anonymisations of quantified ABoxes w.r.t. $\mathcal{EL}$ policies. In: Pan, J.Z., et al. (eds.) ISWC 2020. LNCS, vol. 12506, pp. 3–20. Springer, Cham (2020). https://doi.org/10.1007/978-3-030-62419-4_1
10. Baader, F., Peñaloza, R.: Axiom pinpointing in general tableaux. J. Log. Comput. **20**(1), 5–34 (2010). https://doi.org/10.1093/logcom/exn058

11. Baader, F., Peñaloza, R., Suntisrivaraporn, B.: Pinpointing in the description logic $\mathcal{EL}$. In: Calvanese, D., et al. (eds.) Proceedings of the 2007 International Workshop on Description Logics (DL 2007), Brixen-Bressanone, near Bozen-Bolzano, Italy, 8–10 June 2007. CEUR Workshop Proceedings, vol. 250 (2007). https://ceur-ws.org/Vol-250/paper_16.pdf
12. Calvanese, D., De Giacomo, G., Lembo, D., Lenzerini, M., Rosati, R.: DL-Lite: tractable description logics for ontologies. In: Veloso, M.M., Kambhampati, S. (eds.) Proceedings, The Twentieth National Conference on Artificial Intelligence and the Seventeenth Innovative Applications of Artificial Intelligence Conference, 9–13 July 2005, Pittsburgh, Pennsylvania, USA, pp. 602–607. AAAI Press/The MIT Press (2005). https://www.aaai.org/Papers/AAAI/2005/AAAI05-094.pdf
13. Cóbe, R., Wassermann, R.: Ontology repair through partial meet contraction. In: Booth, R., Casini, G., Klarman, S., Richard, G., Varzinczak, I.J. (eds.) Proceedings of the International Workshop on Defeasible and Ampliative Reasoning, DARe 2015, co-located with the 24th International Joint Conference on Artificial Intelligence (IJCAI 2015), Buenos Aires, Argentina, 27 July 2015. CEUR Workshop Proceedings, vol. 1423 (2015). https://ceur-ws.org/Vol-1423/DARe-15_2.pdf
14. Greiner, R., Smith, B.A., Wilkerson, R.W.: A correction to the algorithm in Reiter's theory of diagnosis. Artif. Intell. **41**(1), 79–88 (1989). https://doi.org/10.1016/0004-3702(89)90079-9
15. Hieke, W., Kriegel, F., Nuradiansyah, A.: Repairing $\mathcal{EL}$ TBoxes by means of countermodels obtained by model transformation. In: Homola, M., Ryzhikov, V., Schmidt, R.A. (eds.) Proceedings of the 34th International Workshop on Description Logics (DL 2021) part of Bratislava Knowledge September (BAKS 2021), Bratislava, Slovakia, 19–22 September 2021. CEUR Workshop Proceedings, vol. 2954 (2021). https://ceur-ws.org/Vol-2954/paper-17.pdf
16. Horrocks, I., Kutz, O., Sattler, U.: The even more irresistible $\mathcal{SROIQ}$. In: Doherty, P., Mylopoulos, J., Welty, C.A. (eds.) Proceedings, Tenth International Conference on Principles of Knowledge Representation and Reasoning, Lake District of the United Kingdom, 2–5 June 2006, pp. 57–67. AAAI Press (2006). https://www.aaai.org/Papers/KR/2006/KR06-009.pdf
17. Kalyanpur, A., Parsia, B., Sirin, E., Cuenca-Grau, B.: Repairing unsatisfiable concepts in OWL ontologies. In: Sure, Y., Domingue, J. (eds.) ESWC 2006. LNCS, vol. 4011, pp. 170–184. Springer, Heidelberg (2006). https://doi.org/10.1007/11762256_15
18. Kazakov, Y.: $\mathcal{RIQ}$ and $\mathcal{SROIQ}$ are harder than $\mathcal{SHOIQ}$. In: Brewka, G., Lang, J. (eds.) Principles of Knowledge Representation and Reasoning: Proceedings of the Eleventh International Conference, KR 2008, Sydney, Australia, 16–19 September 2008, pp. 274–284. AAAI Press (2008). https://www.aaai.org/Papers/KR/2008/KR08-027.pdf
19. Kriegel, F.: Constructing and Extending Description Logic Ontologies Using Methods of Formal Concept Analysis. Doctoral thesis, Technische Universität Dresden, Dresden, Germany (2019). https://nbn-resolving.de/urn:nbn:de:bsz:14-qucosa2-360998
20. Kriegel, F.: Optimal fixed-premise repairs of $\mathcal{EL}$ TBoxes (extended version). LTCS-Report 22-04, Chair of Automata Theory, Institute of Theoretical Computer Science, Technische Universität Dresden, Dresden, Germany (2022). https://doi.org/10.25368/2022.321

21. Küsters, R. (ed.): Non-Standard Inferences in Description Logics. LNCS (LNAI), vol. 2100. Springer, Heidelberg (2001). https://doi.org/10.1007/3-540-44613-3
22. Lam, J.S.C., Pan, J.Z., Sleeman, D.H., Vasconcelos, W.W.: A fine-grained approach to resolving unsatisfiable ontologies. In: 2006 IEEE/WIC/ACM International Conference on Web Intelligence (WI 2006), 18–22 December 2006, Hong Kong, China, pp. 428–434. IEEE Computer Society (2006). https://doi.org/10.1109/WI.2006.11
23. Lutz, C., Piro, R., Wolter, F.: Enriching $\mathcal{EL}$-concepts with greatest fixpoints. In: Coelho, H., Studer, R., Wooldridge, M.J. (eds.) ECAI 2010–19th European Conference on Artificial Intelligence, Lisbon, Portugal, 16–20 August 2010, Proceedings. Frontiers in Artificial Intelligence and Applications, vol. 215, pp. 41–46. IOS Press (2010). https://doi.org/10.3233/978-1-60750-606-5-41
24. Lutz, C., Seylan, I., Wolter, F.: An automata-theoretic approach to uniform interpolation and approximation in the description logic $\mathcal{EL}$. In: Brewka, G., Eiter, T., McIlraith, S.A. (eds.) Principles of Knowledge Representation and Reasoning: Proceedings of the Thirteenth International Conference, KR 2012, Rome, Italy, 10–14 June 2012. AAAI Press (2012). https://doi.org/10.5555/3031843.3031877
25. Lutz, C., Wolterinst, F.: Conservative extensions in the lightweight description logic $\mathcal{EL}$. In: Pfenning, F. (ed.) CADE 2007. LNCS (LNAI), vol. 4603, pp. 84–99. Springer, Heidelberg (2007). https://doi.org/10.1007/978-3-540-73595-3_7
26. Ma, Y., Peñaloza, R.: Towards parallel repair: an ontology decomposition-based approach. In: Bienvenu, M., Ortiz, M., Rosati, R., Šimkus, M. (eds.) Informal Proceedings of the 27th International Workshop on Description Logics, Vienna, Austria, 17–20 July 2014. CEUR Workshop Proceedings, vol. 1193, pp. 633–645 (2014). https://ceur-ws.org/Vol-1193/paper_61.pdf
27. Matos, V.B., Guimarães, R., Santos, Y.D., Wassermann, R.: Pseudo-contractions as gentle repairs. In: Lutz, C., Sattler, U., Tinelli, C., Turhan, A.-Y., Wolter, F. (eds.) Description Logic, Theory Combination, and All That. LNCS, vol. 11560, pp. 385–403. Springer, Cham (2019). https://doi.org/10.1007/978-3-030-22102-7_18
28. Nikitina, N., Rudolph, S.: ExpExpExplosion: uniform interpolation in general $\mathcal{EL}$ terminologies. In: De Raedt, L., et al. (eds.) ECAI 2012–20th European Conference on Artificial Intelligence. Including Prestigious Applications of Artificial Intelligence (PAIS-2012) System Demonstrations Track, Montpellier, France, 27–31 August 2012. Frontiers in Artificial Intelligence and Applications, vol. 242, pp. 618–623. IOS Press (2012). https://doi.org/10.3233/978-1-61499-098-7-618
29. Reiter, R.: A theory of diagnosis from first principles. Artif. Intell. **32**(1), 57–95 (1987). https://doi.org/10.1016/0004-3702(87)90062-2. (See the erratum [14])
30. Scharrenbach, T., Grütter, R., Waldvogel, B., Bernstein, A.: Structure preserving TBox repair using defaults. In: Haarslev, V., Toman, D., Weddell, G.E. (eds.) Proceedings of the 23rd International Workshop on Description Logics (DL 2010), Waterloo, Ontario, Canada, 4–7 May 2010. CEUR Workshop Proceedings, vol. 573 (2010). https://ceur-ws.org/Vol-573/paper_17.pdf
31. Schlobach, S.: Diagnosing terminologies. In: Veloso, M.M., Kambhampati, S. (eds.) Proceedings, The Twentieth National Conference on Artificial Intelligence and the Seventeenth Innovative Applications of Artificial Intelligence Conference, 9–13 July 2005, Pittsburgh, Pennsylvania, USA, pp. 670–675. AAAI Press/The MIT Press (2005). https://www.aaai.org/Papers/AAAI/2005/AAAI05-105.pdf

32. Schlobach, S., Cornet, R.: Non-standard reasoning services for the debugging of description logic terminologies. In: Gottlob, G., Walsh, T. (eds.) IJCAI-03, Proceedings of the Eighteenth International Joint Conference on Artificial Intelligence, Acapulco, Mexico, 9–15 August 2003, pp. 355–362. Morgan Kaufmann (2003). https://ijcai.org/Proceedings/03/Papers/053.pdf
33. Schlobach, S., Huang, Z., Cornet, R., van Harmelen, F.: Debugging incoherent terminologies. J. Autom. Reason. **39**(3), 317–349 (2007). https://doi.org/10.1007/s10817-007-9076-z
34. Tena Cucala, D., Cuenca Grau, B., Horrocks, I.: Pay-as-you-go consequence-based reasoning for the description logic $\mathcal{SROIQ}$. Artif. Intell. **298**, 103518 (2021). https://doi.org/10.1016/j.artint.2021.103518

Health and Habit: An Agent-based Approach

Veronika Kurchyna[1,2](✉), Stephanie Rodermund[2], Jan Ole Berndt[1], Heike Spaderna[2], and Ingo J. Timm[1,2]

[1] German Research Center for Artificial Intelligence (DFKI), Kaiserslautern, Germany
{veronika.kurchyna,jan_ole.berndt,ingo.timm}@dfki.de
[2] Universität Trier, 54293 Trier, Germany
{rodermund,spaderna}@uni-trier.de

Abstract. Data-driven models with weak theoretical foundations for the examination of interventions and concepts to improve and maintain health lack explainability of results and suggestions. The use of agent-based models is a possible approach to remedy this issue. Modelling behaviour and the formation of habits using established theoretical psychological frameworks is a way of improving the utilisation of agent-based models when researching health-related questions. This paper proposes a concept implementing the Health Action Process Approach and the Social Cognitive Learning Theory to model the process of behaviour change within the Beliefs-Desires-Intentions Framework. The concept illustrates how an agent workflow can incorporate these psychological models and explain how social influence contributes to the formation of habits.

Keywords: BDI · Health psychology · Cognitive modeling

1 Introduction

Agent-based models (ABM) are a popular means of modeling and simulating systems across various contexts. These range from self-organising machines in a factory [1] to people transmitting diseases during a pandemic [2]. When it comes to health-related questions, ABMs can offer decision support by simulating the impact of different strategies, giving additional justification to ideas and proposals while reducing strain on potentially affected populations [3]. The motivation for the usage of simulation to examine health behaviour is based on the importance of research into successful methods of encouraging positive changes in health behaviour. Both classical studies using surveys and experiments as well as simulations are used to explore different means of prompting change [4]. Frequent areas of study include exercise and heart disease prevention [5,6], reduction of smoking [7], improving diet [8] and encouraging weight loss in overweight patients [9–11]. A variety of different approaches and suggestions to encourage behaviour change exist. How can researchers decide which

R. Bergmann et al. (Eds.): KI 2022, LNAI 13404, pp. 131–145, 2022.
https://doi.org/10.1007/978-3-031-15791-2_12

idea should be discarded or pursued? What ideas are useful for specific populations? Evidence suggests that explicitly integrating behaviour change theories in intervention design [12] and tailoring interventions to the needs of target groups is more promising than broad delivery [13]. However, models often reduce the complexity of behaviour to outcomes (such as calorie consumption and expenditure [10]) without examining the underlying processes or simplifying them to phenomenons spreading through a network. Explicitly modelling the mechanisms that lead to change is a way of reducing the opaqueness of simulation results and justifying the outcomes of a simulation based on the acceptance of the underlying model.

This paper proposes a theoretical foundation for the modeling of the cognitive processes that are activated when individuals choose to change behaviours and the way conscious attempts turn into habitualized behaviour. The concept introduces two psychological theories into an agent-based model based on the Beliefs-Desires-Intentions Architecture (BDI) [14]. Schwarzer's *Health Action Process Approach*(HAPA) [13] explains the cognitive mechanisms that lead to behaviour change and the way it transitions into habits. Additionally, Bandura's *Social Cognitive Learning Theory*(SCLT) [15] is an explanatory framework for the way agents influence each other, elevating the idea of beliefs spreading through networks into a more nuanced way of viewing social influences. In this paper, the concepts are introduced briefly before examining how these theories are connected and how they may be transferred into a BDI-architecture. Additionally, a concept for an agent workflow implementing these theories is proposed. To illustrate how these concepts can be used for a real-life use case, the mobility of heart failure patients in stationary care has been implemented with promising results in terms of result plausibility.

2 Basics of Behavioural Theory and Agent Modeling

This section offers a small overview of the three key concepts used in this work: the Health Action Process Approach of behaviour change, Social Cognitive Learning Theory and habits. Additionally, a brief refresher on the concept of the BDI architecture in the context of agent-based modelling is presented.

2.1 Health Action Process Approach: From Wanting to Doing

Schwarzer [13] describes a way of expressing behaviour change both in different stages as well as a continuous process. This approach showcases how the behaviour of people evolves and how to classify different people along their journey, allowing interventions tailored to the different needs people experience as their behaviour and mental factors evolve, displayed in Fig. 1. An important aspect in this formalisation is the emphasis on deliberation, planning and active decision-making, which later transforms into habitual behaviours. Throughout the process, external barriers and resources, such as social support, financial means or geographic location, are acknowledged as contributing factors in the

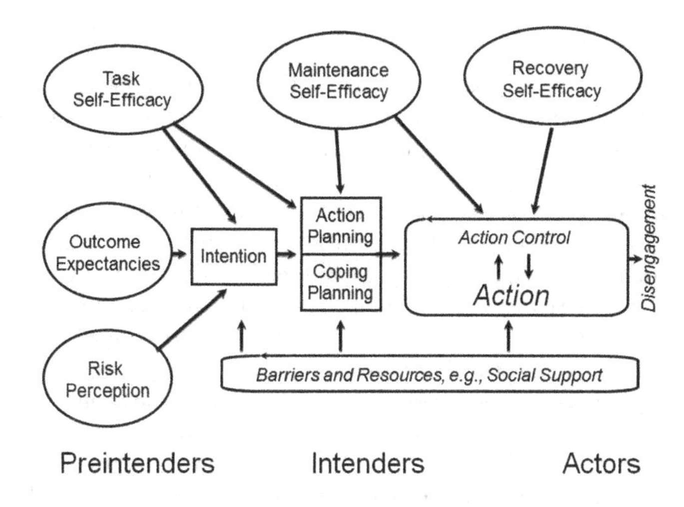

Fig. 1. The health action process approach [13]

formation of intentions and execution of planned behaviours. However, due to its focus on practical application for intervention design, such external factors are largely omitted from considerations, since individuals will generally have no means of changing their circumstances. Instead, there is a strong focus on the mental factors that can be leveraged to explain behaviour change. People are divided into three categories according to the specific stage of their behaviour change process. *Preintenders* have no interest in behaviour change - either due to lack of risk perception, no faith in the outcome of a recommended behaviour or lack of self-efficacy [15], which is a core concept in his approach. It denotes people's belief in their ability to do something and moderates their willingness to attempt actions. Thus, risk perception, outcome expectation and self efficacy moderate how people develop the intention to change their behaviours. Once an intention is set, people start forming plans to implement the new behaviours and how to cope with failures to adherence. In this stage, persons are considered to be *intenders*, at which point information delivery won't offer additional benefits [13]. Finally, they pass into action and become *actors*, motivated by successful maintenance or discouraged by failed attempts. As such, self-efficacy is a governing variable across the different stages of change: the belief in being able to do something, to continue doing it and to resume if stopped for any reason. Within a framework of targeted interventions, action control through healthcare providers can serve as an additional support in maintaining new habits. Given the model's root in intervention design, disengagement is an additional component of this system: once conscious choices have passed into habitualized behaviours, individuals will disengage from targeted interventions. The HAPA-Model has been shown to be an effective means of describing and classifying individuals for different health-related contexts such as obesity reduction [9], dental care [16], drunk driving [17] or cancer screening [18].

2.2 Social Learning: If You Can Do It, Maybe I Can Do It Too

While the HAPA-Model explains how mental factors are transferred into action, it relies on external factors to cause changes to these internal states and opinions before any deliberation takes place. This work proposes the inclusion of the Social Cognitive Learning Theory [15] as an approach to introducing dynamics to the mental state of individuals. This concept describes the way people learn from each other and are influenced by their environments. In classical behavioural theory, complex mental mechanisms may be reduced to stimuli and reaction, comparable to simple reactive agents in ABMs [19]. SCLT places emphasis on human cognition as intermediary between experienced stimuli, displayed reactions and observed outcomes. Is the stimulus perceived at all? Is the outcome that results from an action perceived, understood and associated with the cause? Thus, a cognitive filter is introduced to all perceptions. More importantly, humans don't just learn from their own experiences, but also role models provided by others: observing the actions of another and learning of the outcome leads to possible adjustment of beliefs and opinions. However, this is not a simple imitation. During the cognitive process, *Social Modelling* [20] is employed. People compare themselves to others, introducing yet another filter to perceptions: *is there enough similarity between me and another for my observations of their behaviours to apply to me?*. Similarity may be physical characteristics such as age, sex, and physique or subjective assessments, such as perceived status or personal relationship [21]. Therefore, SCLT introduces a means of adding dynamics to a closed system like the HAPA-model: by observing others successfully implementing a change or suffering the consequences of inaction, the cognitions governing the process of decision-making are altered.

2.3 Forming Habits

Verplanken & Aarts [22] present an extensive review on the state of research on habits and how they relate to behaviour, which is the base for this brief overview of major aspects relevant to this work. Generally, a habit is understood as a largely automatic execution of an action based on certain cues, such as time, weather, location, company or other situations [23]. Such habitual behaviours are exercised without much thought and require little to no mental planning and evaluation. The paradigm of past behaviour being the main predictor for future behaviour [24] holds true across a wide range of publications throughout the past decades. In the absence of habits, implementation intentions gain emphasized importance and have been proven to be reliable predictors for actual behaviour. As behaviours are repeated, the cognitive aspects lose importance and habits are formed. Thus, the more often an action has been repeated in the past, the more likely it will be performed in the future [22].

2.4 Beliefs-Desires-Intentions Architecture: Assembling Cognition

Agents, as used in multiagent-based systems, represent autonomous entities in a simulation [25]. In the context of social simulation, agents represent individuals with different characteristics, goals and actions. The BDI-architecture is an approach to structure the modelling human-like agents capable of complex decisions [14] and is one of the most established approaches to agent modeling [26].

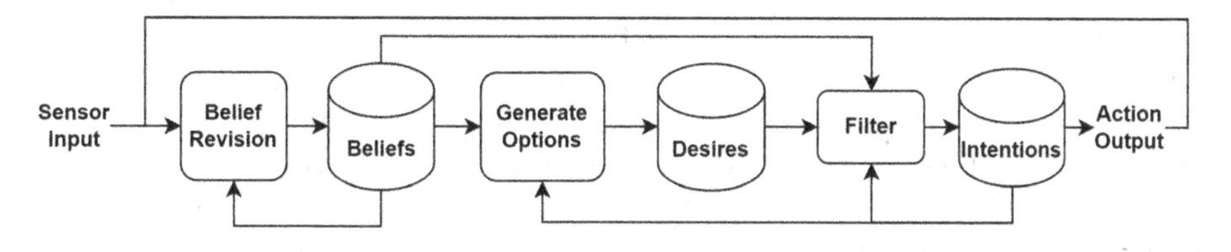

Fig. 2. Workflow of an agent in BDI architecture [27]

It has been used successfully in various contexts, such as emergency evacuations [34], negotiation between agents [35] or reactions to disasters such as landslides [36]. Terms such as 'beliefs' or 'intentions' have been mentioned before, hinting at the intuitive nature of the BDI-architecture: cognition is assembled from these three components and imitated in a simulation. Figure 2 [27] displays the general process of agent decisions. Beliefs describe the agent's knowledge of itself and its environment - such as sensor input and other observations of the environment as well as internal states such as opinions, moods, and other aspects. Based on sensor input, these beliefs are revised and updated in each step. Beliefs determine possible actions and desires an agent may have. Desires range from simple target states to complex, mutually exclusive goals that turn decisions into optimisation problems. These desires act as a filter to possible actions - agents will act rationally in so far that they will pursue the fulfillment of desires and discard non-contributing actions. Finally, intentions express the plans of an agent - be it as simple as the next step to complex plans spanning multiple actions over a span of time. These intentions become actions, which influence the environment in return. An important characteristic of the BDI architecture is its cyclical nature, in which the sensing of the environment, evaluation of beliefs and adjustment of desires and plans is a recurrent process.

3 Bringing Everything Together: Concept Proposal

This section explains how key concepts of the HAPA-model can be represented in a BDI-architecture and how SCLT can be used to introduce belief changes in the system. A concept including both partial concepts is presented to offer a coherent workflow which describes agent behaviour and habit formation.

3.1 Translating HAPA to BDI

Section 2 has shown a strong overlap between the HAPA-model and the BDI-architecture, both in terms of ideas as well as terms that are used to describe them. The classification of agents is simplified into two categories: pre-intenders on the one hand and intenders and actors grouped together on the other hand. Due to the cognitive nature of the model, beliefs are the main focus of this mapping. Barriers and resources, setting the frame a person is acting in, are beliefs about an agent's environment. Mental factors, such as outcome expectation and different types of self-efficacy, are beliefs agents hold about themselves. These beliefs are crucial for the desires of the agents. It is important to note that, while both concepts use the term *intention*, different ideas are expressed. The intentions, as understood by the HAPA model, actually denote the agent's desires, while the intentions, as used by BDI, actually relate to the planning and actions of the agent. Given the practical background of the HAPA model, most of its key features relate to beliefs in the context of BDI: while desires such as a wish for better health, social needs or external constraints such as social support, financial means and environment cannot be influenced directly or easily by healthcare professionals or other methods of intervention delivery. However, attitudes, opinions, knowledge and emotions are more accessible to targeted influences, which makes them focal points of the mechanisms explicitly included.

3.2 Introducing Influence

The proposed mapping produces a largely static system: without a change of beliefs due to external influences, no action will be taken. Thus, SCLT is introduced to create dynamics within the model. Figure 3 displays a concept for the way social observations influence agent beliefs. The graphic covers two different cases: agents acting and experiencing the outcomes and agents observing and evaluating. Agent 1 performs an action, leading to an outcome, which is evaluated and leads to an adjustment of beliefs - this may be a reinforcement of outcome expectation or an increase in confidence. Agent 2, in the meantime, observes and compares themselves to the other based on observable or known traits. If sufficient similarity is found, the observation is relevant and beliefs may be adjusted accordingly. If Agent 2 sees no similarity no learning takes place. Thus, a subjective measure of similarity is necessary for learning - otherwise, an observation or experience will be considered not applicable and has no effect.

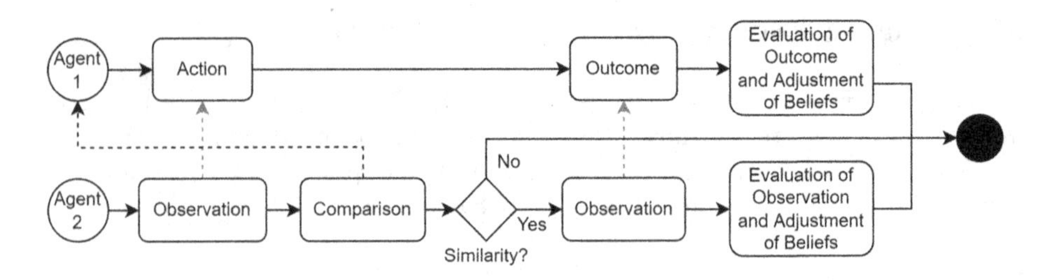

Fig. 3. Observation and evaluation of beliefs in experienced actions and outcomes

3.3 Forming Intentions and Habits

The aim is to simulate how people establish long-term habits. As such, their behaviour and planning must extend beyond short-term moods and experiences. Thus, the proposed workflow takes habits, cognition and social influence into account to show how health behaviour changes over time. An important feature of this concept is the possibility of relapse - if agents have no success with their plans, they adjust their attitudes accordingly and may lose their motivation altogether. Agents follow a loop of behaviour, forming plans based on their current state, including goals, available resources and other aspects specific to a model. Plans are stacks of tasks which are executed sequentially. Once a plan, usually representing a time period of fixed length, such as a week or a month, is finished, agents evaluate the outcome of the plan and adjust their mental attitudes accordingly before forming a new plan. During the execution of a task, described in pseudocode in Algorithm 1, agents follow a series of decision rules. Variables are introduced to save mental states and express their changes, summarised in Table 1. The main concepts, adherence, goals and efficacy are altered by variable magnitudes and upon passing variable thresholds, depending on the context of an implementation. First, agents determine whether a task can be executed at all. This unassuming condition can mean a variety of sub-conditions, such as availability of resources and other preconditions as well as the habits of the agent. Low adherence makes failure more likely. In a case of failure, adh_{temp} is reduced.

Given the long term nature of habits, a single failure is merely a setback, so adh_{temp} is a temporary summary of adherence in this plan cycle. The agent chooses an alternative action depending on available resources, current

Table 1. Variables used in the pseudocode descriptions in Algorithm 1 and Algorithm 2

Variable	Explanation
adh	Adherence: strength of habit in following past plans
adh_{change}	Magnitude of changes to long term adherence
adh_{temp}	Temporary adherence to the current plan
$adh_{modifier}$	Modifier to the adherence changes
$goal$	Expression of an agent goal, depending on the implemented context
$goal_{progress}$	Current progress towards increasing the magnitude of a held goal
$goal_{modifier}$	Modifier to the magnitude of progress towards a goal increase
$goal_{threshold}$	Threshold $goal\_progress$ must reach before a goal increase takes place
$decide_{threshold}$	Threshold that must be passed during deliberation before new goal
$efficacy$	Self-efficacy, held belief in own abilities to perform a desired behaviour
$efficacy_{change}$	Rate at which *efficacy* increases in case of success
$efficacy_{min}$	Minimum value *efficacy* must have to maintain a goal
$expectation$	Expression of the agent's favourable outcome expectation

Algorithm 1. Attempt Task Execution

```
if Task executable then
    ExecuteTask
    adh_modifier ← 1
    if Task ∈ TargetBehaviours then
        if Task ∈ OriginalPlan then
            if TargetBehaviour ∈ AgentGoals then
                adh_modifier ← n ∈ (0, 1)
            else
                goal_progress+ = (goal_modifier × past_behaviour)
            end if
        else
            adh_temp+ = (adh_change × adh_modifier)
        end if
    else
        if (TargetBehaviour ∨ Results) ∈ ObservableEnvironment then
            if Similarity (self, other_agent) > threshold then
                adjust (expectation, observation, similarity)
                adjust (efficacy, observation, similarity)
            end if
        end if
    end if
else
    adh_temp− = adh_change
    Plan.add (choose_alternative ())          ▷ Execute the other task instead
end if
```

mood and needs. This model differentiates between spontaneous implementation of desirable behaviour and planned actions. The lowering of the modifier $adh_{modifier}$ for ad-hoc choices ensures that planned actions contribute more towards forming habits than opportunistic spontaneity. Spontaneous implementations of unplanned target behaviours still contribute towards $goal_{progress}$, moderated by past behaviour. If the behaviour isn't part of the targets, agents examine whether these behaviours (or their results) are observable in the environment. This includes direct communication with others reporting their experience or agents performing these actions in the vicinity. If sufficient similarity is found between an agent, the observation and its source, relevant mental factors are adjusted. Once all tasks have been performed, an evaluation takes place, described in Algorithm 2. This subprocess consist of two branches, depending on whether the desired target behaviour is part of the goals.

If the agent planned desirable behaviours and has a positive score for adh_{temp}, self-efficacy, $goal_{progress}$ and adherence increase. Additionally, if $goal_{threshold}$ is passed, the goal increases. Since this means new planning, adherence is reset, as the agent now adjusts to altered behaviour. If adh_{temp} dropped below its initial value, an agent failed to uphold their plans and loses adherence and efficacy. If $goal_{progress}$ drops into negatives and efficacy below its threshold, the agent

Algorithm 2. Evaluate Plan Outcome

if $TargetBehaviour \in AgentGoals$ **then**
  if $adh_{temp} \geq 1$ **then**
    $efficacy+ = efficacy_{change}$
    $goal_{progress}+ = goal_{modifier}$
    $adh+ = (adh_{change} \times adh_{temp})$
    if $goal_{progress} \geq goal_{threshold}$ **then**
      $increase\_goal\,()$
      $adherence \leftarrow 1$
    end if
  else
    $efficacy- = efficacy_{change}$
    $adh- = adh_{change}$
    $goal_{progress}- = goal_{modifier}$
    if $(goal_{progress} < 0) \wedge (efficacy < efficacy_{min})$ **then**
      $reduce\_goal\,()$
    end if
  end if
else
  $adh- = adh_{change}$
  $efficacy- = efficacy_{change}$
  if $(efficacy + expectation + resources) \geq decide_{threshold}$ **then**
    $goal_{progress}+ = goal_{modifier}$
  end if
  if $goal_{progress} > goal_{threshold}$ **then**
    $increase\_goal\,()$
    $adh+ = adh_{change}$
  end if
end if
$adh_{temp} \leftarrow 0$

reduces their goals. Alternatively, if the target behaviour is not planned, residual adherence and self-efficacy from past success erode. If the combination of efficacy, expectations and available resources surpasses $decide_{threshold}$, a goal is added to the set of existing goals. With the introduction of adh_{temp} and the usage of modifiers to regulate beliefs, behavioural changes are a long-term process as habits develop and strengthen over time. Success (or lack thereof) lead to changes in beliefs such as self-efficacy and outcome expectations, reinforced by experience. A challenge not previously addressed is the calibration of parameters. For some use cases, such as exercise, correlates between concepts and behaviours have been measured empirically [38]. Yet even with some aspects quantified, most variables still need experimental configuration. Therefore, this structure offers a guide on how mechanisms of behaviour change can be achieved, though it omits or simplifies several aspects whose applicability depend on the context.

4 Discussion and Use Cases

The proposed concept combines the two major psychological models presented in Sect. 2. During interaction with other people or from observation, beliefs change to create the dynamics necessary to allow for agents that would be classified as pre-intenders to become intenders and, finally, actors. A major aspect is the interplay between the deliberative nature of the HAPA-model, which places emphasis on plans and evaluation, and the habitualisation of behaviours, which renders execution of tasks easier as habits strengthen. Different concepts regarding health behaviour can be addressed with this architecture. Besides, the availability of resources, such as energy for exercise or accessible vegetarian options for food choices, is an additional factor that will vary across models.

One possible context is increasing the amount of exercise agents perform - observing agents of similar physique exercising or interacting with agents whose health has worsened due to lack of activity can lead towards a change of behaviour, which improves health and allows for easier execution in the future [39]. Due to the possibility of short-term observable results, exercise and diet are a topic easily represented in such a context [40]. Other behaviours, such as vegetarianism [41], less alcohol consumption [42] or reduction of smoking [43], do not have obvious short-term outcomes - in such a situation, social influences gain more importance. As a result, the combination of different psychological theories lends to a higher degree of reusability of this concept. While contexts with observable short-term effects can be portrayed well with the HAPA-model alone, the addition of SCLT ensures that social factors, such as support or perceived norms, can be included to influence long-term decision-making in absence of short-term feedback and behavioural outcomes. Therefore, this combined approach is well suited as a base for the expression of different issues as ABMs.

4.1 Proof of Concept: Exercise in Heart Failure Patients

With exercise being a major means of preventing heart disease and improving the wellbeing of diseased patients [5], the mobility of heart failure patients is an exemplary use case for which a model was implemented using this concept in a simplified form in NetLogo. Figure 4 illustrates the general sequence of agent behaviours. The observed target behaviour is patients meeting recommended exercise quotas - about three units of moderate exercise each week. Since any amount of exercise is preferable to none, this target is pursued by agents through their exercise goal, in which their personal target may be below guideline recommendations at first. This desire is contrasted with social needs, energy and personal well-being. These desires are addressed by different actions: spending time alone to recover, socializing with others to fill social needs, exercise or spend time in a public area where others may be observed. Agents' days are structured into four time slots during which actions may be chosen - as such, a week consists of 28 actions. These actions are planned in advance, using a simplified planning method: agents choose a random plan which incorporates the target number of exercise units corresponding to their current goals. This plan is executed sequentially, as described in Algorithm 1.

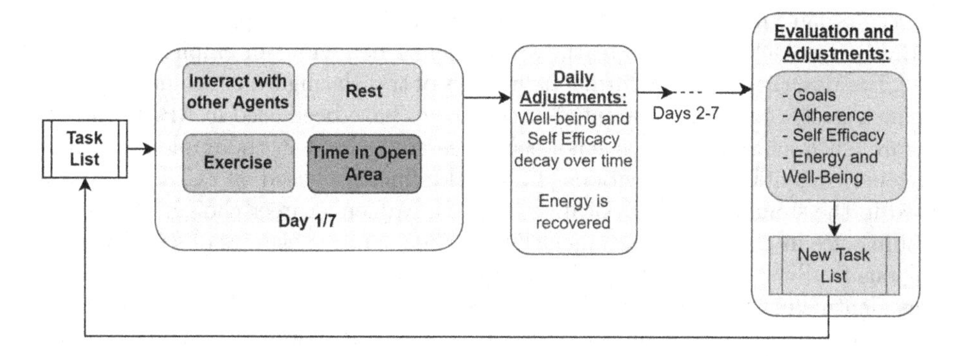

Fig. 4. Observation and evaluation of beliefs in experienced actions and outcomes

Personal beliefs and variables such as wellbeing, past behaviour, self efficacy, outcome expectation and current energy moderate the performance of tasks and their outcomes. This also allows for the inclusion of mechanisms of evaluation in social interactions. Agents with strong adherence and positive health outcomes are more convincing than agents whose past behaviour or current health does not back up their confidence. In this model, social modeling is not actively included - agents, residents of a stationary care facility, see enough similarity to be influenced by observation and communication. To further increase the realism of the simulation, the detrimental effects of isolation and sedentary behaviour were included through the diminishing returns of resting - the more often one chooses to rest within a short period of time, the less beneficial effects will be gained while attributes such as physical wellbeing and self-efficacy slowly decay over time with no exercise to counteract the natural decline observed in the elderly. During the model runs, agents pick up exercise and drop back out in case of failure, repeating the established pattern that individuals who have been physically active in the past are most likely to remain active [24]. To verify the model's ability to produce realistic behaviour, a calibration was performed with the objective of an average of 20% [44] of agents meeting the recommended amount of exercise. The parameters that were calibrated related to the change rate for different variables as expressed in Table 1, the strength of social influence during interactions and the rate at which physical and mental variables decayed. Regardless of whether a group size of 10, 25, 50 or 75 agents was considered, the recommended parameter configuration remained stable, indicating that this implementation and its parameters scale well in terms of behavioural stability. Besides meeting the target behaviour quotes, the configuration also matched the patterns of correlations observed in empirical research [38]. As such, this implementation is promising in terms of its ability to reproduce realistic behaviour.

4.2 Related Works: Behaviour, Social Learning and Habit in ABMs

The integration of psychological theories is not a novel question [28]. Given the fact that theory-driven models can compete with data-driven models in terms of results [30], models focused on theory are an important approach. However, ABMs often lack important features that define behaviour in empirical studies [29]. Besides, there exist no formal definitions of translating concepts into models. Some aspects of theories presented in this paper have been used in ABM context before, such as SCLT, which has been successfully used in the implementation of generic agents for interventions [4]. Social influences based on SCLT were also used in the simulation of physical activity [6], women's career choices [31], leadership roles in disaster management [32] or the cognitive aspects of socially contagious fertility [33]. Another field in which the importance of behaviour change was identified is sustainability. ABMs have been used successfully to examine the interplay between behaviour and cultural factors [45] as well as social contagion of habits [46]. Other models, such as the acceptance of behaviour-changing interventions [47], have applied ABMs for questions of behaviour change. However, models often lack explicit mechanisms of habits and habit formation. The novelty of the approach presented in this work lies in the active deliberation on goals and habits and the explicit inclusion of habit loss not introduced from social sources. While some cognitive theories find application in ABMs [28], social factors are more represented.

5 Conclusion

To increase the explainability and validity of simulations of interventions for health behaviour, a concept for the cognitive process in behaviour change and habit formation is presented. This concept offers an explanation and basic implementation to support the development of models that examine health-related questions. This is a guide for the inclusion of the HAPA-model and a simplified take on SCLT, given that many details will be largely specific to individual use cases. Thus, we present a 'skeleton' that can set the general flow of actions and decision-making in an agent. The major downside of this concept is its greatest strength too, *generality*. With the range of aspects to be considered for each possible use case, researchers and developers will face implementation efforts for the numerous decisions that are made on a case-by-case base. Key questions such as similarity assessment, exact attitude adjustment, determining whether a task can be executed and mechanisms of observation and planning are charged with open questions of design choice. Still, the process of determining possible cognitive theories and how to interleave them is simplified by offering this architecture as baseline. This explanation how habits form using agent-based models is promising due to the flexibility of modeling and range of questions that can be addressed. Thus, as a planned future work, the authors intend to develop and improve ABMs for different health questions using this approach, such as the model presented in Sect. 4.1. In the context of this use case, validation against empirical data from accompanying studies is planned as well as further validation of the implementation using sensitivity analysis. Additionally, a model of other health-related questions may be presented to demonstrate another use case with this architecture to prove the generality of this concept. Given that exercise serves as an example with well-observable short-term benefits, a counterpart with less obvious outcomes, such as smoking cessation or vegetarian diet, might be chosen to present another type of problem this architecture can portray.

References

1. Chiagunye, T.: Markov chain approach to agent based modelling (ABM) of an industrial machine operation control. J. Multi. Eng. Sci. Technol. **2**, 4 (2015)
2. Lorig, F., et al.: Agent-based social simulation of the covid-19 pandemic: a systematic review. J. Artif. Soc. Soc. Simul. **24**(3), 5 (2021)
3. Timm, I., Spaderna, H., Rodermund, S., Lohr, C., Buettner, R., Berndt, J.O.: Designing a randomized trial with an age simulation suit-representing people with health impairments. Healthcare **9**, 27 (2021)
4. Fife-Schaw, C., et al.: Simulating behaviour change interventions based on the theory of planned behaviour: impacts on intention and action. Br. J. Soc. Psychol. **46**(04), 43–68 (2007)
5. Cattadori, G., et al.: Exercise and heart failure: an update: exercise and heart failure. ESC Heart Fail. **5**, 12 (2017)
6. Mollee, J.S., van der Wal, C.N.: A computational agent model of influences on physical activity based on the social cognitive theory. In: Boella, G., Elkind, E., Savarimuthu, B.T.R., Dignum, F., Purvis, M.K. (eds.) PRIMA 2013. LNCS (LNAI), vol. 8291, pp. 478–485. Springer, Heidelberg (2013). https://doi.org/10.1007/978-3-642-44927-7_37
7. Villanti, A.C., et al.: Smoking-cessation interventions for U.S. young adults: updated systematic review. Am. J. Prev. Med. **59**(1), 123–136 (2020)
8. Zhang, D., et al.: Impact of different policies on unhealthy dietary behaviors in an urban adult population: an agent-based simulation model. Am. J. Public Health **104**, 05 (2014)
9. Baldensperger, L., et al.: Physical activity among adults with obesity: testing the health action process approach. Rehabil. Psychol. **59**, 01 (2014)
10. Beheshti, R., et al.: Comparing methods of targeting obesity interventions in populations: an agent-based simulation. SSM - Popul. Health **3**, 211–218 (2017)
11. Giabbanelli, P., et al.: Modeling the influence of social networks and environment on energy balance and obesity. J. Comput. Sci. **3**(03), 17–27 (2012)
12. Allan, V., et al.: The use of behaviour change theories and techniques in research-informed coach development programmes: a systematic review. Int. Rev. Sport Exerc. Psychol. **11**(1), 47–69 (2018)
13. Schwarzer, R., et al.: Mechanisms of health behavior change in persons with chronic illness or disability: the health action process approach (HAPA). Rehabil. Psychol. **56**(08), 161–70 (2011)
14. Rao, A., George, M.: BDI agents: from theory to practice. In: Proceedings of the First International Conference on Multi-Agent Systems (ICMAS-95), pp. 312–319 (1995)
15. Bandura, A.: Lernen am Modell: Ansatze zu einer sozial-kognitiven Lerntheorie. Klett (1976)

16. Smith, S., et al.: Social-cognitive predictors of parental supervised toothbrushing: an application of the health action process approach. Br. J. Health Psychol. **26** (2021). https://doi.org/10.1111/bjhp.12516
17. Wilson, H., et al.: Self-efficacy, planning, and drink driving: applying the health action process approach. Health Psychol. **35** (2016). https://doi.org/10.1037/hea0000358
18. Pourhaji, F., et al.: Application of the health action process approach model in predicting mammography among Iranian women (2020). https://doi.org/10.21203/rs.3.rs-80108/v1
19. Castelfranchi, C., Werner, E. (eds.): MAAMAW 1992. LNCS, vol. 830. Springer, Heidelberg (1994). https://doi.org/10.1007/3-540-58266-5
20. Bandura, A.: The power of observational learning through social modeling. In: Stenberg, R., Fiske, S.T., Foss, D.J. (eds.). Scientists Making a Difference, pp. 235–239 (2016)
21. Andsager, J., et al.: Perceived similarity of exemplar traits and behavior effects on message evaluation. Commun. Res. **33**, 3–18 (2006)
22. Verplanken, B., Aarts, H.: Habit, attitude, and PlannedBehaviour: is habit an empty construct or an interesting case of goal-directed automaticity? Eur. Revi. Soc. Psychol. **10**(1), 101–134 (1999)
23. James, W.: The Principles of Psychology, vol. 1, Macmillan, London (1890)
24. Aarts, H., et al.: Predicting behavior from actions in the past: repeated decision making or a matter of habit? J. Appl. Soc. Psychol. **28**, 1355–1374 (1998)
25. Gilbert, N., Troitzsch, K.G.: Simulation for the Social Scientist. Open University Press, USA (2005)
26. Cardoso, R.C., Ferrando, A.: A review of agent-based programming for multi-agent systems. Computers **10**(2), 16 (2021)
27. Weiss, G.: Multiagent Systems. The MIT Press, Cambridge (2013)
28. Jager, W.: Enhancing the realism of simulation (EROS): on implementing and developing psychological theory in social simulation. J. Artif. Soc. Soc. Simul. **20**, 14 (2017)
29. Mercuur, R., et al.: Integrating social practice theory in agent-based models: a review of theories and agents. IEEE Trans. Comput. Soc. Syst. **7**(5), 1131–1145 (2020)
30. Taghikhah, F., et al.: Where does theory have it right? A comparison of theory-driven and empirical agent based models. J. Artif. Soc. Soc. Simul. **24**, 4 (2020)
31. Mozahem, N.: Social cognitive theory and women's career choices: an agent-based model simulation. Comput. Math. Organ. Theor. **10**, 2020 (2020)
32. Wu, C., et al.: Emergence of informal safety leadership: a social-cognitive process for accident prevention. Prod. Oper. Manag. **30**(11), 4288–4305 (2021)
33. Berndt, J.O., Rodermund, S., Timm, I.J.: Social contagion of fertility: an agent-based simulation study. In: Winter Simulation Conference, pp. 953–964 (2018)
34. Miller, T., Oren, N., Sakurai, Y., Noda, I., Savarimuthu, B.T.R., Cao Son, T. (eds.): PRIMA 2018. LNCS (LNAI), vol. 11224. Springer, Cham (2018). https://doi.org/10.1007/978-3-030-03098-8
35. Barolli, L., Javaid, N., Ikeda, M., Takizawa, M. (eds.): CISIS 2018. AISC, vol. 772. Springer, Cham (2019). https://doi.org/10.1007/978-3-319-93659-8
36. Sugiarto, V., et al.: Modeling agent-oriented methodologies for landslide management. J. Inf. Technol. Comput. Sci. **4**(2), 193–201 (2019)
37. Adam, C.: Emotions: from psychological theories to logical formalization and implementation in a BDI agent, July 2007

38. Courneya, K.S., Hellsten, L.-A.M.: Personality correlates of exercise behavior, motives, barriers and preferences: an application of the five-factor model. Pers. Individ. Differ. **24**(5), 625–633 (1998)
39. Piepoli, M., et al.: Experience from controlled trials of physical training in chronic heart failure. Protocol and patient factors in effectiveness in the improvement in exercise tolerance. Eur. Heart J. **19**(3), 466–475 (1998)
40. James, J., Annesi, P.D.: Relationship of perceived health and appearance improvement, and self-motivation, with adherence to exercise in previously sedentary women. Eur. J. Sport Sci. **4**(2), 1–13 (2004)
41. Singer, P.: Utilitarianism and vegetarianism. Filosofia Unisinos. **17**, (2016). https://doi.org/10.4013/fsu.2016.172.17
42. Coomber, K., et al.: Awareness and correlates of short-term and long-term consequences of alcohol use among Australian drinkers. Aust. NZ J. Public Health. **41**(3), 237–242 (2017)
43. Dalum, P., et al.: A cluster randomised controlled trial of an adolescent smoking cessation intervention: short and long-term effects. Scand. J. Public Health **40**(2), 167–76 (2012)
44. Keadle, S., et al.: Prevalence and trends in physical activity among older adults in the United States: a comparison across three national surveys. Prev. Med. **89**, 05 (2016)
45. Kaaronen, R.O., Stelkovsii, N.: Cultural evolution of sustainable behaviors: pro-environmental tipping points in an agent-based model. One Earth **2**(1), 85–97 (2020)
46. Klein, M., et al.: Contagion of habitual behaviour in social networks: an agent-based model. In: International Conference on Privacy, Security, Risk and Trust and 2012 International Conference on Social Computing, pp. 538–545 (2012)
47. Jensen, T., et al.: Agent-based assessment framework for behavior-changing feedback devices: spreading of devices and heating behavior. Technol. Forecast. Soc. Change **98**, 105–119 (2015)

Knowledge Graph Embeddings with Ontologies: Reification for Representing Arbitrary Relations

Mena Leemhuis[1](✉), Özgür L. Özçep[1], and Diedrich Wolter[2]

[1] University of Lübeck, Lübeck, Germany
{leemhuis,oezcep}@ifis.uni-luebeck.de
[2] University of Bamberg, Bamberg, Germany
diedrich.wolter@uni-bamberg.de

Abstract. Knowledge graph embeddings offer prospects to integrate machine learning and symbolic reasoning. Learning algorithms are designed that map constants, concepts, and relations to geometric entities in a real-valued domain $\mathbb{R}^n$. By identifying logics that feature these geometric entities as their model, one is able to achieve a tight integration of logic reasoning with machine learning. However, interesting description logics are more expressive than current knowledge graph embeddings, as description logics allow concept definitions using arbitrary relations, such as non-functional relationships and partial ones. By contrast, geometric models of relations used so far in knowledge graph embeddings such as translations, rotations, or linear functions can only represent total functional relationships. In this paper we describe a new geometric model of the description logic $\mathcal{ALC}$ based on cones that exploits reification combined with linear functions to represent arbitrary relations. While this paper primarily describes reification in context of a particular model for $\mathcal{ALC}$, the proposed reification technique is general and applicable with other ontology languages and knowledge graph embeddings.

Keywords: Knowledge-graph embedding · Ontologies · Explainable AI

1 Introduction

Knowledge graph embeddings (KGEs) offer prospects of a true integration of machine learning and symbolic reasoning, given that models acquired by means of machine learning can also serve as models in a logic sense. Until now, several properties of prominent ontology languages for representing non-trivial concepts are beyond what can be grounded in machine learning models. In this paper we develop an approach using *reification* to advance the expressiveness of relations in KGEs. In knowledge graph embeddings, concepts are commonly represented as geometric entities (e.g., balls [9], boxes [14], or cones [2,13,18]), constants as points, and relations as geometric operations. TransE [3] continues to be the classical reference for a knowledge graph embedding of relations, drawing its charm from a simple geometrical representation of relations that can be learned efficiently. Indeed, TransE represents relations as vector translations, and hence embedding a triple $(s\ R\ o)$ (stating that a subject s stands in

R. Bergmann et al. (Eds.): KI 2022, LNAI 13404, pp. 146–159, 2022.
https://doi.org/10.1007/978-3-031-15791-2_13

relation R to an object o, also written $R(s, o)$) into a continuous space is easily integrated with a loss function used for learning the embedding. The downside of this simple representation is TransE's limited expressivity [11]: only binary relations that are functional in their first argument can be modeled. Such considerations on expressivity have lead to many other embedding approaches that rely on more complex representations of relations. Most of them are in the tradition of TransE—some more such as TransH [17], TransR [10] that rely on representing relations as translation, and some less, relying on other, more involved geometrical operations such as SimplE [11] or Rescal [12].

Still, all these approaches share the limitation of being restricted to relations that are total and functional. Important features like partiality or non-functionality of relations cannot be modeled correctly. What we mean by correct modeling is not going beyond acceptable performance in some combinations of datasets and tasks, but to give a proper logic-like Tarskian semantics to relations. By doing so, one does not only pave a more solid theoretical foundation but also establishes the basis for KGEs associated with a background ontology which states axiomatically constraints on the entities, concepts, and relations to be embedded. For example, a background theory may state that certain concepts are mutually exclusive (e.g., `familyMovie` and `horrorMovie` are disjoint) or that some elements of a concept are related (e.g., `hasChild` as a partial, non-functional relation on the concept class `human`). In [7] this logic-like representation is expressed as the suggestion to represent (subjects and objects in triples) as vectors (this is represented as arbitrary n-ary relations as subsets of the n-cartesian product over the embedding space). An obvious downside of that approach is the high increase in dimensionality required with adverse effects on learning. In this paper we take a middle-road: We insist on representing concepts (unary relations) as sets of vectors but allow representations of arbitrary binary relations in mathematically well-behaved operations.

In this paper we propose the idea of *reification* to be applied in KGEs to represent relations. We adopt the idea of relying on matrix multiplication to represent relations previously used in KGEs, but we rewrite relations into equivalent structures which allows us to model arbitrary relations, including partial and non-functional ones. The idea of reification is to represent relations as objects in the embedding space. In our case, e.g., a triple $(a\ R\ b)$ would be represented by an object $c_{R(a,b)}$ and functions stating that its "arguments" are a and b: $\pi_{1,R}(c_{R(a,b)}) = a$, $\pi_{2,R}(c_{R(a,b)}) = b$. The only relations $\pi_{i,R}$ that have to be represented and learnt, then, are functional relations, namely projections of triples to its subject and its object. In context of an approach to cone-based embeddings we are able to show that the set of triples $c_{R(a,b)}$ for a particular relation R forms a well-behaved object too, namely a cone itself. In order to achieve partiality, we develop a semantics that allows the projections π_1, π_2 to project pairs (a, b) outside the domain, thus representing non-existence.

While reification is a well-known approach in logic modeling, the technical challenge tackled in this paper is to develop a geometric model that can link machine learning (by using feasible ingredients such as convex sets and simple geometric operations) and ontologies (by defining a model for a logic). While we believe the idea of reification to be compatible with a range of approaches in KGEs, we have opted in this paper to extend an approach using linear functions as projections similar to TransR

[10] as building blocks for relations and convex cones [13] which have already been shown to support full concept negation in background knowledge. Taken together, we can give a feasible geometric model for the well-known description logic $\mathcal{ALC}$ and thereby advance expressivity of KGEs.

In summary, the key contribution of this paper is to give an embedding of $\mathcal{ALC}$-ontologies based on generalized cones with a novel interpretation of relations based on reification having the ability of representing partial and non-functional relations.

The remainder of this paper is structured as follows. In Sect. 2 we introduce notation and summarize relevant properties of ontologies. Section 3 presents the proposed reification approach for cones. In Sect. 4 we show how a geometric model for the description logic $\mathcal{ALC}$ can be constructed and discuss its properties. Section 5 presents related works. The paper concludes with a brief résumé and outlook.

2 Preliminaries

In order to define embeddings of knowledge graph triples with respect to background knowledge, we first introduce a suitable language to model such background knowledge, usually referred to as ontology. Description logics (DLs) [1] are formal languages tailored towards representing ontologies and they thus present themselves as a basis. DLs provide a clear distinction between factual knowledge (expressed in the so-called abox) and terminological knowledge (expressed in the so-called tbox). In context of KGEs, the abox provides specific data instances and the tbox provides background knowledge. We note that one may be interested in semantics beyond classical DLs and as we will discuss later this is indeed possible. For example, one may be interested in some settings to account for partial information, say there may be elements that are neither known to be members of a concept C nor of its negation $C^{\perp}$. This may be accomplished by choosing the appropriate semantics. To keep our approach general, we first describe semantics for a very general orthologic and then refine it to the classic semantics of the well-known and widely used DL $\mathcal{ALC}$.

2.1 Ortholattice and Orthologic

In short, ortholattices are structures similar to Boolean algebras but with fewer properties, e.g., no distributivity.

An *(algebraic) ortholattice* is a partially ordered set L with functions defined on it, namely a structure $(L, \wedge, \vee, \cdot^{\perp}, 0, 1)$ fulfilling the following properties:

- $a \vee a = a$, $a \wedge a = a$. (idempotence)
- $a \vee b = b \vee a$, $a \wedge b = b \wedge a$. (commutativity)
- $(a \vee b) \vee c = a \vee (b \vee c)$, $(a \wedge b) \wedge c = a \wedge (b \wedge c)$ (associativity)
- $a \vee (a \wedge b) = a$, $a \wedge (a \vee b) = a$ (absorption)
- $a \wedge 0 = 0$, $a \vee 0 = a$, $a \vee 1 = 1$, $a \wedge 1 = a$
- $a^{\perp\perp} = a$ (double negation elimination)
- $0 = a \wedge a^{\perp}$ (intuitionistic absurdity)
- $(a \vee b)^{\perp} = a^{\perp} \wedge b^{\perp}$, $(a \wedge b)^{\perp} = a^{\perp} \vee b^{\perp}$ (De Morgan)

Intuitively, $\cdot^{\perp}$ represents negation, more precisely called orthocomplement in ortho-logics, and partial order $\leq$ on L corresponds to concept inclusion $\sqsubseteq$ in the ontology language. Logics defined on ortholattices (as opposed to Boolean algebras) are called *orthologics*. Classical logics like propositional logics are also orthologics, albeit ones that satisfy additional, stronger properties.

2.2 Background Logic

As the basic logical syntax we consider that of $\mathcal{ALC}$ [1]. The syntax of $\mathcal{ALC}$ promises to provide operators to express non-trivial background knowledge. This language is neither trivial nor too complex to distract from developing the main points in this paper. The $\mathcal{ALC}$ syntax rests on a DL vocabulary $\mathcal{V}$ given by a set of constants N_c, a set of role names (binary relation symbols) N_R, and a set of concept names N_C. The set $conc(\mathcal{V})$ of $\mathcal{ALC}$ concepts (concept descriptions) over $N_C \cup N_R$ is described by the grammar

$$C \longrightarrow A \mid \bot \mid \top \mid \neg C \mid C \sqcap C \mid C \sqcup C \mid \exists R.C \mid \forall R.C$$

where $A \in N_C$ is an atomic concept, $R \in N_R$ is a role symbol, and C stands for arbitrary concepts. An *ontology* $\mathcal{O}$ is defined as a pair $\mathcal{O} = (\mathcal{T}, \mathcal{A})$ of a *terminological box (tbox)* $\mathcal{T}$ and an *assertional box (abox)* $\mathcal{A}$. A tbox consists of *general inclusion axioms* $C \sqsubseteq D$ ("C is subsumed by D") with concept descriptions C, D. For ease of notation, we write $C = D$ instead of $C \sqsubseteq D$ and $D \sqsubseteq C$. An abox consists of a finite set of *assertions*, i.e., facts of the form $C(a)$ or of the form $R(a, b)$ for $a, b \in N_c$. We define the notion of an interpretation as usual:

Definition 1. *A structure* $(\Delta, \cdot^{\mathcal{I}})$ *is called an* interpretation $\mathcal{I}$ *for a given* $\mathcal{ALC}$ *vocabulary of constants, concept and role symbols* $\mathcal{V} = N_c \cup N_C \cup N_R$ *iff* Δ*, the so-called* domain, *is a set and* $\cdot^{\mathcal{I}}$ *is the* denotation function *defined for all* $b \in N_c, A \in N_C, R \in N_R$ *and concepts* C, D *over* $\mathcal{V}$ *such that the following conditions are fulfilled:*

$$\begin{array}{ll}
b^{\mathcal{I}} \in \Delta, & (C \sqcap D)^{\mathcal{I}} = C^{\mathcal{I}} \cap D^{\mathcal{I}}, \\
A^{\mathcal{I}} \subseteq \Delta, & (C \sqcup D)^{\mathcal{I}} = C^{\mathcal{I}} \cup D^{\mathcal{I}}, \\
R^{\mathcal{I}} \subseteq \Delta \times \Delta, & (\neg C)^{\mathcal{I}} = \Delta \setminus C^{\mathcal{I}}, \\
\top^{\mathcal{I}} = \Delta, & (\exists R.C)^{\mathcal{I}} = \{x \in \Delta^{\mathcal{I}} \mid \text{There is } y \in \Delta^{\mathcal{I}} \text{ s.t.} \\
\bot^{\mathcal{I}} = \emptyset, & \qquad (x, y) \in R^{\mathcal{I}} \text{ and } y \in C^{\mathcal{I}}\}, \\
 & (\forall R.C)^{\mathcal{I}} = \{x \in \Delta^{\mathcal{I}} \mid \text{For all } y \in \Delta^{\mathcal{I}}: \\
 & \qquad \text{If } (x, y) \in R^{\mathcal{I}},\ y \in C^{\mathcal{I}}\}
\end{array}$$

An interpretation $\mathcal{I}$ *models* a GCI $C \sqsubseteq D$, for short $\mathcal{I} \models C \sqsubseteq D$, iff $C^{\mathcal{I}} \subseteq D^{\mathcal{I}}$. An interpretation $\mathcal{I}$ *models* an ABox axiom $C(a)$, for short $\mathcal{I} \models C(a)$, iff $a^{\mathcal{I}} \in C^{\mathcal{I}}$ and it models an ABox axiom of the form $R(a, b)$ iff $(a^{\mathcal{I}}, b^{\mathcal{I}}) \in R^{\mathcal{I}}$. An interpretation is a *model of an ontology* $(\mathcal{T}, \mathcal{A})$ iff it models all axioms appearing in $\mathcal{T} \cup \mathcal{A}$.

We now discuss embeddings of cones in $\mathbb{R}^n$ before turning our attention to $\mathcal{ALC}$.

3 Cone Embedding

Our approach is based on a geometric interpretation function $\mathcal{I}$ that represents concepts and relations as *convex cones* in $\mathbb{R}^n$. Cones satisfy the property that if x, y are inside a cone, then $\lambda x + \mu x$, $\lambda, \mu \geq 0$ is also inside that cone. We make use out of this property to construct the reification. We adopt the approach of [13] in using *polarity* $(\cdot)^{\circ}$ derived from the scalar product in $\mathbb{R}^n$ to construct a negation of concepts. The polar of a cone C, written C°, is defined as the set $\{x \in \mathbb{R}^n | \forall y \in C.x^T \cdot y \leq 0\}$, i.e., the set of all vectors being rotated at least 90 degrees away from any element of C. For all points neither belonging to a cone C nor to its polar C°, no statement about membership to concepts C, $\neg C$ can be made – the model is thus capable of representing uncertainty and thus is able to cope with the open world assumption.

Definition 2. *A* convex cone *is a set* $C \subseteq \mathbb{R}^n$ *with the property* $\forall x, y \in C.\forall \lambda, \mu \in \mathbb{R}.(\lambda \geq 0 \wedge \mu \geq 0) \rightarrow \lambda x + \mu y \in C$. *For readability, we refer to convex cones simply as cones. We define* H_m *as the* m*-dimensional hyperoctant cone* $H_m \subset \mathbb{R}^m$ *generated by* m *vectors* $\{(1 0 \cdots 0)^T, (0 1 0 \cdots 0)^T, (0 0 1 0 \cdots 0)^T, \ldots, (0 \cdots 0 1)^T\}$.

Closed convex cones are closed under set intersection, so $\cap$ is a meet operator $\wedge$ wrt. $\leq$ but not closed under set union. Instead they have to be closed up by the conic hull operator. The *conic hull* of a set b, for short $ch(b)$, is the smallest convex cone containing b. So, we can define the join operation $\vee$ by $a \vee b = ch(a \cup b)$. Considering $\mathbb{R}^n$ as the largest lattice element 1 and the empty set as the smallest lattice element 0 makes the resulting structure a bounded lattice.

The polarity operator for closed convex cones fulfills properties of orthocomplement. Hence the set of all closed convex cones (over $\mathbb{R}^n$) forms an ortholattice. As de Morgan's laws hold in any ortholattice, one gets in particular the following characterization of the conic hull: $ch(a \cup b) = (a^{\circ} \cap b^{\circ})^{\circ}$. We denote the set of all closed convex cones in $\mathbb{R}^n$ by $\mathcal{C}_n$. Then the following fact holds: For any $n \geq 1$, $\mathcal{C}_n$ is an ortholattice.

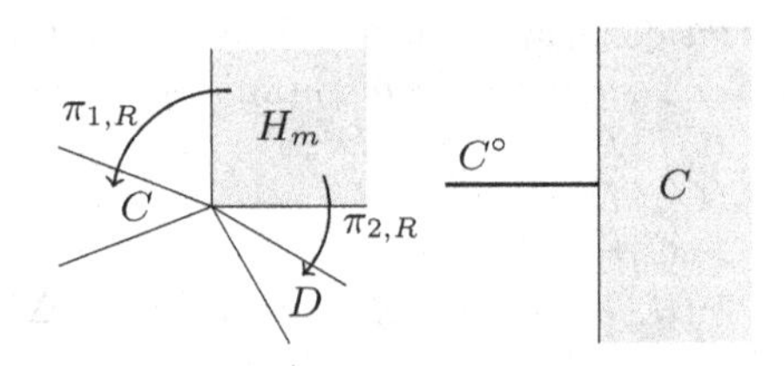

Fig. 1. Left: Reification of relation R is based on linear functions $\pi_{1,R}$, $\pi_{2,R}$ that project relation concept H_m to its arguments. Right: Illustration for reification requiring H_m with $m > n$.

We now define reification as illustrated in Fig. 1 left to relate two concepts C, D. Relations are represented like concepts, i.e., by convex sets of specific geometric shape, and projections π_1, π_2 are introduced that link the embedded relations with the corresponding concepts. The main advantage of this over previous attempts is that the use of projections allows non-functional and partial relations to be represented. Approaches representing relations by geometric transformations in the embedding space such as TransE [3] are attractive as they also do not require further dimensions to be introduced, yet at the severe cost of being only able to represent functional relations, i.e., any x is always related to exactly one y. Several work aimed to remedy this severe limitation, but no general cure is possible when relying on a single geometric transformation function.

Definition 3. *We say a* reification *of a binary relation R between two cones C, D is given by a hyperoctant H_m and two linear functions $\pi_{1,R}$ and $\pi_{2,R}$ given as matrices $M \in \mathbb{R}^{n\times m}$.*

Let us now complement the geometric model. A cone interpretation for $\mathcal{ALC}$ maps symbols and formulae to cones in $\mathbb{R}^n$. The definition is as usual for interpretations, except that we exclude the origin $\vec{0}$ from the domain. This creates a convenient way for projections in the reification of relations to map subspaces to a well-defined 'nirvana' $\vec{0}$ whenever a mapping to the empty set is required. For example, the formula $\forall R.\top \sqsubseteq \bot$ saying that nothing is reachable by role R can elegantly be represented by setting $\pi_{2,R}$ to $\mathbf{0} \in \mathbb{R}^{n\times n}$, the projection to $\vec{0}$.

Definition 4. *A* cone interpretation *for a given $\mathcal{ALC}$ vocabulary $\mathcal{V} = N_c \cup N_C \cup N_R$ of constants, concept and role symbols is a structure $(\Delta, \cdot^{\mathcal{I}})$ where $\Delta = \mathbb{R}^n \setminus \{\vec{0}\}$ for some $n \in \mathbb{N}$ and $\cdot^{\mathcal{I}}$ is the* denotation function *defined for all $b \in N_c, A \in N_C, R \in N_R$ and concepts C, D over $\mathcal{V}$ such that the following conditions are fulfilled:*

$b^{\mathcal{I}} \in \Delta$, $\quad (C \sqcap D)^{\mathcal{I}} = C^{\mathcal{I}} \cap D^{\mathcal{I}}$,

$A^{\mathcal{I}} \in \mathcal{C}_n$, $\quad (\neg C)^{\mathcal{I}} = C^{\circ}$,

$R^{\mathcal{I}} \subseteq \Delta \times \Delta$, $\quad (C \sqcup D)^{\mathcal{I}} = (\neg(\neg C \sqcap D))^{\mathcal{I}}$,

$\top^{\mathcal{I}} = \Delta$, $\quad (\exists R.C)^{\mathcal{I}} = \pi_{1,R}\left(\pi_{2,R}^{-1}(C^{\mathcal{I}}) \cap H_m\right)$

$\bot^{\mathcal{I}} = \emptyset$, $\quad$ *where H_m, $\pi_{1,R}$, $\pi_{2,R}$ are a reification of relation R and $\pi_{1,R}\left(\pi_{2,R}^{-1}(C^{\mathcal{I}}) \cap H_m\right)$ is a convex cone.*

$(\forall R.C)^{\mathcal{I}} = (\neg\exists R.\neg C)^{\mathcal{I}}$

The notion of a cone interpretation being a model *(of an abox, tbox, ontology) is defined in the same way as for classical interpretations according to Definition 1.*

We now show that arbitrary $\mathcal{ALC}$ knowledge bases $\mathrm{KB} = (A, T)$ consisting out of abox A and tbox T are representable by a cone interpretation. First, we define how relations on the abox level are modeled.

Definition 5. *Given an $\mathcal{ALC}$ vocabulary with concept symbols $\mathcal{C}$, constant symbols $\mathcal{A}$, and role symbols $\mathcal{R}$, and $\mathcal{ALC}$ knowledge base* KB, *we say that the roles in* KB *are representable if there is a geometric interpretation $(\Delta, \mathcal{I})$ that is a model of* KB *and* KB $\models R(a, b)$ *if and only if $b \in \pi_{1,R}(\pi_{2,R}^{-1}(a) \cap H_m)$ for some hyperoctant H_m.*

Proposition 1. *A cone interpretation of concepts maps all concept descriptions to closed convex cones.*

Proof. This is clear for atomic concepts, intersection, and for the polar operator. Disjunction is defined by de Morgan via intersection and polarity. But this is the conic hull, hence a mapping to a closed convex cone. Linear mappings also preserve cones, as they distribute over arbitrary linear combinations (not only those with positive scalars). For the existential, being a convex cone is enforced directly by the definition.

Note that enforcing closed convex cones for the embedding of existentials is not a strong constraint. Taking the null vector into account one can show that the inverse preserves cones: Let X be a cone and M be a linear mapping as used in reification. Let $v \in M^{-1}[X]$, then $M(v) = w \in X$. Then for $\lambda > 0$ due to linearity $M(\lambda v) = \lambda M(v) = \lambda w$ and $\lambda w \in X$ due to the fact that X is a cone. Let $v, v' \in M^{-1}[X]$, then $M(v) = w \in X, M(v') = w' \in X$ (for some w, w'). Now $M(v + v') = M(v) + M(v') = w + w' \in X$, so $v + v' \in M^{-1}[X]$.

Reification employs matrix multiplications like several previous approaches, but it employs an 'in-between stop' at H_m which is the central trick to represent 1-to-k relations by making π_1 a k-to-1 mapping. Let us consider a simple example shown in Fig. 1 to see that a stop H_m is necessary and that even $m > n$ may be necessary.

Example 1. We consider the cone C generated by vectors $\{(0\,1)^T, (0\,-1)^T, (1\,0)^T\}$ in $\mathbb{R}^2$ shown in Fig. 1 right. Its negation given by the polarity operator C° is the cone generated by $\{(-1\,0)^T\}$. Now consider background knowledge $\exists R.C = \top$ saying that any entity is reachable from C by means of relation R. $\top$ is interpreted as $\mathbb{R}^2 \setminus \{\vec{0}\}$ and it requires four independent rays $\lambda_i c_i$, $\lambda_i > 0$, $c_i \in C$ to span $\mathbb{R}^2$, more than offered by C. It requires at least H_4 to achieve the desired mapping:

$$\pi_{1,R} := \begin{pmatrix} 0 & 0 & 1 & -1 \\ 1 & -1 & 0 & 0 \end{pmatrix}, \pi_{2,R} := \begin{pmatrix} 0 & 0 & 1 & 0 \\ 1 & -1 & 0 & 0 \end{pmatrix}$$

For any point $\vec{x} = (x_1 \cdots x_4)^T$ in H_4 we have $\pi_2(\vec{x}) = x_1 \begin{pmatrix} 0 \\ 1 \end{pmatrix} + x_2 \begin{pmatrix} 0 \\ -1 \end{pmatrix} + x_3 \begin{pmatrix} 1 \\ 0 \end{pmatrix}$ and since $x_1, \ldots, x_4 \geq 0$ we have $\pi_2(\vec{x}) \in C$ and $\pi_1(\vec{x})$ covers $\mathbb{R}^2$ for $\vec{x} \in H_4$. In general, H_m with $m > n$ is required when concept C is a sub-space of lower dimensionality than the concept it is related to.

Let us discuss a more involved example showcasing the ability to model complex relationships.

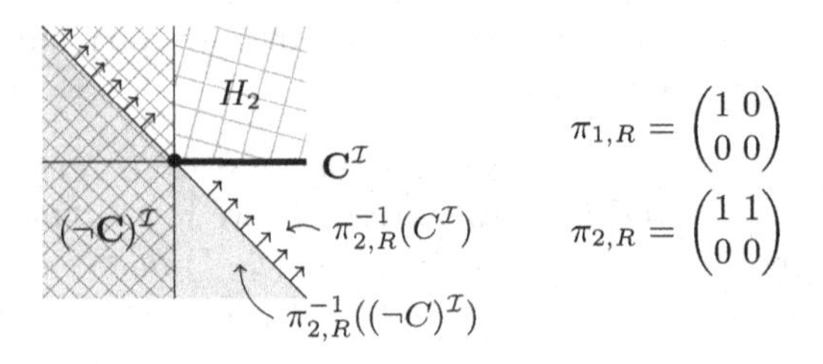

Fig. 2. Example for the construction of a geometric model of a tbox consisting of $\exists R.C = C$ and $\exists R.\neg C = \bot$

Example 2. Consider concept $C^\mathcal{I}$ represented as the positive x-axis, the complement $(\neg C)^\mathcal{I}$ is the negative halfspace in $\mathbb{R}^2$ shown in Fig. 2. This model fulfills the two tbox axioms $\exists R.C = C$ and $\exists R.\neg C = \bot$. The first axiom is fulfilled, as $\pi_{2,R}^{-1}(C^{\mathcal{I}})$ is the region marked with arrows in the figure. The intersection with the region of possible relational facts H_2 results in H_2, the upper right quadrant. This is mapped to $C^{\mathcal{I}}$ by $\pi_{1,R}$. As $\pi_{2,R}^{-1}(\neg C)^{\mathcal{I}}$ does not intersect with H_2, $\exists R.\neg C = \bot$ is valid.

Therefore, partiality is obviously given. To show non-functionality, the instances need to be considered. Assume $c^{\mathcal{I}} = (\lambda\, 0)^T$ for an arbitrary $\lambda > 0$. $\pi_{1,R}^{-1}(c^{\mathcal{I}}) = \{(\lambda\, \mu)^T \mid \mu \geq 0\}$ and $\pi_{2,R}\pi_{1,R}^{-1}(c^{\mathcal{I}}) = \{(\lambda + \mu\, 0)^T \mid \mu \geq 0\}$. Thus, a $c^{\mathcal{I}} = (1\, 0)^T$ has a relation to all $b^{\mathcal{I}} = (\gamma\, 0)^T$ for $\gamma \geq 1$.

A property of our approach, besides its non-distributivity of $\cap$ over $\cup$, is its non-distributivity of $\exists$ over $\vee$, meaning $(\exists R.(C \sqcup D))^{\mathcal{I}} \neq (\exists R.C^{\mathcal{I}} \sqcup \exists R.D)^{\mathcal{I}}$. Despite this is not classical, e.g. different from $\mathcal{ALC}$-semantics, it may be quite useful in modeling: Assume a binary relation E is introduced to model whether a person is examined to have a specific disease. Thus, by asserting $E(\mathit{person}, \mathit{disease})$ medical knowledge about a person at a specific point in time is reported. Now, it might be the case that an examination is not exact and thus results in the knowledge that the person could have disease A or disease B. However, assuming $\exists E.(A \sqcup B) = \exists E.A \sqcup \exists E.B$ would result in the conclusion that at this stage of examination it is already known which exact disease the person has. However, this exact specification was presumed not to be possible, and therefore, the instance representing the person should be placed in the embedding of $\exists E.(A \sqcup B)$ but neither in the embedding of $\exists E.A$ nor in the embedding of $\exists E.B$ because for both of them there is no justification in the examination. Thus, a lack of distributivity can be helpful to bridge gaps in semantics.

4 Distributive Embedding

The approach described so far leads to a possibility of expressing relational knowledge in general orthologics, which may be relevant for some applications as we have argued above. However, many knowledge bases consider stronger orthologics, i.e., expect distributivity to hold, which include all classical logics.

One prominent example is $\mathcal{ALC}$ with classical semantics. Here one requires the ortholattice also to be a Boolean algebra. In fact, classical $\mathcal{ALC}$-tboxes are characterized according to [15] by the fact that the existential is a strong operator, i.e., the existential quantifier with classical $\mathcal{ALC}$ semantics satisfies the following two properties: $(\exists R.\bot)^{\mathcal{I}} = (\bot)^{\mathcal{I}}$ and $(\exists R.(C \sqcup D))^{\mathcal{I}} = (\exists R.C \sqcup \exists R.D)^{\mathcal{I}}$.

Therefore, to adapt our approach to $\mathcal{ALC}$, distributivity of $\sqcap$ over $\sqcup$ and distributivity of $\exists$ over $\sqcup$ must be achieved. The first property can be met by restricting cones to so-called axis-aligned cones (al-cones) as introduced in [13] since geometric models based on al-cones are distributive. Al-cones have a finite basis and their generating vectors only consists out of components $+1$, -1, and 0. The second property can be met by restricting the modeling of relations in form of the role distributivity property.

Definition 6. *Role distributivity property RDP: if* $x \in (C \sqcup D)^{\mathcal{I}}$ *and* $\pi_{2,R}^{-1}(x) \cap H_m \neq \bot^{\mathcal{I}}$, *then it exists* $x_c \in (C)^{\mathcal{I}}$ *and* $x_d \in (D)^{\mathcal{I}}$ *with* $x = x_c + x_d$ *and* $\pi_{2,R}^{-1}(x_c) \subseteq H_m$ *and* $\pi_{2,R}^{-1}(x_d) \subseteq H_m$.

Each two concepts C and D must fulfill the RDP to regain distributivity of the $\exists$-operator.

Proposition 2. *$\exists R.(C \sqcup D) = \exists R.C \sqcup \exists R.D$ is valid if RDP is fulfilled.*

Proof. $\exists R.C \sqcup \exists R.D \sqsubseteq \exists R.(C \sqcup D)$ holds in any case. Therefore, it is sufficient to show $\exists R.(C \sqcup D) \sqsubseteq \exists R.C \sqcup \exists R.D$. Therefore, for all $y \in (\exists R.(C \sqcup D))^{\mathcal{I}}$ it needs to follow that $y \in (\exists R.C \sqcup \exists R.D)^{\mathcal{I}}$. Let $y \notin (\exists R.C)^{\mathcal{I}}$, $y \notin (\exists R.D)^{\mathcal{I}}$, as trivial in the other cases. Therefore, $y = \pi_{1,R}(\pi_{2,R}^{-1}(x_c + x_d) \cap H_m)$ for a $x_c \in C$ and a $x_d \in D$. With linearity of $\pi_{2,R}$ it follows that $y = \pi_{1,R}((\pi_{2,R}^{-1}(x_c) + \pi_{2,R}^{-1}(x_d)) \cap H_m)$ and $y \in \exists R.C \sqcup \exists R.D$ means $y = \pi_{1,R}((\pi_{2,R}^{-1} x_c \cap H_m) + (\pi_{2,R}^{-1} x_d \cap H_m))$. With RDP follows equality.

Having this property, it is possible to show that all $\mathcal{ALC}$ knowledge bases are representable by a geometric interpretation based on al-cones.

Proposition 3. *All $\mathcal{ALC}$ knowledge bases are representable by a geometric interpretation.*

Proof. We show that an al-cone interpretation of a $\mathcal{ALC}$ knowledge base without roles, i.e., only considering the Boolean part, can be extended to a model for roles as well.

Since $\mathcal{ALC}$ features the finite model property we may assume that the geometric model is finite and represents all facts following from a given knowledge base KB. Hence assume all concepts to be represented by cones and all constants by vectors in $\mathbb{R}^n$. We write $a^{\mathcal{I}_B}$ to refer to the vector obtained for constant a in the Boolean embedding and we write $a^{\mathcal{I}}$ for its embedding we seek to construct.

We iteratively construct a geometric model from a Boolean geometric model based on al-cones and a corresponding $\mathcal{ALC}$ model by processing role after role in a two-step process. Initially, we initialize $\cdot^{\mathcal{I}}$ by setting $\cdot^{\mathcal{I}}$ to the Boolean-only model $\cdot^{\mathcal{I}_B}$. In the first step, we consider a role R with $|\{(a,b)|\text{KB} \models R(a,b)\}| = m$ and assume $R = \{(a_1,b_1),\ldots,(a_m,b_m)\}$ in the finite model. We extend the dimensions of our model from n to $n(m+1)$ by cloning the components of all vectors $\vec{x}$ that generate some concept. Let 0_k denote k consecutive zero components in a vector, then we can describe the modification of the embedding $c^{\mathcal{I}}$ for any constant c as follows:

$$\phi(c) = \sum_{i=1,\ldots,m,c=a_i \vee c=b_i} (0_{n \cdot i}\,(c^{\mathcal{I}})^T\,0_{m-i})^T \tag{1}$$

$$c^{\mathcal{I}} \leftarrow \begin{cases} \phi(c) & \phi(c) \neq 0_{n(m+1)} \\ ((c^{\mathcal{I}})^T\,0_{n \cdot m})^T & \text{otherwise} \end{cases} \tag{2}$$

Note that $\phi(c) \neq 0_{n(m+1)}$ occurs exactly if there is at least one a_i or b_i with $c = a_i$ or $c = b_i$. Doing so, we separate all entities in $\text{dom}(R) \cup \text{Img}(R)$ that occur in the model. In particular, we achieve that $\lambda a_i^{\mathcal{I}} \in \text{dom}(R)$, $\lambda > 0$ if and only if $a_i^{\mathcal{I}} \in \text{dom}(R)$ and likewise for $\text{Img}(R)$. We repeat the process for all roles.

In the second step, we need to construct the reification of any role R which can be done as follows. Assume again $R = \{(a_1,b_1),\ldots,(a_p,b_p)\}$ and then define a reification based on hyperoctant H_p embedded in the model using projections

$$\pi_1 = \left[a_1^{\mathcal{I}} \cdots a_p^{\mathcal{I}}\right], \pi_2 = \left[b_1^{\mathcal{I}} \cdots b_p^{\mathcal{I}}\right],$$

where $[\cdots]$ represents a matrix composed out of column vectors. By construction of $a_i^{\mathcal{I}}, b_i^{\mathcal{I}}$, we have $c_{R(a,b)} = (0_{i-1}\, 1\, 0_{n-1})^T \in H_l$ which corresponds to $R(a_i, b_i)$ since $\pi_1(c_{R(a,b)}) = a_i^{\mathcal{I}}$ and $\pi_2(c_{R(a,b)}) = b_i^{\mathcal{I}}$. It thus follows $\pi_1(H_p) \supseteq \mathrm{dom}(R)$ and $\pi_2(H_p) \supseteq \mathrm{Img}(R)$, respectively. Also by construction, for any $c \notin \mathrm{dom}(R)$ we have $c^{\mathcal{I}} \notin \pi_1(H_p)$ since $a_i^{\mathcal{I}}$ and $c^{\mathcal{I}}$ reside in mutually exclusive sub-spaces according to (2).

We note that this proof, albeit constructive, is of theoretical nature since it exploits a large amount of dimensions for H_p to ease the construction.

It is not only possible to represent each $\mathcal{ALC}$ knowledge base with such a geometric interpretation, it is also possible to interpret each geometric model based on the axis-aligned cones introduced in the above proposition as an $\mathcal{ALC}$-ontology.

Proposition 4. *A geometric interpretation based on al-cones fulfilling RDP, where $\pi_{1,R}(\pi_{2,R}(C) \cap H_m)$ maps to an al-cone for each half-axis C, represents an $\mathcal{ALC}$ knowledge base.*

Proof. A geometric interpretation without considering roles is shown in [13]. Therefore, it is sufficient to show that the relational part also fulfills the restrictions of $\mathcal{ALC}$. $\exists R.\bot = \bot$ is fulfilled by construction. The distributivity of $\exists$ over $\vee$ is ensured by RDP, as shown in Proposition 2.

As each half-axis is mapped to an al-cone, because of linearity of π, each concept is mapped to a union of al-cones, which is still an al-cone. Also because of linearity, it is ensured that the properties needed for roles, e.g. $\exists R.C \sqsubseteq \exists R.\top$ are fulfilled. The negation of $\exists R.C$ is given by polarity (as it would not be a geometric model otherwise). Therefore, the resulting geometric model represents a $\mathcal{ALC}$ knowledge base.

5 Related Work

The approach presented in this paper is a contribution to recent efforts on combining knowledge representation (KR) and machine learning (ML). Roughly, those approaches use ML algorithms to learn an ontology or to exploit the ontologies as constraint specifications in order to get more accurate models or in order to optimize statistical models. Our work and many of the recent KGE approaches (see below) tackle the problem of building accurate models in the sense that these are compatible with the background knowledge expressed in an ontology. But there is also relevant work outside of the KGE community which incorporates ontologies into standard statistical models. An example is the approach of Deng and colleagues [5] in which pairwise conditional random fields are optimized by incorporating knowledge of the background as additional factors.

Earlier approaches to knowledge graph embedding—including TransE [3]—were motivated by efficient learning algorithms, hence resolving the expressivity vs. feasibility dilemma strictly in favor of feasibility. For example, consider the notion of "full expressivity" in [11] which only states that an approach is able to differentiate between all class members and non-members of a concept. In those approaches—including the well-known TransE [3]—heads and tails of KGE triples are represented as real-valued vectors and relations are represented as vectors, matrices or tensors, i.e., simple geometric operations. In many occasions, the geometric operations lead to relations that are functional, total or are constrained by other means. But the resulting simple mathematical operations (for representing relations) provide not much expressivity from a KR point of view. Even in later approaches, e.g. [16], functionality is not dealt logically but rather by relying on thresholds. In order to illustrate our point, consider an object represented by a vector x. A relation R is represented in [16] by a rotation M_R. The vector $y := M_R x$ gives only some "prototypical" object to which x stands in R-relation. Other objects y' to which x might stand in R-relation are given by $\|M_R x - y'\| \leq \lambda$ for some threshold λ. In particular, this means that all objects to which x is R-related are close to $M_R x$. In consequence, x can not be related to some objects y' and y'' that are quite different in that they belong to complementary concepts, $y' \in C^{\mathcal{I}}$ and a $y'' \in (\neg C)^{\mathcal{I}}$.

Table 1. Comparison of approaches for embedding with the approach of this paper in bold font

Geometrical structure		Logic	Concept lattice	Negation	Approach/Reference
Concepts	Relations				
Convex sets	Pairs	Quasi-chained Datalog$^{\pm}$	Distr	Atomic	[7]
Hyperspheres	Translation	$\mathcal{EL}^{++}$	Distr	Atomic	[9]
Axis-aligned Cones	Pairs	Rank-restricted $\mathcal{ALC}$	Distr	Full Boolean	[13]
Axis-aligned Cones	**Cones**	$\boldsymbol{\mathcal{ALC}}$	**Distr**	**Full Boolean**	**This paper**
Closed subspaces in Hilbert space	Pairs	Minimal Quantum Logic	Orthomodular	Orthonegation	[6]
Hyperbolic cones	Rotation	Logics for taxonomies	Distrib	Atomic (?)	[2]
Cartesian products of 2D-cones	Rotation + volume change	FOL queries (?) (without ∀)	Distrib	Negation as failure	[18]

In the following we discuss only those KGE approaches that explicitly mention the kind of geometries used for embedding and the logic that characterizes them (see Table 1). Table 1 considers in particular the question how concepts and roles are embedded, whether distributivity of $\sqcap$ over $\sqcup$ is fulfilled, and what kind of negation is expressed. We note that there are good reasons for considering non-distributive logics for the investigation of concept hierarchies as discussed in [4]. Non-distributive logics are investigated thoroughly by Hartonas in [8].

[7] identify a fragment of existential Datalog (fulfilling the quasi-chainedness property) as an appropriate logic for arbitrary convex regions in euclidean spaces. [9] finds a correspondence for hyperspheres and the lightweight description logic $\mathcal{EL}$. [13] identifies axis-aligned cones as an appropriate geometrical class for embedding concepts of the semi-descriptive logic $\mathcal{ALC}$. While [7,9] do not allow for full negation of concepts to be represented, [13] define negation for the model of axis-aligned cones that uses polarity, which possibly gives rise to some interesting logic structure. On the other hand, in [13] binary relations are allowed to be arbitrary pairs of vectors, whereas [7] models also relations (of any arity) by convex regions. The approach of this paper shares the property with [13] of providing full (Boolean) negation. But our approach deviates from [13] in the interpretation of roles—with consequences at three columns of the table: Our approach does not consider arbitrary set of pairs as possible embeddings of roles. In [13], this generality is possible by restricting the quantifier rank of the concepts in the ontology. In contrast, our approach interprets roles by reifying concepts, that are allowed to be (arbitrary) cones. This also allows handling arbitrary (non quantifier-restricted) $\mathcal{ALC}$ ontologies as background knowledge.

In all three approaches the expressible concept hierarchy fulfills distributivity of conjunction over disjunction. The approach of [6] considers minimal quantum logic which does not fulfill distributivity but (only) a weakening: orthomodularity. Relations are handled in [6] by doubling the dimension of space where the concepts are embedded and by treating $R(a, b)$ as a vector $[a\ b]^T$ in this higher-dimensional space.

The approach of [2] uses hyperbolic cones for modelling relation hierarchy graphs and to grasp properties that follow by traversing the edges. The exact logic captured by this approach is not clear (to us), as the authors allow next to subclass relations also part-of relations. Quantifiers, negation (and other Boolean operators) are not handled explicitly in this approach. One can think of antinomies being used in the hierarchy—but this rather correspond to atomic negation. Hence we describe the logic as taxonomic.

The approach of [18] also uses the idea of [13] to handle negation of concepts by using cones. But they do not consider negation as polarity, but negation as set-complement in 2D and cartesian products to embed concepts. Relations in [18] are handled by rotating the support point and changing the volume of a cone. The authors claim to embed FOL queries. Interestingly, they exclude the universal quantifier $\forall$ form their considerations. Given the fact that $\forall$ is dual to $\exists$ via negation we consider this as a sign that negation is not treated in its full expressivity. In particular, they cannot fully account for de Morgan rules since negation as used there is a form of negation as failure.

6 Conclusions and Outlook

Algorithms involving computations over some declarative specification of the world have to trade-off between expressivity and feasibility. Feasibility of embeddings has been traditionally favored over expressivity, because many works are governed by practical implementations. Current investigations now try to push the expressivity envelope and to strive for a better alignment between expressivity provided by an embedding and the expressivity required for sound representation of some domain knowledge. Achieving a true alignment of the geometric structures determined in learning methods with logical models is necessary to exploit embeddings in hybrid AI approaches, in particular with reasoning beyond link prediction. This paper shows how the idea of reification can be applied to knowledge graph embeddings and presented the first geometric model of full $\mathcal{ALC}$ which is based on feasible structures previously employed in knowledge graph embedding, namely convex sets (cones) and linear functions (matrix multiplication). Our approach is not tailored to $\mathcal{ALC}$ but may be useful to a much larger family of orthologics. As this paper has been taken the second roadway of pushing forward expressivity in geometric models, future work will aim to complement these fundamental findings with a learning method to acquire an embedding with reification of roles.

Acknowledgements. The research of Mena Leemhuis and Özgür L. Özçep is funded by the BMBF-funded project SmaDi. Diedrich Wolter acknowledges support by Technologieallianz Oberfranken and BMBF (Dependable Intelligent Software Lab).

References

1. Baader, F.: Description logic terminology. In: Baader, F., Calvanese, D., McGuinness, D., Nardi, D., Patel-Schneider, P. (eds.) The Description Logic Handbook, pp. 485–495. Cambridge University Press (2003)
2. Bai, Y., Ying, R., Ren, H., Leskovec, J.: Modeling heterogeneous hierarchies with relation-specific hyperbolic cones. In: Proceedings 35th Annual Conference on Neural Information Processing Systems (NeurIPS 2021). arXiv:2110.14923 (2021)
3. Bordes, A., Usunier, N., García-Durán, A., Weston, J., Yakhnenko, O.: Translating embeddings for modeling multi-relational data. In: Burges, C.J.C., Bottou, L., Ghahramani, Z., Weinberger, K.Q. (eds.) Advances in Neural Information Processing Systems 26: 27th Annual Conference on Neural Information Processing Systems 2013. Proceedings of a Meeting Held 5–8 December 2013, Lake Tahoe, Nevada, United States, pp. 2787–2795 (2013). http://papers.nips.cc/paper/5071-translating-embeddings-for-modeling-multi-relational-data
4. Conradie, W., Palmigiano, A., Robinson, C., Wijnberg, N.: Non-distributive logics: from semantics to meaning. arXiv e-prints arXiv:2002.04257, February 2020
5. Deng, J., et al.: Large-scale object classification using label relation graphs. In: Fleet, D., Pajdla, T., Schiele, B., Tuytelaars, T. (eds.) ECCV 2014. LNCS, vol. 8689, pp. 48–64. Springer, Cham (2014). https://doi.org/10.1007/978-3-319-10590-1_4
6. Garg, D., Ikbal, S., Srivastava, S.K., Vishwakarma, H., Karanam, H., Subramaniam, L.V.: Quantum embedding of knowledge for reasoning. In: Wallach, H., Larochelle, H., Beygelzimer, A., Alché-Buc, F., Fox, E., Garnett, R. (eds.) Advances in Neural Information Processing Systems, vol. 32. Curran Associates, Inc. (2019)
7. Gutiérrez-Basulto, V., Schockaert, S.: From knowledge graph embedding to ontology embedding? An analysis of the compatibility between vector space representations and rules. In: Thielscher, M., Toni, F., Wolter, F. (eds.) Proceedings of KR 2018, pp. 379–388. AAAI Press (2018)
8. Hartonas, C.: Reasoning with incomplete information in generalized Galois logics without distribution: the case of negation and modal operators. In: Bimbó, K. (ed.) J. Michael Dunn on Information Based Logics. OCL, vol. 8, pp. 279–312. Springer, Cham (2016). https://doi.org/10.1007/978-3-319-29300-4_14
9. Kulmanov, M., Liu-Wei, W., Yan, Y., Hoehndorf, R.: EL embeddings: geometric construction of models for the description logic EL++. In: Proceedings of IJCAI 2019 (2019)

10. Lin, Y., Liu, Z., Sun, M., Liu, Y., Zhu, X.: Learning entity and relation embeddings for knowledge graph completion. In: Proceedings of the Twenty-Ninth AAAI Conference on Artificial Intelligence, AAAI 2015, pp. 2181–2187. AAAI Press (2015)
11. Mehran Kazemi, S., Poole, D.: SimplE embedding for link prediction in knowledge graphs. arXiv e-prints arXiv:1802.04868, February 2018
12. Nickel, M., Tresp, V., Kriegel, H.P.: A three-way model for collective learning on multi-relational data. In: Proceedings of the 28th International Conference on International Conference on Machine Learning, ICML 2011, Omnipress, USA, pp. 809–816 (2011). http://dl.acm.org/citation.cfm?id=3104482.3104584
13. Özçep, Ö.L., Leemhuis, M., Wolter, D.: Cone semantics for logics with negation. In: Bessiere, C. (ed.) Proceedings of IJCAI 2020, pp. 1820–1826 (2020). ijcai.org
14. Ren, H., Hu, W., Leskovec, J.: Query2box: reasoning over knowledge graphs in vector space using box embeddings. In: International Conference on Learning Representations (2020)
15. Schild, K.: A correspondence theory for terminological logics: preliminary report. In: Proceedings of IJCAI 1991, pp. 466–471 (1991)
16. Sun, Z., Deng, Z.H., Nie, J.Y., Tang, J.: RotatE: knowledge graph embedding by relational rotation in complex space. arXiv e-prints arXiv:1902.10197, February 2019
17. Wang, Z., Zhang, J., Feng, J., Chen, Z.: Knowledge graph embedding by translating on hyperplanes. In: Proceedings of the AAAI Conference on Artificial Intelligence, vol. 28, no. 1, June 2014. https://ojs.aaai.org/index.php/AAAI/article/view/8870
18. Zhang, Z., Wang, J., Jiajun, C., Shuiwang, J., Feng, W.: ConE: cone embeddings for multi-hop reasoning over knowledge graphs. In: Proceedings 35th Annual Conference on Neural Information Processing Systems (NeurIPS 2021) (2021)

Solving the Traveling Salesperson Problem with Precedence Constraints by Deep Reinforcement Learning

Christian Löwens(✉), Inaam Ashraf, Alexander Gembus, Genesis Cuizon, Jonas K. Falkner, and Lars Schmidt-Thieme

University of Hildesheim, Hildesheim, Germany
{loewensc,ashraf,gembus,cuizon}@uni-hildesheim.de
{falkner,schmidt-thieme}@ismll.uni-hildesheim.de
https://uni-hildesheim.de

Abstract. This work presents solutions to the Traveling Salesperson Problem with precedence constraints (TSPPC) using Deep Reinforcement Learning (DRL) by adapting recent approaches that work well for regular TSPs. Common to these approaches is the use of graph models based on multi-head attention layers. One idea for solving the pickup and delivery problem (PDP) is using heterogeneous attentions to embed the different possible roles each node can take. In this work, we generalize this concept of heterogeneous attentions to the TSPPC. Furthermore, we adapt recent ideas to sparsify attentions for better scalability. Overall, we contribute to the research community through the application and evaluation of recent DRL methods in solving the TSPPC. Our code is available at https://github.com/christianll9/tsppc-drl.

Keywords: Deep Reinforcement Learning · Traveling salesperson problem with precedence constraints · Heterogeneous attention

1 Introduction

The Traveling Salesperson Problem (TSP) is an NP-hard problem. Many practically relevant operations research problems are formulated as variants of the TSP. In this work, we focus on the TSP with precedence constraints (TSPPC), a variation of the TSP that enforces special ordering constraints, i.e., node i has to be visited before node j. Similar to the regular TSP, the TSPPC can be applied to many practically relevant problems, such as scheduling, routing, process sequencing, etc.

There has been a great deal of attention to the TSP and also work on the TSPPC, discussed in more detail in Sect. 2. In this work, we solve the TSPPC with Deep Reinforcement Learning (DRL) methods. Despite a large number of contributions to the TSP and the rising number of works that applies DRL methods to combinatorial optimization problems, to the best of our knowledge,

R. Bergmann et al. (Eds.): KI 2022, LNAI 13404, pp. 160–172, 2022.
https://doi.org/10.1007/978-3-031-15791-2_14

there do not exist any DRL approaches to the TSPPC yet. Due to the NP-hard nature of the problem, it is not feasible to obtain optimal solutions within a reasonable computational time for large-size problems. Therefore, applying DRL is desirable, as Machine Learning models offer a fast inference time in addition to the ability to generalize and be used on different problem settings.

Building on the approach of Li et al. [12] we adapt their model to cope with precedence constraints. For that, we modify their heterogeneous attention layers to deal with precedence constraints as well as chains of precedence constraints. This enables the model to learn the constraints intrinsically. Furthermore, we sparsify the attentions and show that this not only leads to a better run time performance but also increases the overall performance of the model (cf. Sect. 4).

The rest of the paper is structured as follows. Section 2 highlights some of the most important related work. In Sect. 3, the problem setting is defined, while Sect. 4 describes our methodology. Our experimental setup and results are presented in Sects. 5 and 6, respectively. In Sect. 7, we end with some concluding remarks and thoughts on future work.

2 Related Work

The TSPPC has been dealt with in a more generalized form in the field of operations research. It is termed Sequential Ordering Problem (SOP), which is another name for the asymmetric TSPPC. The SOP was initially presented as an operations research problem by [5], who proposed a heuristic method to solve it. The first work to introduce an algorithm to the SOP for exact solutions [1] formulates it as a mixed-integer linear programming problem and uses a branch-and-cut algorithm. Later several exact branch-and-bound algorithms were proposed [10,15,17], improving the results of [1]. In [9] the authors improve on their previous work [17] by enhancing the lower bound method for the branch-and-bound approach. One of the best heuristic solvers for the traveling salesperson problem is the Lin-Kernighan traveling salesman heuristic (LKH) algorithm [8]. LKH-3 is an extension of the LKH algorithm for solving constrained TSP problems including SOPs.

The TSPPC and the more general TSP belong to the broader family of Vehicle Routing Problems (VRP). A number of works have applied combinatorial optimization using DRL to solve these problems. [3] produced one of the significant works in this regard. They used the Pointer Network of [19] that consists of an LSTM-based encoder-decoder architecture and applied an actor-critic algorithm. They achieve better results in comparison to the supervised learning of [19] and the heuristics library OR-Tools [7]. Kool et al. [11] adapted the Transformer Model of [18] to solve multiple routing problems. They used multi-head attention layers, where multiple heads are concatenated and then transformed to perform message passing between nodes. Furthermore, they trained the model using the REINFORCE algorithm with a greedy rollout baseline and outperformed several TSP and VRP models, including [3].

[2] and [6] adapt the model from [11] to improve the performance on the Capacitated Vehicle Routing Problem (CVRP) and the CVRP with Time Windows respectively by making the feature embeddings more informative. [13] went one step ahead and utilized a consecutive improvement approach using a model partially based on the Transformer Model of [18]. They developed a DRL-based controller that iteratively refines a random initial solution with an improvement operator. Their model outperformed [11] and other state-of-the-art models.

[14] introduced Graph Pointer Networks based on the Pointer Network of [3]. They combined it with a hierarchical policy gradient algorithm to achieve new state-of-the-art results. Although their model lags behind the attention model of [11] for small-scale TSPs, it outperforms every model as the scale increases. Moreover, they also conducted experiments for TSP with time window constraints showing new state-of-the-art results.

[21] presented a Multi-Decoder Attention Model based on [11]. Instead of focusing on only one policy, their model learns multiple diverse policies and then utilizes a special beam search to pick the best of them. They also introduced an Embedding Glimpse layer to add more embedding information and thus improve each policy. [16] solved the CVRP in a supervised fashion and outperforms [11] and [21] for fixed vehicle costs.

Recently, [12] developed a DRL model using heterogeneous attentions for the pickup and delivery problem (PDP), which is a special case of the TSPPC [4]. Here, every node is either a pickup or a delivery node, and every node is part of exactly one precedence constraint. Their model builds extensively on the attention model of [11] and also uses a greedy rollout baseline together with the REINFORCE algorithm. They conducted experiments using randomly generated data and achieved a smaller total tour length than [11] and OR-Tools. Since the PDP is a special case of the TSPPC, we build on the work of [12] and adapt their model to cope with precedence constraints.

3 Problem Setting

The TSP, at its core, is concerned with finding the shortest route between a given set of nodes X while visiting each node only once and returning to the starting node. In addition, with precedence constraints, the starting point is prescribed. Furthermore, each route has to satisfy given precedence constraints. These constraints generally state something comparable to a visiting order. One node can be subject to many constraints. For example, i has to be visited, before j, k and l. The distance from node i to j is given by the distance matrix D. We want to find an optimal permutation σ over the nodes, such that the total travel length L is minimal. Adapted from [14], we can formulate a TSP with n nodes:

$$\begin{aligned} \min L(\sigma, D) = \min \sum_{t=1}^{n} D_{\sigma_t, \sigma_{t+1}} \\ \sigma_1 = \sigma_{n+1} \\ \sigma_t \in \{1, ..., n\} \\ \sigma_t \neq \sigma_{t'} \quad \forall\, t \neq t' \end{aligned} \tag{1}$$

where t is the current node and $\tau(i)$ returns the ordering of node i in the sequence according to the permutation σ, so that $i = \sigma_{\tau(i)}$. P is the set of all precedence constraints. If i has to precede j this is represented by the pair (i, j). Following this notation we can model the precedence constraints like:

$$\tau(i) < \tau(j) \quad \forall\ (i, j) \in P \tag{2}$$

4 Methodology

The proposed model builds on the work of [11] and [12]. We modify their Transformer model by restricting specific attentions. Similar to [11], each problem instance s can be considered as a graph with n nodes (see Sect. 3) having features x_i, where x_i are the 2D coordinates. Our graph would be fully connected in case of the unconstrained TSP. However, for the TSPPC, we use restricted attentions to let the model learn precedence constraints intrinsically [12]. Given a problem instance s, we sample the solution σ from a stochastic policy $p(\sigma|s)$ determined by our Transformer model [11]:

$$p_\theta(\sigma|s) = \prod_{t=1}^{n} p_\theta(\sigma_t|s, \sigma_{1:t-1}) \tag{3}$$

where the parameters θ describe the model. The encoder generates latent embeddings for all nodes using the coordinates as input features. These embeddings along with context are fed in to the decoder. The decoder works iteratively by decoding one state at a time to build the tour σ.

4.1 Encoder

The 2D input x_i is embedded as d_h-dimensional vector $h_i^{(0)} = W^x x_i + b^x$, where $d_h = 128$, W^x is the weight matrix and b^x is the bias term. Then, multiple stacked multi-head attention layers aggregate the embeddings. The output of the final layer h_i^N is used to compute a graph embedding $\bar{h}^N = \frac{1}{n}\sum_{i=1}^{n} h_i^N$, which can be interpreted as context.

Heterogeneous Attentions. In order to solve the PDP, the concept of heterogeneous attentions was introduced by [12]. In addition to the attentions from n nodes to n nodes, they introduce attentions (1) from every pickup/delivery node to its corresponding delivery/pickup node, (2) from every pickup/delivery node to all pickup nodes, and (3) from every pickup/delivery node to all delivery nodes.

In the TSPPC, one node could have the role of a delivery of multiple pickups or a delivery and a pickup simultaneously. These are the cases that cannot occur in the PDP and therefore are not handled by [12]. To generalize their model to the TSPPC, we speak therefore of predecessors instead of pickups and successors instead of deliveries. Additionally, we restrict certain attentions from predecessors to successors (ps) and from successors to predecessors (sp).

The TSPPC can also include chains of precedence constraints, which is another aspect that makes the TSPPC different from the PDP. For example, we could model a problem as follows: Node k can only be visited after node j, which itself can only be visited after node i. All three nodes (i, j, and k) build a chain. So far, there are no specific attentions from k to i or vice versa. Thus, we add a third kind of heterogeneous attentions to our model, where we restrict attentions from and to all members (mm) of a chain of precedence constraints. We define M as the set of all chains, where every node can be a member of a maximum of one chain. The function $chain(i)$ returns the chain of node i. Figure 1 illustrates the different heterogeneous attentions within the encoder.

Sparse Attentions. Sparsifying a graph can improve the run time performance for large-scale problems. To achieve this, we restrict the attentions between all nodes so that node $i \in X$ can only reach its neighborhood N_i. This adds a fourth kind of heterogeneous attentions to the model, namely attentions from neighbors to neighbors (nn). The black lines in Fig. 1 illustrate the attentions between neighbors for the first three nodes.

The neighborhood is calculated using two different approaches. First, we mask every attention between nodes, where the euclidean distance is larger than a fixed value d_t. The second approach uses the euclidean k-NN algorithm. Here, we mask all attentions to nodes that do not belong to the k nearest neighbors. We discuss both approaches in Sect. 5.

Formalization. In total, we use four kinds of heterogeneous attentions $U = \{nn, ps, sp, mm\}$ within each layer $l \in \{1, ..., N\}$ of our encoder. The kind of attention $u \in U$ is characterized by its own trainable weights for queries, keys and values $(W^{Q_u}, W^{K_u}, W^{V_u})$. Following loosely the notation of [11], we can calculate the values for every node $i \in \{1, \ldots, n\}$ and every layer l:

$$q_i^u = W^{Q_u} h_i, \quad k_i^u = W^{K_u} h_i, \quad v_i^u = W^{V_u} h_i \tag{4}$$

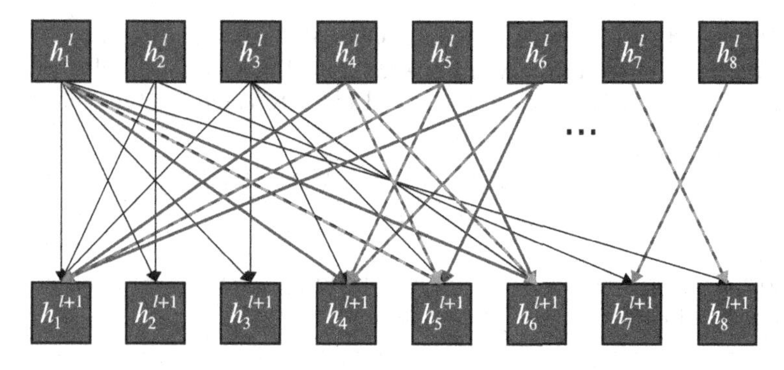

Fig. 1. Heterogeneous attentions between two consecutive multi-head attention layers within the encoder. Black lines show attentions from neighbor-to-neighbor (nn) (for illustration reasons, only the first three nodes were used). Orange lines show attentions from predecessors to successors (ps) and green lines from successors to predecessors (sp). Blue lines show attentions to all other members of the same constraint group (mm).

When calculating the compatibilities c_{ij}^u, we mask some with $-\infty$ to restrict the corresponding attention:

$$c_{ij}^{nn} = \begin{cases} \frac{q_i^{nn\top} \, k_j^{nn}}{\sqrt{d_k}}, & \text{if } j \in N_i \\ -\infty, & \text{otherwise} \end{cases} \tag{5}$$

$$c_{ij}^{ps} = \begin{cases} \frac{q_i^{ps\top} \, k_j^{ps}}{\sqrt{d_k}}, & \text{if } (i,j) \in P \\ -\infty, & \text{otherwise} \end{cases} \tag{6}$$

$$c_{ij}^{sp} = \begin{cases} \frac{q_i^{sp\top} \, k_j^{sp}}{\sqrt{d_k}}, & \text{if } (j,i) \in P \\ -\infty, & \text{otherwise} \end{cases} \tag{7}$$

$$c_{ij}^{mm} = \begin{cases} \frac{q_i^{mm\top} \, k_j^{mm}}{\sqrt{d_k}}, & \text{if } chain(i) = chain(j) \wedge \ chain(i) \in M \\ -\infty, & \text{otherwise} \end{cases} \tag{8}$$

To get the the attention weights a_{ij}^u we apply the softmax function on the compatibilities.

$$a_{ij}^u = \text{softmax}(c_i^u)_j \tag{9}$$

The attention weights a_{ij}^u are multiplied with values v_j^u and then added to get the embeddings for each head as:

$$h_i' = \sum_{u \in U} \sum_{j \in n} a_{ij}^u v_j^u \tag{10}$$

We summarize the aforementioned calculations as applying the multi-head attention function MHA with $B = 8$ heads on the embeddings h:

$$\text{MHA}(h)_i = \sum_{b=1}^{B} W_b^O h_{ib}' \tag{11}$$

where W^O is a trainable weight matrix. Overall, each attention layer l consists of an MHA function, a feedforward layer, skipped connections and a batch normalization (BN) function, which is applied twice to calculate the embedding h_i^l of node i:

$$\hat{h}_i^l = \mathrm{BN}^l(h_i^{(l-1)} + \mathrm{MHA}^l(h^{(l-1)})_i) \tag{12}$$

$$h_i^l = \mathrm{BN}^l(\hat{h}_i^l + \mathrm{FF}^l(\hat{h}_i^l)) \tag{13}$$

We follow [11] and use the same graph embedding as the mean of all final node embeddings, which is used by the decoder for context.

$$\bar{h}^N = \frac{1}{n}\sum_{i=1}^{n} h_i^N \tag{14}$$

4.2 Decoder

Our decoder is similar to [12], where it generates a probability vector based on the graph and node embeddings from the encoder. At the beginning, context h^c is created by concatenating the graph embedding and the last node embedding:

$$h^c = Concat(\bar{h}^N, h_{\sigma_{t-1}}^N) \tag{15}$$

where $Concat$ denotes concatenation.
Similarly, the glimpse $h^g = \mathrm{MHA}(W_g^Q h^c, W_g^K h^N, W_g^V h^N)$ is used for information aggregation. With the values $q_{(c)} = W^Q h^g$ and $k_i = W^K h_i^N$, we can compute the compatibility as:

$$\hat{h}^t = C \cdot \tanh(h^t), \tag{16}$$

where

$$h_i^t = \begin{cases} \frac{q_{(c)}^T k_i}{\sqrt{d_k}}, & \text{if } i \notin \sigma_{1:t-1} \wedge pred(i) \subseteq \sigma_{1:t-1} \\ -\infty, & \text{otherwise} \end{cases} \tag{17}$$

Here, $pred(i)$ is a function returning the set of predecessors of node i. All visited nodes and all successors are masked until their corresponding predecessors are visited. C is a hyperparameter used for clipping and is set to 10. Finally, we chose the next node to be visited based on the probability vector calculated as:

$$p(\sigma_t|X, L_{t-1}) = \mathrm{softmax}(\hat{h}^t) \tag{18}$$

Similar to [12], we optimize the loss $\mathcal{L}$ using gradient descent. We use the REINFORCE [20] algorithm with the greedy rollout [11] baseline $b(s)$:

$$\nabla\mathcal{L}(\theta|s) = \mathbb{E}_{p_\theta(\sigma|s)}[(L(\sigma) - b(s))\nabla \log p_\theta(\sigma|s)] \tag{19}$$

5 Experiments

In order to fairly compare whether the models could generalize well across smaller and larger instances than those seen in training, we set up controlled experiments with fixed configurations.

5.1 Training and Datasets

We train on fixed graph sizes TSPPC20 ($n_{train} = 20$), TSPPC50 ($n_{train} = 50$), TSPPC100 ($n_{train} = 100$), with $|P| = 0.33n$ precedence constraints. Each TSPPC instance consists of n nodes sampled uniformly in the unit square $S = [x_i]_{i=1}^{n}$ and $x_i \in [0,1]^2$. Each model is trained for 150 epochs with 102,400 TSPPC samples each, which are randomly generated for each epoch with the batch sizes of 512 (TSPPC20, TSPPC50) and 256 (TSPPC100) and a validation size of 10,240.

5.2 Model Adaptations and Sparsification

To evaluate the usefulness of our adaptations, we train different versions of our model. First, we use the original model by Kool et al. [11] and adapt only the masking within the decoder so that all tours are feasible solutions for the TSPPC. Additionally, we train a model with and without attentions to the precedence chain members (mm). All these three models are trained on a dense graph without sparsification (i.e., $d_t = \infty \wedge k = \infty$).

Moreover, we try two approaches of sparse attentions. First, by restricting every attention between nodes, where the Euclidean distance is larger than a particular threshold value d_t. Threshold values are chosen to be $d_t \in \{0.3, 0.5, 0.7, 0.9\}$. The second approach uses the euclidean k-NN algorithm, whereby all attentions to nodes, which do not belong to the $k \in \{5, 10, 20\}$ nearest neighbors, are restricted.

5.3 Evaluation and Baselines

All models are evaluated for problem sizes $n \in \{20, 50, 100, 150, 200\}$. We compare them against the simple Nearest Neighbor heuristic and the LKH-3 algorithm.

LKH-3 is a powerful, near-optimal solver for the TSPPC. Other work in the field of Reinforcement Learning often uses this solver and reports its results as the best-found solution, but consistently presents slow inference times [3,11,14]. Although not reported specifically, this suggests the use of LKH-3's standard parameter setting without variation. We analyzed the influence of LKH-3's *maxtrials* parameter on its cost-effectiveness. It can be observed that the standard setting of maxtrials = 10,000 is not the most efficient choice. Based on this finding, we report differently parametrized versions of LKH-3 in Sect. 6.

6 Results

We compare our best performing models for the problem sizes $n = 20$ and $n = 50$ with the aforementioned baselines in Table 1. It is worth noting that the results of the sampling approaches come quite close to those of the LKH-3. The default configuration of LKH-3 (with maxtrials = 10,000) has undoubtedly

Table 1. Average tour length and run time in seconds evaluation of the best performing models vs. baselines of 1,000 TSPPC samples for the problem sizes $n = 20$ and $n = 50$. Note that bold figures in each problem size represent the lowest tour length among baselines and models.

Method	$n = 20$			$n = 50$		
	Obj .	Gap	Time	Obj.	Gap	Time
LKH-3 (maxtrials = 1)	4.48	3.23%	0.041	7.13	9.86%	0.078
LKH-3 (maxtrials = 10)	4.36	0.46%	0.047	6.71	3.39%	0.090
LKH-3 (maxtrials = 10 k)	**4.34**	**0.00%**	1.160	**6.49**	**0.00%**	4.007
Nearest Neighbor	5.33	22.81%	**0.001**	8.61	32.67%	**0.005**
Kool et al. [11] (greedy, masked)	5.12	17.97%	0.007	8.41	29.58%	0.021
Ours (greedy, dense)	4.65	7.14%	0.008	7.76	19.57%	0.028
Ours (greedy, $k = 5$)	4.61	6.22%	**0.005**	7.66	18.03%	**0.012**
Ours (sampling = 1 k, $k = 5$)	4.40	1.38%	0.034	7.15	10.17%	0.085
Ours (sampling = 10 k, $k = 5$)	**4.38**	**0.92%**	0.123	**7.07**	**8.94%**	0.483

the best performance in finding the shortest tour but also has the longest run time across all problem sizes. Nevertheless, by using a different configuration, LKH-3 is still able to compete against all sampling approaches in tour length and run time. On the other hand, the Nearest Neighbor baseline gives the worst tour length performance but the fastest run times. Furthermore, we can see that using heterogeneous attentions can achieve better results compared to a masking approach based on the model by [11].

The sparse model (where $k = 5$) outperforms its dense counterpart not only in terms of run time but also achieves a shorter average tour length. The reason for this might be that sparsification forces the model to search for optimal follow-up nodes in the proximity of the current node. The hypothesis that the optimal next node lies close to the current node intuitively makes sense. By enforcing this assumption through sparsification, the model focuses on the most promising next candidate nodes, thus achieving better performance.

To evaluate the ability to generalize, we show the results of our models for different problem sizes in Table 2. Comparing TSPPC20, TSPPC50, and TSPPC100 models, the common observation is that models trained at a particular problem size perform better when evaluated on the same problem size they are trained on. Furthermore, TSPPC50 models tend to scale better when evaluated on larger problem sizes (i.e., $n = 100$, 150, and 200). Focusing on TSPPC50 models, the sparse k-NN $k = 5$ model outperforms all other dense and sparse models in all problem sizes they are evaluated on. Moreover, it is also the only model that beats the Nearest Neighbor baseline at $n = 200$, indicating the scalability of this sparse model to larger problem sizes. Note that the results for other model parameters can be found in appendix A Table 3.

Table 2. Comparison of average tour length and run time (in seconds) between multiple models vs. baselines on the evaluation of 1,000 TSPPC samples at each varying problem size. Bold figures represent the lowest tour length among baselines and models.

Method	$n = 20$		$n = 50$		$n = 100$		$n = 150$		$n = 200$	
	Obj.	Time	Obj.	Time	Obj.	Time	Obj.	Time	Obj.	Time
Baselines										
LKH-3 (maxtrials = 1)	4.48	0.041	7.13	0.078	10.49	0.096	13.27	0.175	15.75	0.274
LKH-3 (maxtrials = 10)	4.36	0.047	6.71	0.090	9.57	0.126	12.03	0.204	14.15	0.321
LKH-3 (maxtrials = 10 k)	**4.34**	1.160	**6.49**	4.007	**8.93**	12.160	**11.99**	22.928	**12.42**	36.086
Nearest Neighbor	5.33	0.001	8.61	0.005	12.31	0.022	15.16	0.065	17.54	0.124
TSPPC20										
Greedy Evaluation										
Kool et al. [11] (masked)	5.12	0.007	8.51	0.021	12.93	0.077	16.66	0.257	20.05	0.345
Ours (dense, without mm)	4.63	0.008	7.99	0.026	13.50	0.093	19.72	0.267	26.38	0.444
Ours (dense)	4.65	0.008	7.92	0.028	13.43	0.101	18.85	0.285	24.18	0.505
Ours ($d_t = 0.5$)	4.63	0.008	7.80	0.023	12.46	0.079	16.91	0.240	20.99	0.364
Ours ($k = 5$)	4.61	0.005	7.86	0.015	12.75	0.049	17.09	0.103	21.09	0.176
Ours ($k = 20$)	4.61	0.006	10.35	0.020	26.37	0.057	40.00	0.120	53.29	0.193
Sampling (1000)										
Ours ($d_t = 0.5$)	4.41	0.041	7.32	0.087	13.18	0.246	20.07	0.507	27.32	0.858
Ours ($k = 5$)	4.40	0.034	7.26	0.085	12.41	0.247	17.67	0.519	22.90	0.825
Ours ($k = 20$)	4.41	0.027	9.56	0.086	29.33	0.246	50.82	0.482	72.01	0.781
Sampling (5000)										
Ours ($d_t = 0.5$)	4.39	0.066	7.18	0.232	12.83	0.894	19.07	1.741	25.23	3.119
Ours ($k = 5$)	**4.38**	0.078	7.16	0.272	12.20	0.803	17.38	1.737	22.54	3.015
Ours ($k = 20$)	4.39	0.063	9.27	0.254	28.53	0.852	49.76	1.784	70.68	3.127
TSPPC50										
Greedy Evaluation										
Kool et al. [11] (masked)	5.27	0.007	8.41	0.021	12.45	0.077	15.70	0.257	18.44	0.345
Ours (dense, without mm)	5.09	0.008	7.76	0.026	11.52	0.093	14.94	0.267	18.19	0.444
Ours (dense)	5.00	0.008	7.76	0.028	11.50	0.101	14.85	0.285	17.99	0.505
Ours ($d_t = 0.5$)	5.00	0.008	7.70	0.023	11.46	0.079	14.85	0.240	18.10	0.364
Ours ($k = 5$)	4.95	0.004	7.66	0.012	11.30	0.050	14.54	0.108	17.47	0.175
Ours ($k = 20$)	5.02	0.005	7.76	0.019	11.71	0.062	15.65	0.122	19.48	0.202
Sampling (1000)										
Ours ($d_t = 0.5$)	4.67	0.030	7.16	0.080	10.83	0.238	14.61	0.553	18.51	0.863
Ours ($k = 5$)	4.64	0.027	7.15	0.085	10.83	0.234	14.60	0.531	18.41	0.845
Ours ($k = 20$)	4.70	0.031	7.23	0.086	11.31	0.276	16.18	0.494	21.49	0.806
Sampling (5000)										
Ours ($d_t = 0.5$)	4.62	0.059	7.14	0.268	10.79	0.819	14.58	1.720	18.53	3.001
Ours ($k = 5$)	4.61	0.073	**7.09**	0.250	10.68	0.827	14.38	1.749	18.16	3.037
Ours ($k = 20$)	4.66	0.069	7.17	0.246	11.14	0.804	15.92	1.694	21.16	3.256
TSPPC100										
Greedy Evaluation										
Ours ($d_t = 0.5$)	5.18	0.004	7.87	0.020	11.32	0.072	14.22	0.159	16.91	0.299
Ours ($k = 5$)	5.28	0.004	8.01	0.015	11.41	0.047	14.45	0.106	17.26	0.175
Ours ($k = 20$)	5.35	0.006	7.94	0.020	11.30	0.058	14.27	0.123	17.02	0.194
Sampling (1000)										
Ours ($d_t = 0.5$)	4.71	0.027	7.29	0.078	10.63	0.232	13.84	0.524	17.05	0.827
Ours ($k = 5$)	4.81	0.031	7.49	0.087	10.77	0.236	13.94	0.507	17.13	0.815
Ours ($k = 20$)	4.81	0.031	7.35	0.077	10.68	0.237	13.95	0.482	17.33	0.840
Sampling (5000)										
Ours ($d_t = 0.5$)	4.66	0.056	7.21	0.253	**10.51**	0.852	**13.65**	1.727	**16.84**	2.993
Ours ($k = 5$)	4.76	0.057	7.41	0.264	10.67	0.792	13.79	1.724	16.94	2.950
Ours ($k = 20$)	4.76	0.058	7.27	0.260	10.55	0.846	13.77	1.738	17.10	2.943

During our experiments, we observed unusual behavior regarding the scalability of some of our models. For instance, the TSPPC20 k-NN $k = 20$ model in Table 2 exhibits significantly worse results when scaled to higher problem sizes. TSPPC20 with 20 nearest neighbors means that the model has access to all nodes of the graph during training, which essentially makes it a dense model. Thus, it should also scale in a similar way as the dense model. We hypothesize that the reason behind the poor scalability is the fact that the model is practically trained as a dense model and then evaluated with sparse attentions for larger problem sizes. Hence, we conclude that a model needs to be trained on sparse attentions if it is to be evaluated on sparse attentions.

While the k-NN $k = 5$ model achieved the best performance for both, TSPPC20 and TSPPC50 models, we can see that the threshold-based sparse TSPPC100 model (where $d_t = 0.5$) gives a slightly better tour length than the corresponding k-NN models. Therefore, we would say that no single sparse model works best for all problem sizes. Hence, the sparsity level should be carefully selected since it is not a trivial task. Interestingly, we can see that the effect of sampling methods shrinks when increasing n, because the solution space grows exponentially. For $n = 200$, the greedy approach already beats the sampling of 1000 tours.

7 Conclusion and Future Work

In this paper, we successfully deploy a DRL-based training method using a Transformer model to solve the TSPPC. Furthermore, we sparsify our attentions, achieving not only faster computation time but also a gain in performance. Our model achieves better results than the model from Kool et al. [11] with a simple masking adaptation for the TSPPC. However, the scalability of our model to very large sizes was lacking in our experiments, which can be an aspect to concentrate on in future work.

We analyze the LKH-3 heuristic algorithm and, unlike most other publications, also report results for non-standard LKH-3 settings. This shows that LKH-3, while still having outstanding performance, can be very fast in inference as well.

It can be said that DRL methods applied to the TSPPC have shown promising results, notably when evaluated on the same problem size as seen in training or similar.

Acknowledgment. This work was supported by the German Federal Ministry of Education and Research (BMBF) via the project "Learning to Optimize" (L2O) under grant no. 01IS20013A.

A Appendix

Table 3. Average tour length (and run time in seconds) comparison of all TSPPC20 and TSPPC50 models vs. baselines evaluated on 1,000 TSPPC samples at each varying problem sizes. The number of precedence constraints is fixed at $|P| = 0.33n$. For a better comparison of the effects caused by using different model parameters, we show only the greedy evaluation.

Method	$n = 20$		$n = 50$		$n = 100$		$n = 150$		$n = 200$	
	Obj.	Time	Obj.	Time	Obj.	Time	Obj.	Time	Obj.	Time
Baselines										
LKH-3 (maxtrials = 1)	4.48	0.041	7.13	0.078	10.49	0.096	13.27	0.175	15.75	0.274
LKH-3 (maxtrials = 10)	4.36	0.047	6.71	0.090	9.57	0.126	12.03	0.204	14.15	0.321
LKH-3 (maxtrials = 10 k)	**4.34**	1.160	**6.49**	4.007	**8.93**	12.160	**11.99**	22.928	**12.42**	36.086
Nearest Neighbor	5.33	0.001	8.61	0.005	12.31	0.022	15.16	0.065	17.54	0.124
TSPPC20										
Greedy Evaluation										
Kool et al. [11] (masked)	5.12	0.007	8.51	0.021	12.93	0.077	16.66	0.257	20.05	0.345
Ours (dense, without mm)	4.63	0.008	7.99	0.026	13.50	0.093	19.72	0.267	26.38	0.444
Ours (dense)	4.65	0.008	7.92	0.028	13.43	0.101	18.85	0.285	24.18	0.505
Ours ($d_t = 0.3$)	4.63	0.006	7.88	0.189	13.20	0.062	19.05	0.178	24.75	0.303
Ours ($d_t = 0.5$)	4.63	0.008	7.80	0.023	12.46	0.079	16.91	0.240	20.99	0.364
Ours ($d_t = 0.7$)	4.64	0.008	7.96	0.026	13.19	0.089	18.28	0.312	23.20	0.445
Ours ($d_t = 0.9$)	4.63	0.008	7.92	0.028	13.08	0.095	17.98	0.277	22.50	0.500
Ours ($k = 5$)	4.61	0.005	7.86	0.015	12.75	0.049	17.09	0.103	21.09	0.176
Ours ($k = 10$)	4.62	0.005	8.13	0.016	14.13	0.050	19.63	0.112	24.74	0.190
Ours ($k = 20$)	4.61	0.006	10.35	0.020	26.37	0.057	40.00	0.120	53.29	0.193
TSPPC50										
Greedy Evaluation										
Kool et al. [11] (masked)	5.27	0.007	8.41	0.021	12.45	0.077	15.70	0.257	18.44	0.345
Ours (dense, without mm)	5.09	0.008	7.76	0.026	11.52	0.093	14.94	0.267	18.19	0.444
Ours (dense)	5.00	0.008	7.76	0.028	11.50	0.101	14.85	0.285	17.99	0.505
Ours ($d_t = 0.3$)	4.97	0.006	7.66	0.188	11.37	0.062	14.76	0.178	17.96	0.302
Ours ($d_t = 0.5$)	5.00	0.008	7.70	0.023	11.46	0.079	14.85	0.240	18.10	0.364
Ours ($d_t = 0.7$)	5.02	0.008	7.72	0.026	11.47	0.089	14.97	0.312	18.28	0.445
Ours ($d_t = 0.9$)	5.01	0.008	7.69	0.028	11.39	0.095	14.66	0.277	17.68	0.500
Ours ($k = 5$)	4.95	0.004	7.66	0.012	11.30	0.050	14.54	0.108	17.47	0.175
Ours ($k = 10$)	5.05	0.005	7.73	0.014	11.61	0.054	15.18	0.110	18.50	0.187
Ours ($k = 20$)	5.02	0.005	7.76	0.019	11.71	0.062	15.65	0.122	19.48	0.202

References

1. Ascheuer, N., Jünger, M., Reinelt, G.: A branch & cut algorithm for the asymmetric traveling salesman problem with precedence constraints. Comput. Optim. Appl. **17**(1), 61–84 (2000)
2. Bdeir, A., Boeder, S, Dernedde, T., Tkachuk, K., Falkner, J. K., Schmidt-Thieme, L. RP-DQN: an application of q-learning to vehicle routing problems. KI **2021**, 3–16 (2021)
3. Bello, I., Pham, H., Le, Q.V., Norouzi, M., Bengio, S.: Neural combinatorial optimization with reinforcement learning. In: ICLR Workshop (2017)

4. Dumitrescu, I., Ropke, S., Cordeau, J.-F., Laporte, G.: The traveling salesman problem with pickup and delivery: polyhedral results and a branch-and-cut algorithm. Math. Program. **121**(2), 269–305 (2010)
5. Escudero, L.F.: An inexact algorithm for the sequential ordering problem. Eur. J. Oper. Res. **37**(2), 236–249 (1988)
6. Falkner, J. K., Schmidt-Thieme, L. Learning to solve vehicle routing problems with time windows through joint attention. CoRR abs/2006.09100 (2020)
7. Google Inc. OR-Tools (2016)
8. Helsgaun, K. An extension of the lin-kernighan-helsgaun tsp solver for constrained traveling salesman and vehicle routing problems. Roskilde: Roskilde University, pp. 24–50 (2017)
9. Jamal, J., Shobaki, G., Papapanagiotou, V., Gambardella, L.M., Montemanni, R.: Solving the sequential ordering problem using branch and bound. In: 2017 IEEE Symposium Series on Computational Intelligence (SSCI), pp. 1–9 (2017)
10. Karan, M., Skorin-Kapov, N.: A branch and bound algorithm for the sequential ordering problem. In: 2011 Proceedings of the 34th International Convention MIPRO, pp. 452–457. IEEE (2011)
11. Kool, W., Hoof, H., Welling, M.: Attention, learn to solve routing problems!. In: International Conference On Learning Representations (2019). https://openreview.net/forum?id=ByxBFsRqYm
12. Li, J., Xin, L., Cao, Z., Lim, A., Song, W., Zhang, J.: Heterogeneous attentions for solving pickup and delivery problem via deep reinforcement learning. IEEE Trans. Intell. Transp. Syst. **23**(3), 2306–2315 (2022)
13. Lu, H., Zhang, X., Yang, S.: A learning-based iterative method for solving vehicle routing problems. In: International Conference on Learning Representations (2020)
14. Ma, Q., Ge, S., He, D., Thaker, D., Drori, I.: Combinatorial optimization by graph pointer networks and hierarchical reinforcement learning (2019)
15. Mojana, M., Montemanni, R., Di Caro, G., Gambardella, L.M., Luangpaiboon, P.: A branch and bound approach for the sequential ordering problem. Lecture Notes Manag. Sci. **4**(1), 266–273 (2012)
16. Thyssens, D., Falkner, J. K., Schmidt-Thieme, L.: Supervised permutation invariant networks for solving the CVRP with bounded fleet size. CoRR. abs/2201.01529 (2022)
17. Shobaki, G., Jamal, J.: An exact algorithm for the sequential ordering problem and its application to switching energy minimization in compilers. Comput. Optim. Appl. **61**(2), 343–372 (2015). https://doi.org/10.1007/s10589-015-9725-9
18. Vaswani, A., et al.: Attention is all you need. In: Guyon, I., Luxburg, U. V., Bengio, S., Wallach, H., Fergus, R., Vishwanathan, S., Garnett, R., (eds.), Advances in Neural Information Processing Systems, vol. 30, pp. 5998–6008 (2017)
19. Vinyals, O., Fortunato, M., Jaitly, N.: Pointer networks. In: Cortes, C., Lawrence, N., Lee, D., Sugiyama, M., Garnett, R., (eds.), Advances in Neural Information Processing Systems, vol. 28, pp. 2692–2700 (2015)
20. Williams, R.J.: Simple statistical gradient-following algorithms for connectionist reinforcement learning. Mach. Learn. **8**(3–4), 229–256 (1992)
21. Xin, L., Song, W., Cao, Z., Zhang, J.: Multi-decoder attention model with embedding glimpse for solving vehicle routing problems. Proc. AAAI Conf. Artif. Intell. **35**(13), 12042–12049 (2021)

HanKA: Enriched Knowledge Used by an Adaptive Cooking Assistant

Nils Neumann[(✉)] and Sven Wachsmuth

Institute for Cognition and Robotics (CoR-Lab), Universitätsstr. 25, 33615 Bielefeld, Germany
{nneumann,swachsmu}@techfak.uni-bielefeld.de

Abstract. Cooking a meal is a challenging and recurring task that requires the consideration of various environmental influences and constraints as well as significant domain knowledge. One way to reduce the complexity is to follow recipes that provide an ordered set of tasks for the preparation of a dish. This concept was already transferred to cooking assistants which present the recipe to the cook while adding further assistance like device control or step-by-step visualization. Although recipes and assistants simplify the cooking process itself, other factors like the available devices or differences in cooking skills are ignored. Aside from that, current assistants are often limited to one recipe at a time, ignoring the regularly occurring requirement to prepare a meal of multiple components. Considering these challenges, we propose our adaptive cooking assistant *HanKA* that considers the individual user skill and environmental kitchen setup, while assisting the cook at the preparation of their freely combined and synchronized recipes. This is achieved through a modular approach consisting of the automatic detection and control of the available devices and user interfaces, the scheduling of multiple recipes based on the distributed knowledge representation, and a deviation management that considers the user experience. Hereby, we created an adaptive cooking assistant that considers various influences that occur in a cooking scenario, resulting in a better assistance for the user.

Keywords: AI applications and innovations · Knowledge representation and reasoning · Planning and scheduling · Human monitoring · Assisted living · Intelligent assistive environments

1 Introduction

Cooking is a complex instrumental activity of daily living [9], whereby cooking skills correlate with healthy eating [7]. Hereby, a layperson has to continuously transfer and coordinate recipe instructions in his individual cooking environment. Due to an increasing digitization of life, multiple cooking assistants were developed that assist a cook in the preparation of a dish. *Cooking Navi* [6] focuses on the synchronization of cooking recipes, *PIC2DISH* [3] generates the cooking instructions from a picture and *MimiCook* [16] projects the instructions

R. Bergmann et al. (Eds.): KI 2022, LNAI 13404, pp. 173–186, 2022.
https://doi.org/10.1007/978-3-031-15791-2_15

on the kitchen counter. While these provide a step-by-step assistance for the cook, they do not provide key characteristics of user assistance systems [11] like context-awareness and adaptation to the assisted person. In the cooking domain, this includes the detection, control, and consideration of the available devices. Although recipes provide step-by-step instructions for the cook, and in some cases customization features like exchanging ingredients [1,2,12], these are often limited to a single component. A completely flexible composition of multiple recipes to an assisted cooking process including an automated sequencing, distributed device control, and dynamic synchronization of preparation steps has not been realized so far. Time estimations in traditional recipes do not consider the skills of the cook and frequently differ greatly from the time actually needed. Similar to *MAMPF* [15], we individually measure and adapt the duration time for each step. Beyond this, we further generalize these time estimates to new recipes and ingredients. Although various assistive systems for the kitchen exist that all successfully assist the cook when preparing dishes, most of these focus on a step-by-step instruction and simplify contextual factors, as shown in Table 1. However, the consideration of these contextual factors would lead to an individualized *in situ* assistance of the user in his regular kitchen environment as an ultimate goal.

Table 1. Functionality provided by cooking assistants: Asterisk (*) means only build-in devices are controlled. The first 5 are research prototypes, the next 2 are available products, followed by our *HanKA* assistant. "Skill adaptation" means that the cooking skills of the user are considered while planning, and does not describe if the system adapts while cooking.

Cooking assistance	Device		Recipe		Skill adaptation
	Control	Detection	Combination	Synchronization	
MimiCook [16]	-	-	-	-	-
CookingNavi [6]	-	-	✓	✓	-
PIC2DISH [3]	-	-	-	-	-
KogniChef [13]	✓	-	-	-	-
MAMPF [15]	✓	-	✓	-	✓
Thermomix [17]	*	-	-	-	-
Cookit [4]	*	-	-	-	-
HanKA	✓	✓	✓	✓	✓

In this paper we present our cooking assistant *HanKA* (German acronym for: action centered coordination of assistance processes) that provides the following contributions: (i) automatic detection and control of available user devices; (ii) progress detection utilizing device interactions; (iii) combination and synchronization of freely combinable recipes; (iv) adaptation of recipes to available devices and cooking skills; (v) utilizing a distributed knowledge base for enriching minimal recipes with general cooking knowledge; (vi) a working cooking assistant that guides the user, monitors the progress, and adjusts to occurring

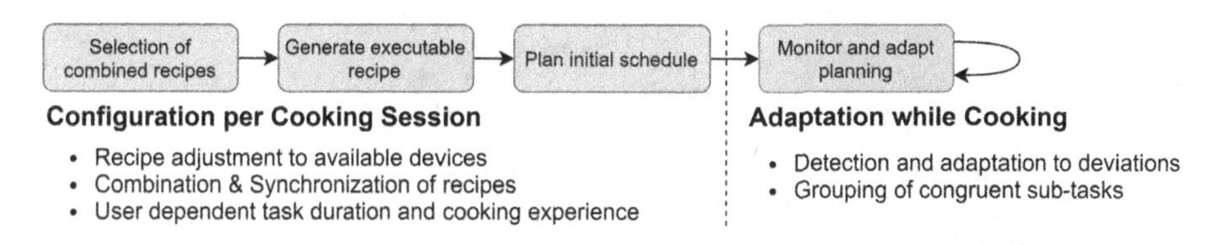

Fig. 1. Visualization of the phases from a cooking session and the assignment of the procession loops with their related adaptations.

deviations. In this way, the cook can concentrate on the cooking task itself and entrusts the scheduling of the tasks and managing of the devices to the assistant.

2 Methodology

In order to create a cooking assistant that provides the previously described flexibility, the provision of multiple adaptations is required (Fig. 1). These adaptations are organized in two different processing loops. The outer loop initiates a configuration at the start of each cooking session and the inner loop repeatedly applies a process adaptation. Before the adaptation concepts are described in detail, a brief summery of the underlying knowledge representation is given.

2.1 Underlying Knowledge Representation

In the following, the concepts of the knowledge representation are briefly described. More details can be found in [14]. The distributed knowledge representation consists of three sources (component recipes, ingredients, *action_templates*) which, when combined, result in the executable recipes for the cooking assistant. These three knowledge sources are linked by the action type of each task in the component recipe (Fig. 2). The hierarchical representations of these knowledge sources allow to reuse and generalize knowledge that is utilized in the proposed concepts. *Action_templates* are abstract actions that are parameterized by the other knowledge sources. They provide all actions executable by the cooking assistant with generalized and action-dependent information: e.g., whether a user and/or device are required for the execution, the urgency to proceed with the following action (connection urgency) and all subordinate (pre/post) actions. The recipe representation contains a set of tasks with a reference to the corresponding *action_template*, the logical dependency between the tasks and all recipe specific information, like ingredients and amount per serving. The ingredient representation for each usable ingredient contains actions executable with it, together with their general parameters (e.g. cooking time). If an information is not defined in the recipe, the missing information is added from the ingredient representation or, if not available for the ingredient, from the *action_template.*

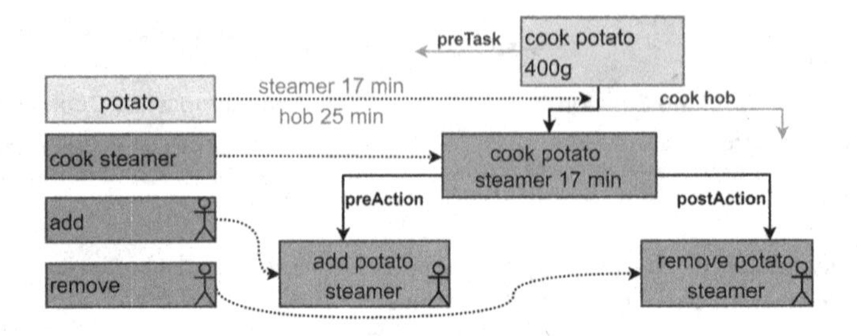

Fig. 2. Exemplary creation of executable recipe steps (orange) for the single recipe representation step *cook potato* (blue), while leaving out the connection to previous tasks or the alternative preparation with a hob. The recipe task *cook potato* is enriched with information (cooking times) from the cook action of the ingredient representation *potato* (green). These combined information parametrize the abstract *action_template cook steamer* to create the executable *cook potato steamer* recipe step (orange). Since the *action_template cook steamer* has (sub-)tasks, these executable steps are also generated with information from the recipe and *action_templates*, visualized by stick figure. (Color figure online)

2.2 Configurations per Cooking Session

The configuration per cooking session mostly takes place while creating the executable representation of the recipes that are selected by the cook. This provides the necessary infrastructure to define the inference processes for any configuration or adaptation of the executable recipe proposed in this paper. This representation contains all information about the available devices and cook as well as all recipes that should be cooked together.

Recipe Adjustment to Available Devices. In order to deal with varying devices in different kitchens, available devices must be discovered and device-specific programs have to be abstracted to functionalities used by the cooking assistant. Describing the required device-independent functionality in the *action_templates* enables the mapping of recipes to devices and the filtering of executable recipes in the context of the available devices. Through this, only recipes with executable tasks on the available devices are planned and executed. Adapting to the devices and utilizing the abstracted device program description from the *action_templates*, enables the automatic adjustment to different setups and allows automated device control in the recipe context.

Combination and Synchronization of Recipes. Enabling the user to freely combine different recipes requires a flexible, automated enriching and scheduling of instruction steps. In this case, further challenges arise, such as, where recipes can be interrupted, which tasks are connected time-critically, which tasks can be parallelized, which tasks require specific devices or the user attention, and how are all component preparations synchronized to be served hot. Therefor, the self descriptive recipe contains all information about the recipe steps that are utilized by the scheduler to arrange the tasks into a temporal order. In order to plan the recipes based on the current setup, the scheduler needs the available devices and information about the cook in addition to the recipes. Based on the cook and device workload for each task, the scheduler can take the occupancy into account. This enables the planner to execute tasks in parallel in regard to the resources, which made it possible to synchronize the recipes. Utilizing the logical connection between tasks and their connection urgency from the recipes, enables the scheduler to know which tasks have to be carried out one after the other and gives an indication where recipes can be interrupted. When alternative methods of preparation are described by a recipe, further options are available for the scheduler. The approach described so far leads to a valid plan, but does not take the synchronization of the serving times into account. Because device tasks that heat an ingredient are critical for serving everything hot, the last one of these for each recipe must end as late as possible for all component recipes.

User Dependent Task Duration and Cooking Experience. The duration of many preparation tasks greatly varies for different cooks. A better prediction will result in less deviations, less re-planning, and a better estimation of the serving time. While storing the last time for a recipe task would be sufficient, this only provides a prediction for the step in the specific recipe. Using a more generalized approach that does not just refer to the recipe but also to general cooking skills, enables the estimation of preparation time for recipes yet not cooked by the user. A solution for this is presented by estimating factors that scale the default task duration appropriately. Such factors can be estimated on three different levels of abstraction utilizing the distributed knowledge representation of recipes, ingredients, and actions. After each cooking session, these three factors are updated for each executed step, as further explained in Sect. 3.3. While some cooking actions scale with improved cooking skills of the user like peeling, others only depend on the recipe task for actions like cooling. Therefore, it is necessary to know which tasks scale with the cooking skill and which not. Since this information is action-dependent, it is provided by the *action_templates*, which define if an action can adapt to the user. As the duration for some tasks scale with the number of servings and some have fixed values, these information are also provided by the *action_template*. By saving the default user duration of the task as a factor, this value is independent of the number of servings and calculated when these are selected.

2.3 Adaptation While Cooking

Based on the initial plan with a complete schedule of tasks, the user starts the cooking process. Although the plan contains only tasks executable with the given devices and uses task durations adapted to the user, this does not ensure that no further adaptations are required. Therefore, a monitoring and re-planning of the cooking process is implemented.

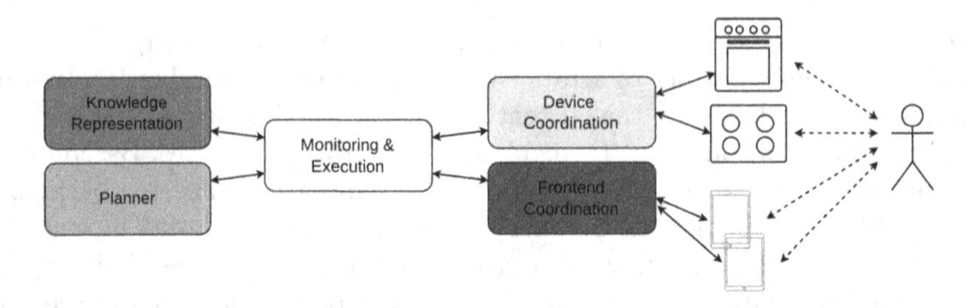

Fig. 3. Component overview of the cooking assistant: The user interacts with the devices and user interfaces which are connected to the corresponding coordination components (3.1, 3.2). These send the interaction commands to the *Monitoring and Execution* (3.4) component, that executes the recipe and detects deviations based on the scheduled task order from the *Planner* (3.5), which is calculated on the enriched knowledge representation from the *Knowledge Representation* (3.3).

Detection and Adaptation to Deviations from the Plan. While cooking a meal, the estimated duration of the tasks can differ from the real duration. Possible reasons are varying attention levels, external distractions or delays from devices. Examples are the missing confirmation of the device execution or a prolonged heating process with larger amounts of water. Therefore, it is necessary to monitor the cooking session and react to occurring deviations. These deviations can be that a task is not executable because a previous task is not finished, or that a running task should have ended but the user or a device has never confirmed it. If a deviation is detected, the discrepancy to the old plan is corrected by re-scheduling critical tasks. By adapting to the current situation, the cook receives a permanently executable plan and is supported until the dish is finished.

Grouping of Congruent Sub-tasks for Compact User Instructions. While experienced cooks profit from compact instructions, beginners require detailed simple cooking instructions. Therefore, we propose an automated strategy to merge congruent cooking instructions. Here, the frequency how often a user cooked the recipe and the appropriate ingredient-action pair beforehand is used to determine if tasks are combined. As the final task sequence is a result of the planning process, the grouping of tasks is considered after planning. Before executing a task, the following tasks in time are inspected if they are merge-able with the current task and if the user has enough experience for the combination. Whether tasks can be combined is decided in regard to the used ingredient, logical connection, action type and device usage. If a combination of tasks is possible, the description, title, and images of the tasks are merged, and both tasks are started. Clustering the task shortly before the execution retains that scheduling can find the best possible solution, while also providing compact user instructions in an adaptable manner.

3 System Description

Creating an adaptive cooking assistant covers a wide set of areas, resulting in thematically appropriate components roughly following *MAPE-K* from autonomic computing [8], as shown in Fig. 3. These components are connected with a customized request/response pattern, based on the *MQTT* protocol [10] and communicate with serialized protocol buffer messages. This results in a well-defined API and clear task areas for each component and allows the decoupling of components. In the following subsections, the individual components, their contribution to the overall system, and their adaptation strategies are described.

3.1 Device Coordination

The *Device Coordination* discovers and provides the available devices, controls them, and detects process events required in the active device tasks from the recipe. To keep it flexible, the component abstracts device-specific information into a device-independent description that is used to describe device tasks inside a recipe through which a mapping between functionalities in the recipe and device specific programs is enabled. In addition to information such as operating mode (e.g. two-sided heat), temperature, and duration, the description contains the general type of actions (*preheat*, *execute*, *add*, *remove*). This is utilized by the individual device controller which interprets the sensor data in the recipe context and tracks the task progress. Through the additional information the device functionality is extended and allows a context-aware interpretation of built-in device sensors. Whereby preheat is finished if the temperature is reached, execute is finished if the required duration is expired and add/remove are finished if other device sensors detect the addition or removal of an ingredient at the time of the task (e.g. door switch sensor for an oven). If the end of a task is detected by the device, it notifies the *Monitoring and Execution*, which results in automatic recipe progress.

3.2 Frontend Coordination

The *Frontend Coordination* is responsible for the coordination and synchronization of the user interfaces and for the generalization of the user input. While the specific user interfaces are not known beforehand for each setup analogous to the available devices, they register via a discovery service to the *Frontend Coordination.* The registered interfaces are connected with and synchronized by the *Frontend Coordination.* Hereby, they implement a defined API to interact with the system that abstracts the user input to commands, relevant for the cooking process. Through this, the assistant can connect to different user interfaces (tablet, speech interface or several of these) available in the concrete setup (e.g. Fig. 4). Via the user interfaces the cook can combine recipes into the desired dish, start and stop the cooking process, add or remove recipes to an active cooking session, and visualize the planned task order. During cooking, the cook is always able to finish any running task, select any valid new task, as well as to preview tasks before they are active.

Fig. 4. Exemplary cooking scenario, where the user confirms the cooking step *cut onion* via a tablet.

3.3 Knowledge Representation

The *Knowledge Representation* provides the executable and combinable recipes by utilizing the distributed knowledge sources (recipe representation, ingredient representation and *action_templates*), as summarized in Sect. 2.1. The enriched recipes provide the required information for scheduling (*Planner*), visualization (*Frontend Coordination*), device controlling (*Device Controller*) and monitoring (*Monitoring and Execution*). As the recipe should be cookable in the individual kitchen, the created executable recipe representation also contains the registered devices with their functionalities from the *Device Coordination* and the registered cook from the *Frontend Coordination.* Based on the available devices, non executable recipes are removed. If an unknown user starts cooking with the assistant, the system creates a user-specific ontology that stores the cooking experience and duration factors for each recipe step in reference to the recipe, ingredient, and action. Due to the different levels of abstraction and the more frequent use of more generalized actions, a weight factor X is added in regard to the referred knowledge source. In the ontology, the cooking skill of the user is saved regarding the recipe step (R $(X = 5)$), the ingredient-action pair (I $(X = 10)$), and the action type itself (A $(X = 20)$). After each cooking session, these three factors (R,I,A) are updated for each executed step (Eq. 1), with the corresponding *X-value* for R,I,A.

$$newFactor(R, I, A) = \frac{X * oldFactor(R, I, A) + \frac{measuredTime}{defaultTime}}{X + 1} \tag{1}$$

$$userTaskDuration = \frac{R + I + A}{3} * taskDuration \tag{2}$$

When a new recipe is generated, the duration for each task that the user can perform is calculated by Eq. 2, with a factor of 1 if no value was saved before. Even though the R value has no impact for recipes the user has never cooked, the I and A values are recipe independent, allowing a prediction for recipes that are unknown to the cook. Through the recipe representation, *Monitoring and Execution* also receives information how often the cook executed certain recipes and utilized ingredients in order to merge tasks into more compact instructions.

3.4 Monitoring and Execution

The *Monitoring and Execution* component connects all components of the cooking assistant. However, the main task is the monitoring of the cooking process and the execution of the active tasks. Therefore, the execution of tasks and the detection of deviations is split into multiple conditions and actions that are repeated in an adaptation loop. While the *Monitoring and Execution* component detects the deviations, the adjustment of the task schedule is carried out by the *Planner* component.

- **Task removal:** Checks if the *Device* or *Frontend Coordination* finished a task and removes it from the active tasks.
- **Active task verification:** Checks if an active task should have been ended but is still running. If a task is still running with an end time in the past it is extended.
- **Task start:** If the time for a new task to run has been reached, it is verified that there are no resource conflicts resulting from an extended step. It is further checked whether all logically connected previous tasks are finished. When all tasks are finished and no resource conflicts are found, the task is added to the active tasks. If not, a re-planning is necessary due to a deviation.
- **Execution triggering:** If the list of active tasks changes, they are send to the *Device-* and *Frontend Coordination* in regard to the required resources. If the cook has experience with the recipe or ingredient, it is checked if the task can be combined with the following tasks, based on the ingredient, task connection and action type of the tasks. If detected they, are combined and communicated as one task with merged content information.
- **Planning:** If any of the previous steps detects a deviation, a planning request with the actual state, available devices and cook is send to the *Planner*. The solved response is used as the new plan for execution.

While the initial combination and synchronization of recipes is performed beforehand, the combination of recipes can be changed while cooking, e.g. removing, replacing or adding a component recipe. In this case, the appropriate action is executed and the new recipe combination in regard to the progress of the already running recipes is synchronized. This enables an even greater adaptability of the cooking process for the cook.

3.5 Planner

The scheduling is considered as an optimization problem, where the domain, rules and moves are defined and implemented in a way that the optaplanner [5] engine can be utilized to combine and synchronize the recipes with the available

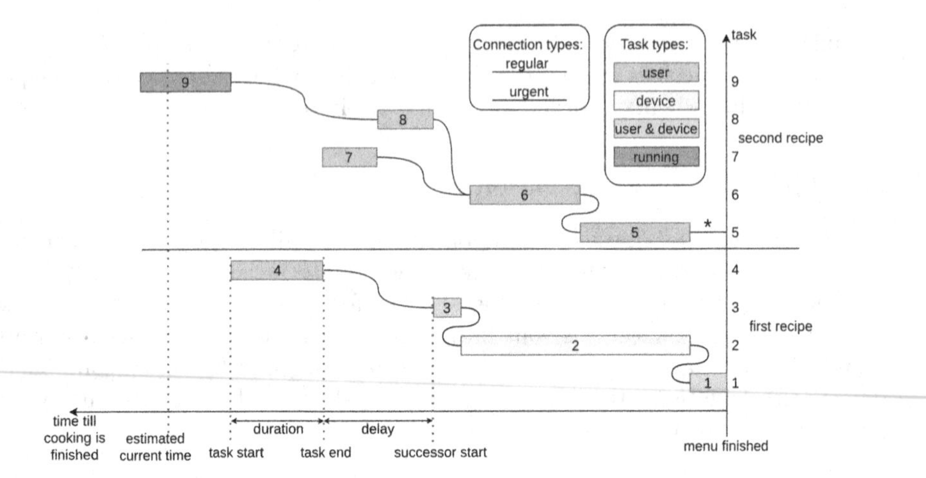

Fig. 5. Simplified planning structure with two recipes and nine tasks: The end time of a task is the earliest successor start time together with a variable delay, changeable by the planner. The task 4 has a delay to his successor task (3) as the user attention is required for tasks 7/8. In contrast, the user tasks 5/6 are parallel to the device task 2, because it only needs a device and no user. The urgent connection between 1/2 and 2/3 urges the planner to prevent a delay. For the synchronization of the recipes, the planner minimizes the delay (*). Since it is a planning request in a running state, task 9 is executed by the user at the moment. Therefore, if the task is moved, the time till cooking is finished also changes, resulting in a changed serving time.

devices. Hereby, each planning request is stateless and utilizes the cooking state of the request, which can only have small deviations for re-planning requests. The main planning entity of the domain is the task. Each task is assigned to a recipe. Since the planner synchronizes the recipes, all recipes have to be finished around the same time. Therefore, the synchronization point of our planning problem is the serving time, with the serving time defining the origin and the time axis representing the time till the cooking process is finished, as shown in Fig. 5. Utilizing this time representation, the first two optimizing criteria are the difference between the finished menu and the actual time (cook as fast as possible) and the distance from the last recipe step to the finished menu (synchronized serving). The logical dependency of the tasks is modeled as a built-in constraint, because any deviation from the logical order results in an invalid plan. Therefore, the tasks are connected as a chain where the end of a task is the earliest start time for all its successor tasks. This includes a delay that is used as a planning variable, allowing the planner to change the time between tasks. Using the time representation with the chain connection creates the advantage that if a deviation occurs in the cooking process, only the ends of the chain have to be adjusted, resulting in an easy re-planning. Solving the logical dependency with the built-in constraint, the main constraint remaining is the consideration of the workload for the user and devices. These are assigned based on the functionalities required by the task. In order to not exceed the maximum workload for user and devices,

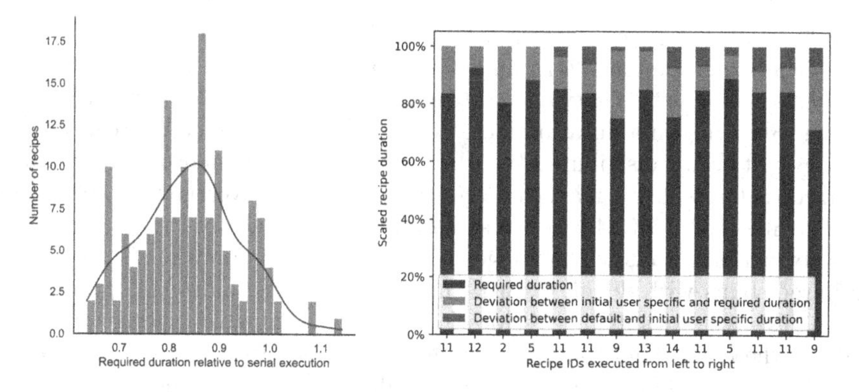

Fig. 6. Left: Distribution of the relative duration for the 153 combinations in regard to their serialized execution duration. Right: User Adaptation over multiple cooking sessions. The single recipes are executed by a new simulated user that validates each user task after 70% of the required duration defined in the base recipe. The device tasks are executed normally. The recipes, with their IDs were executed from left to right. Hereby, 100% is the necessary duration without user adaptation and the blue bar is the required duration the simulated user actually needed. As the system adjusts to the user skills, the deviation between initial plan and required duration (orange) reduces, while the deviation between the initial plan and the required duration without user adaptation increases (green). This happens especially strong if a known recipe is repeated (ID 11) due to the different factor (Sect. 3.3). (Color figure online)

their workload is calculated over all tasks and if the maximum value is exceeded, a hard constraint is detected which results in a penalty that outranks all other scores. In order to ensure that action sequences requiring the same device, e.g., adding multiple ingredients, are still consistently assigned, device groups are formed that treat them as a single compound task. For the creation of a plan that is acceptable for the user, the logical dependency/cohesion is considered as soft constraint. This takes the urgency of the connection into account where delays between tasks with a high urgency result in higher penalties. Nevertheless, tasks with a low urgency connection are still sticked together if possible. Since the planner should not always move all tasks in the chain, a custom move was implemented that moves a single task in the range between the next task and the following task, which expands the possibilities of the planner. Based on the described domain, rules, and scores, the tasks are ordered, deviations are solved, and an executable plan is returned.

4 Evaluation

The flexibility of the *HanKA* cooking assistant is provided by multiple adaptations. Therefore, they are evaluated individually as far as that is possible in a quantitative evaluation. Unless otherwise stated, a kitchen setup consisting of a steamer, oven and hob is used. The evaluation is based on 18 recipes, each consists of 6 to 64 tasks (avg. 19.6). Hereby, 1/3 of the recipes has two options of preparation, e.g. cooking noodles with a hob or steamer. The initial scheduling of a single recipe requires 2474 ms on average. The average planned execution time for a recipe with 4 servings is 46 min. and 39 s. and 38 min. and 15 s. for 1 serving. Resulting in a duration increase of 22.09% on average between 1 and 4 servings, not evenly distributed over the recipes, due to differences in task scaling (min. cooked noodle 0%/max. cabbage turnip 94%). On average the recipes have an idle time of 40.98% (4 serv.) that describes, how much of the time the cook does not perform a task. This ranges from 0% for vegan mayonnaise to 81.69% for burger buns with dough resting time. The idle time indicates optimization possibilities when combining recipes and thus is a valuable scoring criteria to evaluate the recipe combinations. For the evaluation of the recipe combination, all 153 2-pair combinations of the 18 available recipes were calculated with an average initial planning time of 3394 ms. Hereby, the planned cooking duration is reduced to 83.68% of the time that is required to cook these successively, as shown in Fig. 6 (left) for all combinations. The longer duration in 6/153 cases is noticeable and results from changing types of preparation, that take longer but provide a better synchronization at the end. For a single component, the time between heating the last heating task and serving was 129.44 s, while for the combined recipes the average for each component was 229.64 s. In this duration, the user removes the components from the devices and performs the last tasks that are only executable afterwards. The lower idle time of 25.52% for two recipes in contrast to 40.98% for a single recipe indicates a useful combination of the recipes.

The cooking skills are evaluated with a simulated user, visualized in Fig. 6 (right). This user operates the assistant including devices just as a normal user, but takes over the user interface due to the well-defined API. Hereby, the initially estimated duration approaches the required duration over multiple cooking sessions, especially if the recipe was cooked before, but also for completely unknown recipes (e.g. ID 9). Due to varying amounts of user tasks, the possible adaptation varies depending on the recipe. While executing the recipes, the average adaptation time to deviations was 908.27 ms, which happened 24.14 times per cooking process on average.

5 Conclusion

While supporting a cook at the preparation of a meal, cooking assistants and recipes focus mostly on the cooking process itself. Although the cooking process is the most important part, ignoring various other influences prevents the exploitation of the full potential. Therefore, we identified these influences, presented approaches that take these into account, and combined them into our adaptive cooking assistant *HanKA*.

As a result, the cooking assistant combines and synchronizes multiple recipes, and considers the available devices while also controlling these in the context of the task. It considers the cooking skills of the user, adjusts the granularity of steps based on the user experience, and adapts the cooking process in case of deviations. Due to the adaptation of the system, the workload for the cook is reduced whereby the user no longer has to check and control the available devices or merge the recipes by hand. Through this, the cook can focus on the cooking process itself.

References

1. Innit - Your food. Simplified and solved (2021). https://www.innit.com/. Accessed 02 Nov 2021
2. Plantjammer: Dynamic recipes (2021). https://www.plantjammer.com/dynamic-recipes. Accessed 02 Nov 2021
3. An, Y., et al.: PIC2DISH: a customized cooking assistant system. In: Proceedings of the 25th ACM International Conference on Multimedia, pp. 1269–1273. MM 2017, Association for Computing Machinery, NY (2017). https://doi.org/10.1145/3123266.3126490
4. Bosch: Cookit - küchenmaschine mit kochfunktion - bosch hausgeräte (2021). https://www.bosch-home.at/shop/kuechenmaschine-mit-kochfunktion1. Accessed 02 Nov 2021
5. De Smet, G., Open Source Contributors: OptaPlanner User Guide. Red Hat, Inc. or third-party contributors (optaPlanner is an open source constraint solver in Java) (2006). https://www.optaplanner.org
6. Hamada, R., Okabe, J., Ide, I., Satoh, S., Sakai, S., Tanaka, H.: Cooking navi: assistant for daily cooking in kitchen. In: Proceedings of the 13th Annual ACM International Conference on Multimedia, pp. 371–374. MULTIMEDIA 2005, Association for Computing Machinery, NY (2005). https://doi.org/10.1145/1101149.1101228
7. Hartmann, C., Dohle, S., Siegrist, M.: Importance of cooking skills for balanced food choices. Appetite **65**, 125–131 (2013). https://doi.org/10.1016/j.appet.2013.01.016
8. Kephart, J., Chess, D.: The vision of autonomic computing. Computer **36**(1), 41–50 (2003). https://doi.org/10.1109/MC.2003.1160055
9. Lawton, M.P., Brody, E.M.: Assessment of older people: self-maintaining and instrumental activities of daily living. Gerontologist **9**(3_Part_1), 179–186 (1969)
10. Light, R.A.: Mosquitto: server and client implementation of the MQTT protocol. J. Open Source Softw. **2**(13), 265 (2017). https://doi.org/10.21105/joss.00265 https://doi.org/10.21105/joss.00265 https://doi.org/10.21105/joss.00265
11. Maedche, A., Morana, S., Schacht, S., Werth, D., Krumeich, J.: Advanced user assistance systems. Bus. Inf. Syst. Eng. **58**(5), 367–370 (2016). https://doi.org/10.1007/s12599-016-0444-2
12. Müller, G., Bergmann, R.: CookingCAKE: a framework for the adaptation of cooking recipes represented as workflows. In: ICCBR (Workshops), pp. 221–232 (2015)
13. Neumann, A., et al.: "KogniChef": a cognitive cooking assistant. Künstl. Intell. **31**(3), 273–281 (2017). https://doi.org/10.1007/s13218-017-0488-6
14. Neumann, N., Wachsmuth, S.: Recipe enrichment: knowledge required for a cooking assistant. In: Proceedings of the 13th International Conference on Agents and Artificial Intelligence - vol. 2: ICAART, pp. 822–829. INSTICC, SciTePress (2021). https://doi.org/10.5220/0010250908220829

15. Reichel, S., Muller, T., Stamm, O., Groh, F., Wiedersheim, B., Weber, M.: MAMPF: an intelligent cooking agent for zoneless stoves. In: 2011 Seventh International Conference on Intelligent Environments, pp. 171–178 (2011). https://doi.org/10.1109/IE.2011.18
16. Sato, A., Watanabe, K., Rekimoto, J.: MimiCook: a cooking assistant system with situated guidance, pp. 121–124 (2014). https://doi.org/10.1145/2540930.2540952
17. Thermomix: Thermomix® das original | vorwerk thermomix® (2021). https://www.vorwerk.com/de/de/c/home/produkte/thermomix. Accessed 02 Nov 2021

Automated Kantian Ethics: A Faithful Implementation

Lavanya Singh(✉)

Harvard University, Cambridge, MA 02138, USA
lavanyasingh2000@gmail.com

Abstract. As we grant artificial intelligence increasing power and independence in contexts like healthcare, policing, and driving, AI faces moral dilemmas but lacks the tools to solve them. Warnings from regulators, philosophers, and computer scientists about the dangers of unethical AI have spurred interest in automated ethics-i.e., the development of machines that can perform ethical reasoning. However, prior work in automated ethics rarely engages with philosophical literature. Philosophers have spent centuries debating moral dilemmas so automated ethics will be most nuanced, consistent, and reliable when it draws on philosophical literature. In this paper, I present an implementation of automated Kantian ethics that is faithful to the Kantian philosophical tradition. I formalize Kant's categorical imperative in an embedding of Dyadic Deontic Logic in HOL, implement this formalization in the Isabelle/HOL theorem prover, and develop a testing framework to evaluate how well my implementation coheres with expected properties of Kantian ethic. My system is an early step towards philosophically mature ethical AI agents and it can make nuanced judgements in complex ethical dilemmas because it is grounded in philosophical literature. Because I use an interactive theorem prover, my system's judgements are explainable.

Keywords: Automated ethics · Kant · Isabelle · Ai ethics

1 Introduction

AI is making decisions in increasingly important contexts, such as medical diagnoses and criminal sentencing, and must perform ethical reasoning to navigate the world responsibly. This ethical reasoning will be most nuanced and trustworthy when it is informed by philosophy. Prior work in building computers that can reason about ethics, known as automated ethics, rarely capitalizes on philosophical progress and thus often cannot withstand philosophical scrutiny. This paper presents an implementation of philosophically faithful automated ethics.

Faithfully automating ethics is challenging. Representing ethics using constraint satisfaction [20] or reinforcement learning [1] fails to capture most ethical theories. For example, encoding ethics as a Markov Decision Process assumes that ethical reward can be aggregated, a controversial idea [47]. Even once ethics

R. Bergmann et al. (Eds.): KI 2022, LNAI 13404, pp. 187–208, 2022.
https://doi.org/10.1007/978-3-031-15791-2_16

is automated, context given to the machine, such as the description of an ethical dilemma, plays a large role in determining judgements.

I implement automated Kantian ethical reasoning that is faithful to philosophical literature.[1] I formalize Kant's moral rule in Dyadic Deontic Logic (DDL), a logic that can express obligation and permissibility [15]. I implement my formalization in Isabelle, an interactive theorem prover that can automatically generate proofs in user-defined logics [40]. Finally, I use Isabelle to automatically prove theorems (such as, "murder is wrong") in my new logic. Because my system automates reasoning in a logic that represents Kantian ethics, it automates Kantian ethical reasoning. It can classify actions as prohibited, permissible or obligatory with minimal factual background. I make the following contributions:

1. In Sect. 4.1, I formalize a philosophically accepted version of Kant's moral rule in DDL.
2. In Sect. 4.2, I implement my formalization in Isabelle. My system can judge appropriately-represented actions and show the facts used in the proof.
3. In Subsects. 1 and 2 of Sect. 4.2, I use my system to produce nuanced answers to two well-known Kantian ethical dilemmas. Because my system draws on Kantian literature, it can perform sophisticated moral reasoning.
4. In Sect. 4.3, I present a testing framework to evaluate how faithful my system is to philosophical literature. Tests show that my implementation outperforms two other formalizations of Kantian ethics.

2 The Need for Faithful, Explainable Automated Ethics

AI operating in high-stakes environments like policing and healthcare must make moral decisions. For example, self-driving cars may face the following moral dilemma: an autonomous vehicle approaching an intersection fails to notice pedestrians until it is too late to brake. The car can continue on its course, running over and killing three pedestrians, or it can swerve to hit a tree, killing its single passenger. While this example is (hopefully) not typical of the operation of a self-driving car, every decision that such an AI agent makes, from avoiding congested freeways to carpooling, is morally tinged.

Machine ethicists recognize this need and have made theoretical [8,19,26,53] and practical progress in automating ethics [6,16,30,54]. Prior work in machine ethics using deontology [2,4], consequentialism [1,3,17], and virtue ethics [13] rarely engages with philosophical literature, and so misses philosophers' insights. The example of the self-driving car is an instance of the trolley problem [24], in which a bystander watching a runaway trolley can pull a lever to kill one and save three. Decades of philosophical debate have developed nuanced answers to the trolley problem. AI's moral dilemmas are not entirely new, so solutions should draw on philosophical progress. The more faithful that automated ethics is to philosophy, the more trustworthy and nuanced it will be.

[1] Source code can be found at https://github.com/lsingh123/automatedkantianethics.

A lack of engagement with philosophical literature also makes automated ethics less explainable, as seen in the example of Delphi, which uses deep learning to make moral judgements based on a training dataset of human decisions [30]. Early versions of Delphi gave unexpected results, such as declaring that the user should commit genocide if it makes everyone happy [52]. Because no explicit ethical theory underpins Delphi's judgements, we cannot determine why Delphi thinks genocide is obligatory. Machine learning approaches like Delphi often cannot explain their decisions. This reduces human trust in a machine's controversial ethical judgements. The high stakes of automated ethics require explainability to build trust and catch mistakes.

3 Automated Kantian Ethics

I present a faithful implementation of Kantian ethics, a testing framework to evaluate how well my implementation coheres with philosophical literature, and examples of my system performing sophisticated moral reasoning.

I formalize Kant's moral rule in a semantic embedding of Dyadic Deontic Logic in HOL [10,15]. Deontic logic can express obligation, or binding moral requirements and is often an extension of a modal logic. Modal logics include the necessitation operator $\Box$, where $\Box p$ is true at world w if p is true at all worlds that neighbor w [18]. Modal logics also contain operators of propositional logic like $\neg, \wedge, \vee, \rightarrow$. Some deontic logics replace the $\Box$ operator with an obligation operator O. I use Carmo and Jones's Dyadic Deontic Logic (DDL) [15], which uses the dyadic obligation operator $O\{A|B\}$ to represent the sentence "A is obligated in the context B."

Because this work is an early step towards faithful automated ethics, I use Kantian ethics, a theory that is amenable to formalization. I do not argue that Kantian ethics is the best theory, but that it is the most natural to automate.[2] I automate the Formula of Universal Law (FUL), a version of Kant's moral rule which states that moral principles can be acted on by all people without contradiction. For example, if everyone falsely promises to repay a loan, lenders will stop offering loans, so not everyone can act on this principle, so it is prohibited.

Prior work by Benzmüller et al. [10,12] implements an embedding of DDL in HOL using Isabelle. I add the Formula of Universal Law as an axiom to their library. The resulting Isabelle theory can automatically generate proofs in a new logic that has the categorical imperative as an axiom. Because interactive theorem provers are designed to be interpretable, my system is explainable. Isabelle can list the facts used in a proof and construct human-readable proofs. In Sect. 4.2, I use my system to generate sophisticated solutions to two ethical dilemmas. Because my system is faithful to philosophical literature, it produces nuanced judgements.

I also contribute a testing framework that evaluates how well my formalization coheres with philosophical literature. I formalize expected properties of

[2] The full argument is in Appendix A.

Kantian ethics as sentences in my logic and run the tests by using Isabelle to automatically find proofs or countermodels for the test statements. My system outperforms two other attempts at formalizing Kantian ethics [36].

Given an action represented as a sentence in my logic, my system proves that it is morally obligatory, permissible, or prohibited. My system serves as one step towards philosophically sophisticated automated ethics.

4 Details

4.1 Formalizing the Categorical Imperative in DDL

The Formula of Universal Law reads, "act only according to that maxim by which you can at the same time will that it should become a universal law" [31]. To formalize this, I represent willing, maxims, and the FUL in Benzmüller, Farjami, and Parent's [10] semantic embedding of DDL in HOL. This allows me to use quantifiers and worlds[3], which don't exist in the object level of DDL. This detail does not affect the correctness of my implementation, which operates on top of Benzmüller, Farjami, and Parent's implementation of their embedding.[4] Prior work formalizing Gewirth's ethics also adopts this approach [25].

Willing a Maxim. Kantian ethics evaluates "maxims," which are "the subjective principles of willing," or the principles that the agent understands themselves as acting on [31]. I adopt O'Neill's view that a maxim includes the act, the circumstances, and the agent's purpose of acting or goal that the maxim seeks to achieve [42]. The maxim's goal is the end that the agent seeks to achieve by acting on the maxim.

Definition 1 (Maxim). *A circumstance, act, goal tuple (C, A, G), read as "In circumstances C, do act A for goal G."*

For example, one maxim is "When strapped for cash, falsely promise to repay a loan to get some easy money," with goal "to get some easy money." A maxim includes an act and the circumstances[5] under which it should be performed. It must also include a goal because human activity, guided by a rational will, pursues ends that the will deems valuable [31].

I define "willing a maxim" as adopting it as a principle to live by.

[3] The embedding of DDL in HOL operates only over sets of worlds, but I abbreviate singleton sets as a single world to make the presentation clearer.

[4] I reproduce the relevant sections of their embedding, as implemented in Isabelle, in Appendix B.

[5] The inclusion of circumstances in a maxim raises the "tailoring objection" [33,55], under which maxims are arbitrarily specified to pass the FUL. For example, the maxim "When my name is John Doe, I will lie to get some easy money," passes the FUL but should be prohibited. One solution is to argue that the circumstance "when my name is John Doe" is not morally relevant, but this requires defining morally relevant circumstances. The difficulty in determining relevant circumstances and formulating a maxim is a limitation of my system and requires that future work develop heuristics to classify circumstances as morally relevant.

Definition 2 (Willing). *For maxim $M = (C, A, G)$ and actor s,*

$$\text{will}\, M\, s \equiv \forall w\, (C \longrightarrow A\,(s))\, w$$

At all worlds w, if the circumstances hold at that world, agent s performs act A.

If I will the example maxim above about falsely promising to repay a loan, then whenever I need cash, I will falsely promise to repay a loan.

Practical Contradiction Interpretation. My project uses Korsgaard's canonical practical contradiction interpretation of the FUL [22, 34].

The logical contradiction interpretation prohibits maxims that are impossible when universalized. Under this view, falsely promising is wrong because, in the universalized world, the practice of promising would end, so falsely promising would be impossible. This view cannot handle natural acts, like that of a mother killing her crying children so that she can sleep [21, 34]. Universalizing this maxim does not generate a contradiction, but it is clearly wrong. Killing is a natural act, so it can never be impossible so this view cannot prohibit it.

As an alternative to the logical contradiction view, Korsgaard endorses the practical contradiction view, which prohibits maxims that are self-defeating, or ineffective, when universalized. By willing a maxim, an agent commits themselves to the maxim's goal, so they cannot rationally will that this goal be undercut. This can prohibit natural acts like that of the sleep-deprived mother: in willing the end of sleeping, she is willing that she is alive. If all mothers kill all loud children, then she cannot be secure in the possession of her life, because her mother could have killed her as an infant. Willing this maxim thwarts the end that she sought to secure.

Formalizing the FUL. The practical contradiction interpretation interprets the FUL as, "If, when universalized, a maxim is not effective, then it is prohibited." If an agent wills an effective maxim, then the maxim's goal is achieved, and if the agent does not will it, then the goal is not achieved.

Definition 3 (Effective Maxim). *For a maxim $M = (C, A, G)$ and actor s,*

$$\text{effective}\, M\, s \equiv \forall w\, (\text{will}\, (C, A, G)\, s \iff G)\, w$$

A maxim is universalized if everyone wills it. If, when universalized, it is not effective, it is not universalizable.

Definition 4 (Universalizability). *For a maxim M and agent s,*

$$\text{not\_universalizable}\, M\, s \equiv [\forall w\, (\forall p\, \text{will}\, M\, p)\, w \longrightarrow \neg\, \text{effective}\, M\, s]$$

Using these definitions, I formalize the Formula of Universal Law.

Definition 5 (Formula of Universal Law).

$$\forall M, s\, (\forall w\, \text{well\_formed}\, M\, s\, w) \longrightarrow (\text{not\_universalizable}\, M\, s \longrightarrow \forall w\, \text{prohibited}\, M\, s\, w)$$

For all maxims and people, if the maxim is well-formed, then if it is not universalizable, it is prohibited.[6]

Definition 6 (Well-Formed Maxim). *A maxim is well-formed if the circumstances do not contain the act and goal. For a maxim* (C, A, G), *and subject* s,

$$\text{well\_formed}\, (C, A, G)\, s \equiv \forall w\, (\neg(C \longrightarrow G) \wedge \neg(C \longrightarrow A\, s))\, w$$

For example, the maxim "When I eat breakfast, I will eat breakfast to eat breakfast" is not well-formed because the circumstance "when I eat breakfast" contains the act and goal. Well-formedness is not discussed in the literature, but I discovered that if the FUL holds for badly formed maxims, then it is not consistent.[7] The fact that the FUL cannot hold for badly formed maxims is philosophically interesting. Maxims are an agent's principle of action, and badly-formed maxims cannot accurately represent any action. The maxim "I will do X when X for reason X" is not useful to guide action, and is thus the wrong kind of principle to evaluate. This property has implications for philosophy of doubt and practical reason. The fact that I was able to derive this insight using my system demonstrates that, in addition to guiding AI agents, automated ethics can help philosophers make philosophical progress.

4.2 Isabelle/HOL Implementation

I implement my formalization in Isabelle[8], which allows the user to define types, axioms, and lemmas. It integrates with theorem provers [39,43] and counter-model generators [14] to automatically generate proofs.

I use Benzmüller et al.'s implementation of DDL [10]. They define the atomic type i, a set of worlds. Term t is true at set of worlds i if t holds at all worlds in i. I add the atomic type s, which represents a subject or person. I also introduce the type abbreviation $os \equiv s \rightarrow term$, which represents an open sentence. For example, run is an open sentence, and run applied to the subject $Sara$ produces the term $Sara\, runs$, which can be true or false at a world.

I define the type of a maxim to be a (t, os, t) tuple. Circumstances and goals are terms because they can be true or false at a world. In the falsely promising example, the circumstance "when I am strapped for cash" is true in the real world and the goal "so I can get some easy money" is false. An act is an open sentence because whoever wills the maxim performs the action. "Falsely promise to repay a loan" is an open sentence that, when applied to a subject, produces a term, which is true if the subject falsely promises.

---

[6] The definition of prohibition is given in Appendix C and uses the dyadic obligation operator.

[7] See Fig. 10 for the experiment that tests consistency of an implementation.

[8] For further implementation details, see Appendix C.

I add the definitions from Sect. 4.1 as abbreviations, include logical background to simplify future proofs, and add the FUL as an axiom. I use countermodel checker Nitpick [14] to show that my formalization of the FUL does not hold in DDL, so adding it as an axiom will strengthen the logic. After I add the FUL as an axiom, I use Nitpick to find a satisfying model, demonstrating that the logic is consistent. The results of these experiments are in Figs. 9 and 10.

Application: Lies and Jokes. I demonstrate my system's power on two ethical dilemmas. First is the case of joking. Many of Kant's critics argue that his prohibition on lies includes lies told in the context of a joke. Korsgaard [35] responds by arguing that there is a crucial difference between lying and joking: lies involve deception, but jokes do not. The purpose of a joke is amusement, which does not rely on the listener believing the story told. Given appropriate definitions of lies and jokes, my system shows that jokes are permissible but lies are not. Because my system is faithful to philosophical literature, it can perform nuanced reasoning, demonstrating the value of faithful automated ethics.

```
lemma lying_prohibited:
  assumes "m ≡ (c::t, a::os, g::t)"
  assumes "∀w. ∀s. well_formed m s w"
—‹Initial technical set-up: m is a well-formed maxim composed of some circumstances, act, and goal.›
  assumes "lie m"
—‹m is a maxim in which the action requires knowingly uttering a falsehood and the goal requires
that someone believe this falsehood.›
  assumes "∀t w. ((∀p. utter_falsehood p t w) ⟶ (∀p. ¬ (believe p t) w))"
—‹The convention of trust assumption that if everyone utters false statement t, then no one will
believe t. This simple assumption encodes the common sense knowledge that human communication
involves an implicit trust, and that when this trust erodes, the convention of communication begins
to break down and people no longer believe each other.›
  assumes "∀w. c w"
—‹Restrict our focus to worlds in which the circumstances hold. A technical detail. ›
  shows "⊨ (prohibited m me)"
proof -
  have "(∀p w. (W m p) w) ⟶ (⊨ (c → (¬ g)))"
    by (smt assms(1) assms(2) assms(5) case_prod_beta fst_conv old.prod.exhaust snd_conv)
  have "not_universalizable m me"
    by (metis (mono_tags, lifting) assms(1) assms(2) case_prod_beta fst_conv snd_conv)
  thus ?thesis
    using FUL assms(2) by blast
qed
```

Fig. 1. The proof that lying is prohibited. This proof relies on some technical details, a definition of lying, and the convention of trust assumption.

First, I implement the argument that lies are prohibited because they require deception. The goal of a maxim about lying requires that someone believe the lie. This is a thin definition of deception; it does not include the liar's intent. I also assume that if everyone lies about a particular statement, then people will no longer believe that statement. This is the uncontroversial fact that we tend to believe people only if they are trustworthy in a given context. I call this the "convention of trust" assumption. The full proof is in Fig. 1.

Next, I use my system to show that jokes are permissible. Korsgaard notes that the purpose of jokes "is to amuse and does not depend on deception" [35].

The goal of a joke does not require that anyone believe the statement. As in the case of lying, this is a thin definition; it does not involve any definition of humor. With this definition of a joke and with the convention of trust assumption above, my system shows that joking is permissible. The full proof is in Fig. 2.

My system can show that lying is prohibited but joking is not because of its robust conception of a maxim. Because my implementation is faithful to philosophical literature, it is able to recreate Korsgaard's solution to a complex ethical dilemma that philosophers debated for decades. Moreover, the reasoning in this section requires few, uncontroversial common sense facts. The deepest assumption is that, if everyone lies about a given statement, no one will believe that statement. This is so well-accepted that most philosophers do not bother to justify it.

```
lemma joking_not_prohibited:
  assumes "m ≡ (c::t, a::os, g::t)"
  assumes "∀w. ∀s. well_formed m s w"
—‹Initial set-up: m is a well-formed maxim composed of some circumstances, act, and goal.›
  assumes "joke m"
—‹m is a maxim about joking. Precisely, it is a maxim in which the action is to knowingly utter a
falsehood and the goal does not require that someone believe this falsehood.›
  assumes "∀t w. ((∀p. utter_falsehood p t w) ⟶ (∀p. ¬ (believe p t) w))"
—‹The convention of trust assumption.›
  assumes "∀w. c w"
—‹Restrict our focus to worlds in which the circumstances hold. A technical detail. ›
  shows "⊨ (permissible m me)"
  by (smt assms(1) assms(2) assms(3) case_prod_conv)
```

Fig. 2. The proof that joking is permissible. This proof relies on technical assumptions, a definition of joking, and the convention of trust assumption.

```
lemma lying_to_murderer_permissible:
  assumes "⊨ (well_formed murderers_maxim murderer)"
  assumes "⊨ (well_formed my_maxim me)"
—‹Assume that we're working with well-formed maxims.›
  assumes "⊨ (protect_victim → (murderer believes victim_not_home))"
—‹In order for you to protect the victim, the murderer must believe that the victim is not home. ›
  assumes "∀sentence::t. ∀p1::s. ∀p2::s. ∀w::i.
    ((p1 believes (utter_falsehood p2 sentence)) w) ⟶ (¬ (p1 believes sentence) w)"
—‹The convention of belief assumption: if person1 believes that person2 utters a sentence as a
falsehood, then person1 won't believe that sentence.›
  assumes "∀c a g w. (universalized (c, a, g) w) ⟶
           ((person1 believes (person2 believes c)) → (person1 believes (a person2))) w"
—‹The universalizability assumption: if a maxim is universalized, then if person1 believes person2
believes they are in the given circumstances, then person1 believes person2 performs the act. ›
  assumes "∀w. murderer_at_door w"
—‹Restrict our focus to worlds in which the circumstance holds. A technical detail. ›
  shows "⊨ (permissible my_maxim me)"
  using assms(1) assms(6) by auto
—‹The common sense assumptions given are not strong enough to generate a prohibition against
lying to a liar, and are thus unused in this proof. ›
```

Fig. 3. The proof that lying to the murderer is permissible. This proof relies on technical assumptions, specification of the example, the convention of belief assumption, and the universalizability assumption.

Application: Murderer at the Door. My system can also resolve the paradox of the murderer at the door. In this dilemma, murderer Bill knocks on your door asking about Sara, his intended victim. Sara is at home, but you should lie to Bill and say that she is away to protect her. Critics argue that the FUL prohibits you from lying; if everyone lied to murderers, then murderers wouldn't believe the lies and would search the house anyways. Korsgaard resolves this debate by noting that the maxim of lying to a murderer is actually that of lying to a liar. Bill cannot announce his intentions to murder; instead, he "must suppose that you do not know who he is and what he has in mind" [35].[9] Thus, the maxim of lying to the murderer is actually the maxim of lying to a liar.

My system correctly shows that lying to a liar is permissible. Implementing this argument requires formalizing Korsgaard's assumptions. First, she assumes that Bill believes you, so he won't search your house if he thinks Sara isn't home. Second is what the convention of belief assumption: if X thinks Y utters a statement as a lie, X won't believe that statement. For example, if you say that it is raining, but I think that you are lying, I will think that it is sunny. This assumption is almost definitional; if you think someone is lying, you won't believe them. Third, she assumes that if a maxim is universalized, then everyone believes that everyone else wills it. For example, if the falsely promising maxim is universalized, everyone notices that people who are strapped for cash falsely promise to repay loans. This is the heaviest assumption of the three; if you observe that many do X in circumstances C, you will assume that everyone does X in circumstance C. I call this the universalizability assumption.

Using these assumptions, my system proves that lying to a murderer is permissible. The full proof is in Fig. 3. These examples show that, even with uncontroversial assumptions, my system can make nuanced moral judgements.

4.3 Testing Framework

I contribute a testing framework to evaluate how well my implementation coheres with philosophical literature. These tests make "philosophical faithfulness" precise. Each test consists of a sentence in my logic, such as that obligations cannot contradict each other. The rest of the tests are presented in Appendix E.

To run the tests, I prove or refute each test sentence in my logic. Because these tests are derived from moral intuition and philosophical literature, they evaluate how reliable my system is. As I implemented my formalization, I checked it against the tests, performing test-driven development for automated ethics. My testing framework shows that my implementation outperforms DDL with no other axioms added (a control group) and Kroy's [36] prior attempt at formalizing the FUL, which I implement in Isabelle. My implementation outperforms both other attempts. Full test results are summarized in Fig. 4.

[9] Korsgaard assumes that the murderer will lie about his identity in order to take advantage of your honesty to find his victim. In footnote 5 of [35], she accepts that her arguments will not apply in the case of the honest murderer who announces his intentions, so she restricts her focus to the case of lying to a liar. She claims that in the case of the honest murderer, the correct act is to refuse to respond.

Test	Naive	Kroy	Custom
FUL Stronger than DDL	×	✓	✓
Obligation Universalizes Across People	×	✓	✓
Obligations Never Contradict	×	×	✓
Distributive Property for Obligations	×	×	✓
Prohibits Actions That Are Impossible to Universalize	×	×	✓
Robust Representation of Maxims	×	×	✓
Can Prohibit Conventional Acts	×	×	✓
Can Prohibit Natural Acts	×	×	✓

Fig. 4. Table showing which tests each implementation passes. The naive interpretation is raw DDL and the custom formalization is my novel implementation.

5 Future Work

My implementation can evaluate sentences represented in my logic but it is not yet ready for deployment. Like much work in automated ethics [1,30], it uses a rigid representation for its inputs (i.e., sentences in my logic) and outputs (i.e., proof of judgement). A deployment-ready ethics engine requires an input parser to translate moral dilemmas into my logic and an output parser to translate judgements into action. Figure 5 depicts this example ethics engine .

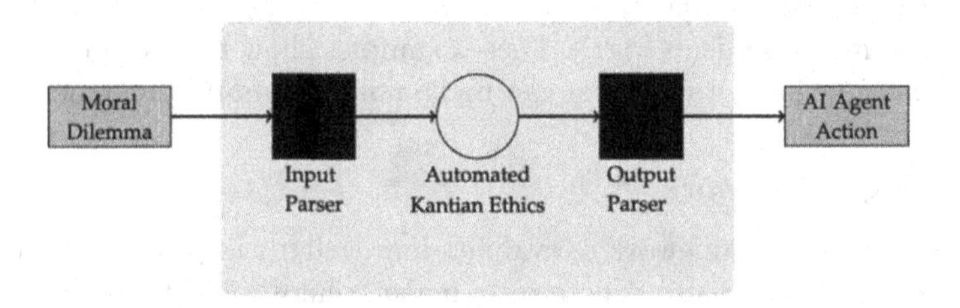

Fig. 5. An ethics engine that passes a moral dilemma through an input parser, applies the automated Kantian ethics test, and then processes the output using an output parser. I contribute the automated Kantian ethics component.

Future work must solve the open problem of translating real-life situations to a structured, logical representation (e.g., a maxim). For example, consider an AI-operated drone deciding whether to bomb a weapons factory, knowing that shrapnel could likely harm civilians. The input parser must translate this potential action into the maxim, "When I am at war, I will bomb a factory next to civilians in order to end the war soon," and evaluate its moral status. Defining a maxim is a central challenge in Kantian ethics because it requires deciding which circumstances are morally relevant, a decision that must be informed by social context.[10] Future work could address this limitation by defining "moral closeness" heuristics or using machine learning to learn maxims.

This work uses Kantian ethics. Like any ethical theory, there are objections to Kantian ethics, such as the assumption of an ideal moral society [35].[11] Moreover, some argue that human ethics cannot apply to AI [49]. The use of Kantian ethics does not impact the central contribution of this work, which is demonstrating that philosophically sophisticated automated ethics is possible.

This work does not address all of AI's ethical harms. Much harm is caused by the decisions that humans make while building AI. For example, biased datasets are responsible for biased algorithms, and automated ethics cannot resolve this problem [23]. This work, like other work in automated ethics, addresses the specific challenge of dynamically resolving the moral dilemmas that AI faces as it navigates the world.

6 Related Work

Automated ethics is a growing field, spurred in part by the need for ethically intelligent AI agents. Tolmeijer et al. surveyed the state of the field of machine ethics [50] and characterized implementations in automated ethics by (1) the choice of ethical theory, (2) implementation design decisions (e.g. logic programming), and (3) implementation details (e.g. choice of logic).

Two branches of automated ethics are top-down and bottom-up ethics. Top-down automated ethics begins with an ethical theory, whereas bottom-up automated ethics learns ethical judgements from prior judgements (e.g., using machine learning to make ethical judgements as in [30]). Bottom-up approaches often lack an explicit ethical theory explaining their judgements, so analytically arguing for or against their conclusions is impossible. Top-down approaches, on the other hand, must be explicit about the underlying ethical theories, and are thus more explainable.

In this paper, I use a top-down approach to formalize Kantian ethics. There is work automating other ethical theories, like consequentialism [1,3] or particularism [7,28]. Kantian ethics is a deontological, or rule based ethic, and there is prior work implementing other deontological theories [2,4,27].

There has been both theoretical and practical work on automating Kantian ethics [37,44]. In 2006, Powers [44] argued that implementing Kantian ethics presented technical challenges, such as automation of a non-monotonic logic, and

[10] Many misconceptions about Kantian ethics arise from misreading social context. For example, critics of Kantian ethics worry that the maxim, "When I am a man, I will marry a man because I want to spend my life with him" fails the universalizability test because if all men marry men, sexual reproduction would stop. Kantians often respond by arguing that the correct formulation of this maxim is, "When I love a man, I will marry him because I want to spend my life with him," which is universalizable. Arriving at this correct formulation requires understanding the social fact that marriage is generally driven by love, not solely by the gender of one's partner.

[11] Philosophers call Kantian ethics an "ideal theory," or one that functions best when everyone behaves morally.

philosophical challenges, like a definition of the categorical imperative. I address the former through my use of Dyadic Deontic Logic, which allows obligations to be retracted as context changes, and the latter through my use of the practical contradiction interpretation. There has also been prior work in formalizing Kantian metaphysics using I/O logic [48].

Kroy [36] presents a formalization of the first two formulations of the categorical imperative, but does not implement it. I implement his formalization of the FUL to compare it to my system. Lindner and Bentzen [9] presented one of the first formalizations and implementations of Kant's second formulation of the categorical imperative. They present their goal as "not to get close to a correct interpretation of Kant, but to show that our interpretation of Kant's ideas can contribute to the development of machine ethics." My work builds on theirs by formalizing the first formulation of the categorical imperative as faithfully as possible. Staying faithful to philosophical literature makes my system capable of making robust and reliable judgements.

The implementation of this paper was inspired by and builds on Benzmüller, Parent, and Farjami's foundational work with the LogiKEy framework for machine ethics, which includes their implementation of DDL in Isabelle [10,12]. The LogiKEy project has been used to study metaphysics [11,32], law [56], and ethics [25], but not Kant's categorical imperative.

7 Conclusion

In this paper, I present an implementation of automated Kantian ethics that is faithful to philosophical literature. I formalize Kantian ethics in Dyadic Deontic Logic, implement my formalization in the Isabelle/HOL theorem prover, and use my system to make nuanced ethical judgements. I also present a testing framework that evaluates how faithful an implementation of automated ethics is to philosophical literature. Tests show that my system outperforms two other implementations of Kantian ethics.

This paper contributes a proof-of-concept system that demonstrates that automating philosophically sophisticated ethics is possible. Ethics is the study of how best to navigate the world, and as AI becomes more powerful and independent, it must be equipped with ethical reasoning. Growing public consciousness about the dangers of unregulated AI is creating momentum in automated ethics; the time is ripe to create usable, reliable automated ethics. This paper is one step towards building computers that can think ethically in the richest sense of the word.

Acknowledgements. I thank Jeremy D. Zucker for contributing a definition of causality; Professor Christoph Benzmüller and Dr. Xavier Parent for helpful discussion and correspondence. I thank my advisors, Professor Nada Amin and Dr. William Cochran, for discussion at every stage of the project and valuable feedback on the manuscript.

A Why Automate Kantian Ethics

T.M. Powers posits that Kantian ethics is an attractive candidate for formalization because of its emphasis on formal rules, which are generally computationally tractable [44]. In this section, I extend this argument and argue that Kantian ethics is more natural to formalize than the two other major ethical traditions, consequentialism and virtue ethics, because it requires little data about the world and is easy to represent to a computer. Given that this work is an early step in philosophically-sophisticated automated ethics, I automated an ethical theory that is amenable to formalization, but application-ready automated ethics may be best served by using a different ethical theory. Full discussion of the benefits and limitations of Kantian ethics is outside the scope of this paper. First I present the challenges of automating consequentialism and virtue ethics, and then I describe how Kantian ethics overcomes these challenges.

A.1 Consequentialism

A consequentialist ethical theory evaluates an action by evaluating its consequences. Some debates in the consequentialist tradition include which consequences matter, what constitutes a "good" consequence, and how we can aggregate the consequences of an action over all the individuals involved [47].

Because consequentialism evaluates the state of affairs following an action, it requires more knowledge about the world than Kantian ethics. Under naive consequentialism, an action is judged by all its consequences. Even if we cut off the chain of consequences at some point, evaluating a single consequence is data-intensive because it requires knowledge about the world before and after the event. As acts become more complex and affect more people, the computational time and space required to calculate and store their consequences increases. Kantian ethics, on the other hand, does not suffer this scaling challenge because it merely evaluate the structure of the action itself, not its consequences. Actions that affect one person and actions that affect one million people share the same representation.

The challenge of representing the circumstances of action is not unique to consequentialism, but is particularly acute in this case. Kantian ethicists robustly debate which circumstances of an action are "morally relevant" when evaluating an action's moral worth.[12] Because Kantian ethics merely evaluates a single action, the surface of this debate is much smaller than the debate about circumstances and consequences in a consequentialist system. An automated consequentialist system must make such judgements about the act itself, the circumstances in which it is performed, and the circumstances following the act. All ethical theories relativize their judgements to the situation in which an act is performed, but consequentialism requires far more knowledge about the world than Kantian ethics.

[12] Powers [44] identifies this as a challenge for automating Kantian ethics and briefly sketches solutions from O'Neill [41], Silber [46], and Rawls [45]. For more on morally relevant circumstances, see Sect. 4.1.

A.2 Virtue Ethics

Virtue ethics centers the virtues, or traits that constitute a good moral character and make their possessor good [29]. For example, Aristotle describes virtues as the traits that enable human flourishing. Just as consequentialists define "good" consequences, virtue ethicists present a list of virtues. Such theories vary from Aristotle's virtues of courage and temperance [5] to the Buddhist virtue of equanimity [38]. An automated virtue ethical agent will need to commit to a particular theory of the virtues, a controversial choice. Unlike Kantian ethicists, who generally agree on the meaning of the Formula of Universal Law, virtue ethicists robustly debate which traits qualify as virtues, what each virtue actually means, and what kinds of feelings or attitudes must accompany virtuous action.

The unit of evaluation for virtue ethics is a person's moral character. While Kantians evaluate the act itself and utilitarians evaluate the act's consequences, virtue ethicists evaluate how good of a person the actor is, a difficult concept to represent to a machine. Formalizing the concept of character appears to require significant philosophical and computational progress, whereas Kantian ethics immediately presents a formal rule to implement.

A.3 Kantian Ethics

Kantian ethics is more natural to formalize than the traditions outlined above because the FUL evaluates the form or structure of an agent's maxim,[13] or principle of action as they themselves understand it. For example, when I falsely promise to repay a loan, my maxim is, "When I am strapped for cash, I falsely promise to repay a loan to make some easy money." Evaluating a maxim has little to do with the circumstances of behavior, the agent's mental state, or other contingent facts; it merely requires analyzing the hypothetical world in which the maxim is universalized. This property not only reduces computational complexity, but it also makes the system easier for human reasoners to interact with. A person crafting an input to a Kantian automated agent needs to reason about relatively simple features of a moral dilemma, as opposed to the more complex features that consequentialism and virtue ethics base their judgements on.[14]

B Relevant Features of the Embedding of DDL in HOL

In this appendix, I present some of the relevant features of Benzmüller, Farjami, and Parent's [10] semantic embedding of DDL in HOL. The semantic

[13] For a more detailed definition of a maxim, see Sect. 4.1.

[14] As is the case with any ethical theory, Kantians debate the details of their theory. I assume stances on debates about the definition of a maxim and the correct interpretation of the Formula of Universal Law. Those who disagree with my stances will not trust my system's judgements. Unlike consequentialism or virtue ethics, these debates are close to settled in the Kantian literature, so my choices are relatively uncontroversial [22].

embedding consists of types, constants, and axioms. Figure 6 introduces the semantic embedding.

```
1  typedecl i <type for a set of worlds>
2  type_synonym t = "(i -> bool)" <a set of DDL formulas>
3      <A set of formulas is defined by its truth value at a set of worlds.>
4  (...)
5  consts ob::"t -> (t -> bool)"
6      <ob (context) (term) is true if term is obligated in this context>
7      <this is the neighborhood function for DDL's neighborhood semantics>
8  (...)
9  axiomatization where
10 (...)
11 and ax_5d: "∀ X Y Z. ((∀w. Y(w) -> X(w)) ∧ ob(X)(Y) ∧ (∀w. X(w)-> Z(w))
12     -> ob(Z)(λ.(Z(w) ∧ ¬ X(w)) ∧ Y(w))"
13     <If some subset Y of X is obligatory in the context X, then in>
14     <a larger context Z, any obligatory proposition must either be>
15     <in Y or in  Z \ X. Expanding the context can't cause>
16     <something unobligatory to become obligatory.>
```

Fig. 6. Example types, constants, and axioms in Benzmüller et al.'s implementation of DDL.

This semantic embedding requires defining a set of worlds. I use this concept heavily in my formalization of the FUL as presented in Sect. 4.1.

In my implementation, I use the syntax of DDL, which Benzmüller et al. define as abbreviations using the semantic axioms above. In Fig. 7, I present some of the most important syntactic symbols and operators for my purposes.

C Additional Implementation Details

In Sect. 4.1, I present my formalization of the FUL in a semantic embedding of DDL in HOL. I then implement this formalization in Isabelle/HOL on top of prior work implementing DDL [10]. The code for my implementation is given in Fig. 8.

D Experimental Figures

Figs. 9 and 10 depict the Nitpick output showing the FUL does not hold in DDL and that the FUL is consistent.

E Additional Tests

Below I present details and philosophical justification for the individual tests in my testing framework.

```
1  abbreviation ddlneg::"t->t" ("¬")
2    where "¬A ≡ λw. ¬A(w)" <DDL contains propositional logic operators>
3  abbreviation ddlbox::"t->t" ("□")
4    where "□ A ≡ λw.∀y. A(y)" <and modal logic operators>
5  abbreviation ddlob::"t->t->t" ("O{_|_}")
6    where "O{B|A} ≡ λw. ob(A)(B)" <the dyadic obligation operator>
7    <O{B|A} can be read as "B is obligatory in the context A">
8
9  <In some cases, a monadic obligation operator suffices.>
10 abbreviation ddltrue::"t" ("⊤")
11   where "⊤ ≡ λw. True"
12 abbreviation ddlob_normal::"t->t" ("O {_}")
13   where "(O {A}) ≡ (O{A|⊤})"
14   <True is the widest context because it holds at all worlds>
15 (...)
```

Fig. 7. Examples of syntactic operators and symbols from Benzmüller et al.'s implementation of DDL.

FUL Stronger Than DDL. The base logic DDL does not come equipped with the categorical imperative built-in. It defines basic properties of obligation, such as ought implies can, but contains no axioms that represent the formula of universal law. Therefore, if a formalization of the FUL holds in the base logic, then it is too weak to actually represent the FUL. The naive control group definitionally holds in DDL but Kroy's formalization does not and neither does my implementation.

Obligation Universalizes Across People. Another property of the Formula of Universal Law that any implementation should satisfy is that obligation generalizes across people. In other words, if a maxim is obligated for one person, it is obligated for all other people because maxims are not person-specific. Velleman argues that, because reason is accessible to everyone identically, obligations apply to all people equally [25,51]. When Kant describes the categorical imperative as the objective principle of the will, he is referring to the fact that, as opposed to a subjective principle, the categorical imperative applies to all rational agents equally [16,31]. At its core, the FUL best handles, "the temptation to make oneself an exception: selfishness, meanness, advantagetaking, and disregard for the rights of others" [30,34]. Kroy latches onto this property and makes it the center of his formalization, which says that if an act is permissible for someone, it is permissible for everyone.[15] While Kroy's interpretation clearly satisfies this property, the naive interpretation does not.

Distributive Property. A property related to contradictory obligations is the distributive property for the obligation operator.[16] The rough English transla-

[15] Formally, $P\{A(s)\} \longrightarrow \forall p.P\{A(p)\}$.
[16] Formally, $O\{A\} \wedge O\{B\} \longleftrightarrow O\{A \wedge B\}$.

tion of $O\{A \wedge B\}$ is "you are obligated to do both A and B". The rough English translation of $O\{A\} \wedge O\{B\}$ is "you are obligated to do A and you are obligated to do B." We think those English sentences mean the same thing, so they should mean the same thing in logic as well. Moreover, if that (rather intuitive) property holds, then contradictory obligations are impossible, as shown in Fig. 11. This property fails in the base logic and Kroy's formalization, but holds in my implementation.

Un-universalizable Actions. Under a naive reading of the Formula of Universal Law, it prohibits lying because, in a world where everyone simultaneously lies, lying is impossible. In other words, not everyone can simultaneously lie because the institution of lying and believing would break down. More precisely, the FUL should show that actions that cannot possibly be universalized are prohibited, because those acts cannot be willed in a world where they are universalized. This property fails to hold in both the naive formalization and Kroy's formalization, but holds in my formalization.

Conventional Acts and Natural Acts. A conventional act like promising relies on a convention, like the convention that a promise is a commitment, whereas a natural act is possible simply because of the laws of the natural world. It is easier to show the wrongness of conventional acts because there are worlds in which these acts are impossible; namely, worlds in which the convention does not exist. For example, the common argument against falsely promising is that if everyone were to falsely promise, the convention of promising would fall apart because people wouldn't believe each other, so falsely promising is prohibited. It is more difficult to show the wrongness of a natural act, like murder or violence. These acts can never be logically impossible; even if everyone murders or acts violently, murder and violence will still be possible, so it is difficult to show that they violate the FUL.

Both the naive and Kroy's interpretations fail to show the wrongness of conventional or natural acts. My system shows the wrongness of both natural and conventional acts because it is faithful to Korsgaard's practical contradiction interpretation of the FUL, which is the canonical interpretation of the FUL [34].

Maxims. Kant does not evaluate the correctness of acts, but rather of maxims. Therefore, any faithful formalization of the categorical imperative must evaluate maxims, not acts. This requires representing a maxim and making it the input to the obligation operator, which neither of the prior attempts do. Because my implementation includes the notion of a maxim, it is able to perform sophisticated reasoning as demonstrated in Sect. 4.2. Staying faithful to the philosophical literature enables my system to make more reliable judgements.

```
1  abbreviation will :: "maxim -> s-> t" ("W _ _")
2    where "will ≡ λ(c, a, g) s. (c -> (a s))"
3    <subject S wills maxim (C, A, G) if, in circumstances C, S performs A.>
4  abbreviation effective :: "maxim->s->t" ("E _ _")
5    where "effective ≡ λ(c, a, g) s. ((will (c, a, g) s) ≡ g)"
6    <a maxim is effective if willing it is a necessary and>
7    <sufficient condition for achieving the goal>
8  abbreviation universalized::"maxim->t" where
9    "universalized ≡ λM. (λw. (∀p. (W M p) w))"
10   <a maxim is universalized at a world if everyone wills it at that world.>
11 abbreviation not_universalizable :: "maxim->s->bool" where
12   "not_universalizable ≡ λM s. ∀w. ((universalized M) -> (¬ (E M s))) w"
13
14 abbreviation prohibited::"maxim->s->t"
15   where "prohibited ≡ λ(c, a, g) s. O{¬ (will (c,a, g) s) | c}"
16   <the unit of evaluation is willing a maxim, not merely the maxim itself>
17 abbreviation permissible::"maxim->s->t"
18   where "permissible ≡ λM s. ¬ (prohibited M s)"
19
20 abbreviation well_formed::"maxim->s->i->bool"
21   where "well_formed ≡ λ(c, a, g). λs. λw. (¬ (c -> g) w) ∧ (¬ (c -> a s) w)"
22 abbreviation FUL
23   where "FUL ≡ ∀M::maxim. ∀s::s. (∀w. well_formed M s w) ->
24         (not_universalizable M s -> ∀ w. (prohibited M s) w )"
25   <the consistent version of the FUL only holds for well-formed maxims>
26
27 abbreviation non_contradictory where
28    "non_contradictory A B c w ≡ ((O{A|c} ∧ O{B|c}) w) -> ¬(A ∧ (B ∧ c)) w -> False)"
29    <obligations are noncontradictory in circumstances>
30    <if their conjunction with the circumstances does not lead to a contradiction>
31
32 axiomatization
33   where no_contradictions:"∀A::t. ∀B::t. ∀c::t. ∀w::i. non_contradictory A B c w"
34   <all obligations must be non-contradictory in all circumstances>
35   and FUL:FUL
```

Fig. 8. Implementation of the FUL

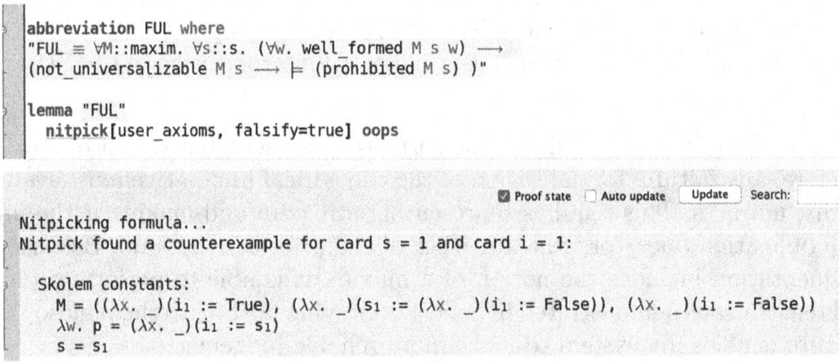

```
abbreviation FUL where
"FUL ≡ ∀M::maxim. ∀s::s. (∀w. well_formed M s w) ⟶
(not_universalizable M s ⟶ ⊨ (prohibited M s) )"

lemma "FUL"
  nitpick[user_axioms, falsify=true] oops
```

```
Proof state   Auto update   Update   Search:

Nitpicking formula...
Nitpick found a counterexample for card s = 1 and card i = 1:

  Skolem constants:
    M = ((λx. _)(i1 := True), (λx. _)(s1 := (λx. _)(i1 := False)), (λx. _)(i1 := False))
    λw. p = (λx. _)(i1 := s1)
    s = s1
```

Fig. 9. Nitpick output showing that the FUL does not hold in DDL.

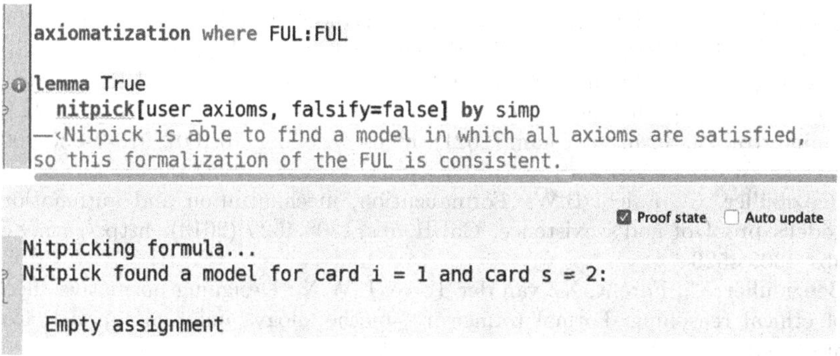

```
axiomatization where FUL:FUL

lemma True
  nitpick[user_axioms, falsify=false] by simp
—‹Nitpick is able to find a model in which all axioms are satisfied,
so this formalization of the FUL is consistent.
```

```
Proof state   Auto update
Nitpicking formula...
Nitpick found a model for card i = 1 and card s = 2:

  Empty assignment
```

Fig. 10. Nitpick model showing that the FUL is consistent.

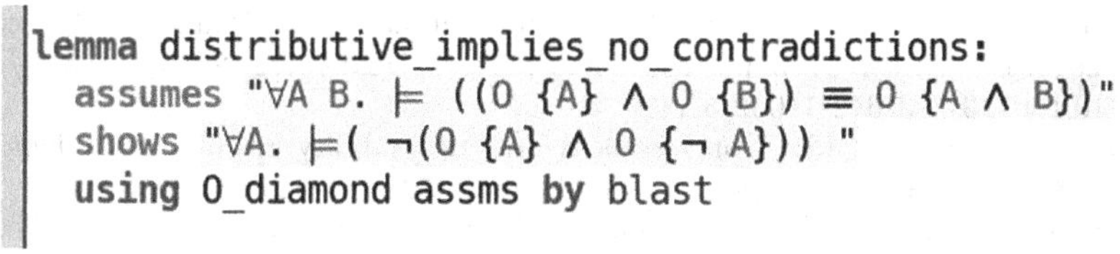

```
lemma distributive_implies_no_contradictions:
  assumes "∀A B. ⊨ ((O {A} ∧ O {B}) ≡ O {A ∧ B})"
  shows "∀A. ⊨( ¬(O {A} ∧ O {¬ A})) "
  using O_diamond assms by blast
```

Fig. 11. The proof that the distributive property implies that contradictory obligations are impossible.

References

1. Abel, D., MacGlashan, J., Littman, M.L.: Reinforcement learning as a framework for ethical decision making. In: AAAI Workshop: AI, Ethics, and Society (2016)
2. Anderson, M., Anderson, S.: Geneth: a general ethical dilemma analyzer **1** (2014)
3. Anderson, M., Anderson, S., Armen, C.: Towards machine ethics (2004)
4. Anderson, M., Anderson, S.L.: Ethel: Toward a principled ethical eldercare robot
5. Aristotle: The nicomachean ethics. Journal of Hellenic Studies **77**, 172 (1951). https://doi.org/10.2307/628662
6. Arkoudas, K., Bringsjord, S., Bello, P.: Toward ethical robots via mechanized deontic logic. AAAI Fall Symposium - Technical report (2005)
7. Ashley, K.D., McLaren, B.M.: A CBR knowledge representation for practical ethics. In: Haton, J.-P., Keane, M., Manago, M. (eds.) EWCBR 1994. LNCS, vol. 984, pp. 180–197. Springer, Heidelberg (1995). https://doi.org/10.1007/3-540-60364-6_36
8. Awad, E., Dsouza, S., Shariff, A., Rahwan, I., Bonnefon, J.F.: Universals and variations in moral decisions made in 42 countries by 70,000 participants. In: Proceedings of the National Academy of Sciences, vol. 117, no. 5, pp. 2332–2337 (2020). https://doi.org/10.1073/pnas.1911517117, https://www.pnas.org/content/117/5/2332
9. Bentzen, M.M., Lindner, F.: A formalization of kant's second formulation of the categorical imperative. CoRR abs/1801.03160 (2018). http://arxiv.org/abs/1801.03160

10. Benzmüller, C., Farjami, A., Parent, X.: Dyadic deontic logic in HOL: faithful embedding and meta-theoretical experiments. In: Armgardt, M., Nordtveit Kvernenes, H.C., Rahman, S. (eds.) New Developments in Legal Reasoning and Logic: From Ancient Law to Modern Legal Systems. Logic, Argumentation & Reasoning, vol. 23. Springer, Cham (2021). https://doi.org/10.1007/978-3-030-70084-3
11. Benzmüller, C., Paleo, B.W.: Formalization, mechanization and automation of gödel's proof of god's existence. CoRR abs/1308.4526 (2013). http://arxiv.org/abs/1308.4526
12. Benzmüller, C., Parent, X., van der Torre, L.W.N.: Designing normative theories of ethical reasoning: Formal framework, methodology, and tool support. CoRR abs/1903.10187 (2019). http://arxiv.org/abs/1903.10187
13. Berberich, N., Diepold, K.: The virtuous machine - old ethics for new technology? (2018)
14. Blanchette, J.C., Nipkow, T.: Nitpick: a counterexample generator for higher-order logic based on a relational model finder. In: Kaufmann, M., Paulson, L.C. (eds.) ITP 2010. LNCS, vol. 6172, pp. 131–146. Springer, Heidelberg (2010). https://doi.org/10.1007/978-3-642-14052-5_11
15. Carmo, J., Jones, A.J.I.: Completeness and decidability results for a logic of contrary-to-duty conditionals. J. Logic Comput. **23**, 585–626 (2013)
16. Cervantes, J.A., Rodríguez, L.F., López, S., Ramos, F.: A biologically inspired computational model of moral decision making for autonomous agents. In: 2013 IEEE 12th International Conference on Cognitive Informatics and Cognitive Computing, pp. 111–117 (2013). https://doi.org/10.1109/ICCI-CC.2013.6622232
17. Cloos, C.: The utilibot project: An autonomous mobile robot based on utilitarianism. AAAI Fall Symposium - Technical report (2005)
18. Cresswell, M., Hughes, G.E.: A New Introduction to Modal Logic. Routledge, Milton Park (1996)
19. Davenport, D.: Moral Mechanisms. Philos. Technol. **27**(1), 47–60 (2014). https://doi.org/10.1007/s13347-013-0147-2
20. Dennis, L., Fisher, M., Slavkovik, M., Webster, M.: Formal verification of ethical choices in autonomous systems. Robot. Auton. Syst. **77**, 1–14 (2016). https://doi.org/10.1016/j.robot.2015.11.01, https://www.sciencedirect.com/science/article/pii/S0921889015003000
21. Dietrichson, P.: When is a maxim fully universalizable? **55**(1–4), 143–170 (1964). https://doi.org/10.1515/kant.1964.55.1-4.143
22. Ebels-Duggan, K.: Kantian Ethics, chap. Continuum, Kantian Ethics (2012)
23. Eubanks, V.: Automating Inequality: How High-Tech Tools Profile, Police, and Punish the Poor. St. Martin's Press, New York (2018)
24. Foot, P.: The problem of abortion and the doctrine of the double effect. Oxford Rev. **5**, 5–15 (1967)
25. Fuenmayor, D., Benzmüller, C.: Formalisation and evaluation of Alan Gewirth's proof for the principle of generic consistency in isabelle/hol. Arch. Formal Proofs (2018). https://isa-afp.org/entries/GewirthPGCProof.html. Formal proof development
26. Gabriel, I.: Artificial intelligence, values, and alignment. Mind. Mach. **30**(3), 411–437 (2020). https://doi.org/10.1007/s11023-020-09539-2
27. Govindarajulu, N.S., Bringsjord, S.: On automating the doctrine of double effect. CoRR abs/1703.08922 (2017). http://arxiv.org/abs/1703.08922
28. Guarini, M.: Particularism and the classification and reclassification of moral cases. IEEE Intell. Syst. **21**(4), 22–28 (2006). https://doi.org/10.1109/MIS.2006.76

29. Hursthouse, R., Pettigrove, G.: Virtue ethics. In: Zalta, E.N. (ed.) The Stanford Encyclopedia of Philosophy. Metaphysics Research Lab, Stanford University, Winter 2018 edn. (2018)
30. Jiang, L., et al.: Delphi: towards machine ethics and norms (2021)
31. Kant, I.: Groundwork of the Metaphysics of Morals. Cambridge University Press, Cambridge (1785)
32. Kirchner, D., Benzmüller, C., Zalta, E.N.: Computer science and metaphysics: a cross-fertilization. CoRR abs/1905.00787 (2019). http://arxiv.org/abs/1905.00787
33. Kitcher, P.: What is a maxim? Philos. Top. **31**(1/2), 215–243 (2003). https://doi.org/10.5840/philtopics2003311/29
34. Korsgaard, C.: Kant's formula of universal law. Pac. Philos. Q. **66**, 24–47 (1985)
35. Korsgaard, C.: The right to lie: kant on dealing with evil. Philos. Publ. Aff. **15**, 325–349 (1986)
36. Kroy, M.: A partial formalization of kant's categorical imperative an application of deontic logic to classical moral philosophy. Kant-Studien **67**(4), 192–209 (1976). https://doi.org/10.1515/kant.1976.67.1-4.192
37. Lin, P., Abney, K., Bekey, G.A.: Robotics, ethical theory, and metaethics: a guide for the perplexed, pp. 35–52 (2012)
38. McRae, E.: Equanimity and intimacy: a buddhist-feminist approach to the elimination of bias. Sophia **52**(3), 447–462 (2013). https://doi.org/10.1007/s11841-013-0376-y
39. de Moura, L., Bjørner, N.: Z3: an efficient SMT solver. In: Ramakrishnan, C.R., Rehof, J. (eds.) TACAS 2008. LNCS, vol. 4963, pp. 337–340. Springer, Heidelberg (2008). https://doi.org/10.1007/978-3-540-78800-3_24
40. Nipkow, T., Paulson, L.C., Wenzel, M.: Isabelle/HOL: A Proof Assistant for Higher Order Logic. Springer-Verlag, Berlin Heidelberg, Berlin (2002)
41. O'Neill, O.: Constructions of Reason: Explorations of Kant's Practical Philosophy. Cambridge University Press, Cambridge (1990). https://doi.org/10.1017/CBO9781139173773
42. O'Neill, O.: Acting on Principle: An Essay on Kantian Ethics. Cambridge University Press, Cambridge (2013)
43. Paulson, L., Blanchette, J.: Three years of experience with sledgehammer, a practical link between automatic and interactive theorem provers (2015). https://doi.org/10.29007/tnfd
44. Powers, T.: Prospects for a kantian machine. IEEE Intell. Syst. **21**(4), 46–51 (2006). https://doi.org/10.1109/MIS.2006.77
45. Rawls, J.: Kantian constructivism in moral theory. J. Philosophy **77**(9), 515–572 (1980). http://www.jstor.org/stable/2025790
46. Silber, J.R.: Procedural formalism in kant's ethics. Rev. Metaphysics **28**(2), 197–236 (1974). http://www.jstor.org/stable/20126622
47. Sinnott-Armstrong, W.: Consequentialism. In: Zalta, E.N. (ed.) The Stanford Encyclopedia of Philosophy. Metaphysics Research Lab, Stanford University, Fall 2021 edn. (2021)
48. Stephenson, A., Sergot, M., Evans, R.: Formalizing kant's rules: a logic of conditional imperatives and permissives. J. Philos. Logic **49** (2019). https://eprints.soton.ac.uk/432344/
49. Tafani, D.: Imperativo categorico come algoritmo. kant e l'etica delle macchine. Sistemi intelligenti, Rivista quadrimestrale di scienze cognitive e di intelligenza artificiale, pp. 377–392 (2021). https://doi.org/10.1422/101195, https://www.rivisteweb.it/doi/10.1422/101195

50. Tolmeijer, S., Kneer, M., Sarasua, C., Christen, M., Bernstein, A.: Implementations in machine ethics. ACM Comput. Surv. **53**(6), 1–38 (2021). https://doi.org/10.1145/3419633
51. Velleman, J.D.: A Brief Introduction to Kantian Ethics, pp. 16–44. Cambridge University Press (2005). https://doi.org/10.1017/CBO9780511498862.002
52. Vincent, J.: The AI oracle of Delphi uses the problems of reddit to offer dubious moral advice (2021)
53. Wallach, W., Allen, C.: Moral Machines: Teaching Robots Right From Wrong. Oxford University Press, London (2008)
54. Winfield, A.F.T., Blum, C., Liu, W.: Towards an ethical robot: internal models, consequences and ethical action selection. In: Mistry, M., Leonardis, A., Witkowski, M., Melhuish, C. (eds.) TAROS 2014. LNCS (LNAI), vol. 8717, pp. 85–96. Springer, Cham (2014). https://doi.org/10.1007/978-3-319-10401-0_8
55. Wood, A.W.: Kant's Ethical Thought. Cambridge University Press, Cambridge (1999)
56. Zahoransky, V., Benzmüller, C.: Modelling the us constitution to establish constitutional dictatorship (2020)

PEBAM: A Profile-Based Evaluation Method for Bias Assessment on Mixed Datasets

Mieke Wilms, Giovanni Sileno(✉), and Hinda Haned

University of Amsterdam, Amsterdam, The Netherlands
mieke.wilms@student.uva.nl, g.sileno@uva.nl

Abstract. Bias evaluation methods focus either on individual bias or on group bias, where groups are defined based on protected attributes such as gender or ethnicity. More generally, however, descriptively relevant combinations of feature values in the data space (*profiles*) may serve also as anchors for biased decisions. This paper introduces therefore a semi-hierarchical clustering method for profile extraction from mixed datasets. It elaborates on how profiles can be used to reveal *historical, representational, aggregation* and *evaluation biases* in algorithmic decision-making models, taking as example the German credit data set. Our experiments show that the proposed profile-based evaluation method for bias assessment on mixed datasets (PEBAM) can reveal forms of bias towards profiles expressed by the dataset that are undetected when using individual- or group-bias metrics alone.

Keywords: Algorithmic fairness · Bias prevention · Bias evaluation · Clustering · Domain analysis

1 Introduction

The wider introduction of machine learning algorithms in decision-making processes feeds an ongoing debate over algorithmic decision-making (ADM). To prevent or correct ADM from taking biased decisions several fairness-aware machine learning algorithms have been proposed [8]. However, these algorithms are not always accessible to practitioners due to their 'black-box' nature [2]; they are highly dependent on the data preprocessing phase [8]; and, at more fundamental level, fairness and bias can be given technical meanings but cannot be captured by one single definition [14,16]. Contemporary bias evaluation methods used for fairness analysis generally focus either on individual bias or on group bias, where groups are defined based on protected attributes such as gender or ethnicity, but this is not without drawbacks. For instance, analysing the German credit dataset—a real world dataset[1] collecting features of loan applicants and a credit risk label *good* or *bad* assigned to them—the group of young individuals

[1] Available at: https://www.kaggle.com/uciml/german-credit.

R. Bergmann et al. (Eds.): KI 2022, LNAI 13404, pp. 209–223, 2022.
https://doi.org/10.1007/978-3-031-15791-2_17

(age below 25) obtains more often a false negative label than the group above 25, hence young individuals are discriminated when applying for a loan [11]. The simplest solution would be to take this sensitive attribute out of consideration, however there are multiple attributes that correlate with the "age" attribute (e.g. "own house" [12]). From a more general standpoint, one may ask whether there exist relevant descriptive combinations of feature values in the data space, that we will call here *profiles*, which may act as anchors for (assessing the presence of) biased decisions. As an additional source of complexity, we need also to take into account that data is commonly presented in form of mixed datasets (i.e. including both categorical and numeric features). Discretization of numeric dimensions, or embedding of categorical dimensions, add further complexity and potentially undesired effects. Given this context, we address the following research questions: *How can profiles be defined? How can we extract profiles from mixed datasets? How can profiles be used to assess biases? How does a profile-based assessment compare with existing individual- or group-based methods?*The goal of this paper is to develop and test a Profile-based Evaluation method for Bias Assessment of algorithmic decision-making on Mixed datasets (PEBAM). Our contribution is twofold: (i) we present an effective and computationally efficient method for profile extraction on mixed datasets based on clustering; (ii) we show how profiles can be utilized to evaluate various forms of biases—most of them associated to trained ADM models. The paper is structured as follows. Section 2 provide a brief overview of relevant concepts. Section 3 presents the proposed methodology. Section 4 elaborates on the experiments and results on the German credit dataset. A note on future work ends the paper.

2 Theoretical Background

Types of Bias. Several types of bias have been identified in the literature (see e.g. the 23 types in [14]), but for the scope of this research we will focus in particular on biases that can arise during a ML-product lifecycle (see e.g. [16]). For instance, during the *data generation* process, we may have: *historical bias*, produced by the world as it is, and occurring even if data is perfectly measured and sampled; *representation bias*, occurring when the training data for the ML model under-represents parts of the population the algorithm will be used on. During the *model building* and *implementation* phases, we may have: *aggregation bias*, arising when a general model is used for all groups, while in reality different groups have a different mapping from input features to labels (e.g. some ethnic groups can have different indicators for a disease than others); *evaluation bias*, occurring when the data on which the model is evaluated is a misrepresentation of the target population. These four types of bias do not cover all possible sources of bias, but they will be used as relevant examples about how to set up a profile-based evaluation.

Bias Evaluation Methods for Algorithmic Fairness. In their extensive literature review, Mehrabi et al. [14] give an overview of the most widely used definitions of fairness within machine learning, providing 3 definitions focused on individual fairness, 6 on group fairness, in which groups are defined by protected attribute classes (e.g. sex, ethnicity, etc.), 1 on subgroup fairness. In this work we will build upon two (group-fairness) measures. The first is *equal opportunity* [10], a criterion for fairness in binary algorithms. Reading the outcome $y = 1$ as the "advantaged" outcome, and A as the protected class attribute, we have:

Definition 1. ***Equal opportunity*** *A binary predictor* $\hat{y}$ *satisfies equal opportunity w.r.t. attribute* A *and ground truth* y *iff:* $Pr\{\hat{y} = 1\, A = 0, y = 1\} = Pr\{\hat{y} = 1\, A = 1, y = 1\}$.

The second is *contextual demographic (dis)parity* (CDD)—based on *conditional (non-)discrimination* by [12]—a measure found to be the most compatible with the decisions of the European Court of Justice on cases of discrimination [18].

Definition 2. ***Conditional Demographic Disparity*** *Let* R *be a given set of attributes,* A_r *be the proportion of people belonging to a protected class in the advantaged group and with attribute* $r \in R$*, and let* D_r *be the proportion of people of protected class in the disadvantaged group with attribute* r*. A decision-making process exhibits conditional demographic disparity iff:* $\forall r \in R : D_r > A_r$

The conditions r in R should be *explanatory* [12], i.e. they should hypothetically explain the outcome even in the absence of discrimination against the protected class (e.g., different salaries between men and women might be due to different working hours). Under this view, R is derived from domain expert knowledge.

Clustering Algorithm for Mixed Data. In ADM one very often has to deal with mixed datasets, i.e. datasets that consist of both categorical and numerical features. Various solutions have been proposed in the literature to the known difficulty to capture distributions on mixed datasets [15]; the present work will rely in particular on *k-medoids clustering* [5]. The main benefit of k-medoids clustering over k-means is that it is more robust to noise and outliers; we also do not have to come up with a measure to compute the mean for categorical features. On the other hand, the k-medoids clustering problem is NP-hard to solve exactly. For this reason, in our work we will make use of the heuristic Partitioning Around Medoids (PAM) algorithm [4].

3 Methodology

PEBAM (*Profile-based Evaluation for Bias Assessment for Mixed datasets*) is a method consisting of three main steps: (1) a profile selection—based on the iteration of clustering controlled by a measure of variability—to extract profiles representative of the target domain from an input dataset; (2) profiles are evaluated in terms of stability over repetitions of extractions; (3) a given ADM classification model is evaluated for bias against those profiles.

3.1 Profile-Selection Based on Clustering

Informally, profiles can be seen as relevant descriptive elements that, as a group, act as a "summary" of the data space. Because individuals sharing to an adequate extent similar attributes should intuitively be assigned to the same profile, clustering can be deemed compatible with a profile selection task. We consider then three different clustering algorithms to implement profile selection: (i) *simple clustering*; (ii) a form of *hierarchical clustering* based on the iteration of the first; and (iii) a novel *semi-hierarchical clustering* method based on adding static and dynamic constraints to control the second. The first two algorithms will be used as baselines to evaluate the third one, and will be described succintly.

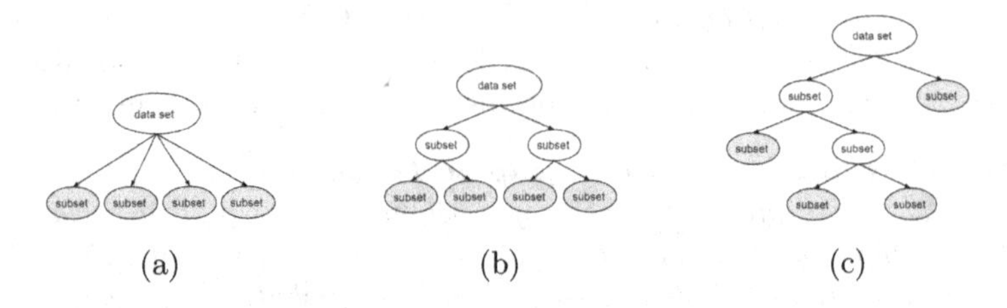

Fig. 1. Schematic overview of the simple clustering algorithm with $k = 4$ (a), the hierarchical clustering algorithm with $k = 2$ and $l = 2$ (b) and a possible outcome of the semi-hierarchical clustering algorithm (c).

The **simple clustering** algorithm consists in the application of the chosen clustering algorithm (in our case, k-medoids, Sect. 2), and results in a flat partition of the sample space (see e.g. Fig. 1a). The challenge is to determine the right number of clusters k. **Hierarchical clustering** consists in the nested iteration of the previous algorithm set with k clusters over l layers (see e.g. Figure 1b). The benefit of this method is that the resulting tree-based structure gives further insight on the basis of which feature clusters are created. The downside is that we have to tune an extra hyperparameter besides k: the number of layers l.

The **semi-hierarchical clustering** algorithm is an automatically controlled version of the second algorithm. It is based on performing iteratively two operations. First, we apply a clustering method with $k = 2$, i.e. at every step we divide input subsets into two new subsets (clusters). Then, we test each new subset to decide whether we should continue clustering by looking at the *variability* of the features it expresses, or at its cardinality. For variability of a feature f w.r.t. a dataset D we intend a measure between 0 and 1 that indicates to what extent the feature f is varying within D. A value close to 0 indicates that f is rather stable over D, and a value close to 1 indicates that f mostly varies in D (and so it is not relevant for describing it). The intuition behind using this measure is that stable (i.e. not varying) features help to discriminate a profile over another one. For instance, coffees are characterized by a black (or dark brown) colour, so the colour feature is very stable to support discriminating coffee from other drinks; in contrast, colour is not a relevant feature to discriminate books from other objects. In general, any feature can then be: (1) an *irrelevant feature*, when the variability of a feature exceeds an upper bound c_u, is deemed not to be a characteristic property of the cluster; (2) a *relevant feature:* when the variability of a feature is smaller than a lower bound c_l, it means that the feature is strongly characteristic of that cluster. When all features expressed by a subset satisfy either case (1) or case (2), or the subset has less than a fixed amount of elements n_{stop}, there is no need to continue further clustering the input subset, and therefore it is selected as a *profile.* The resulting structure is then a binary tree, possibly unbalanced (see Fig. 1c), whose leaves are the selected profiles. The benefit of semi-hierarchical clustering over simple and hierarchical clustering is that we do not have to decide the numbers of clusters and layers in advance, but requires setting the variability thresholds values c_u and c_l, as well as the threshold cluster cardinality n_{stop}.

In quantitative terms, when *numerical features* are not significant, we expect that their distribution should approximate a uniform distribution. Let us assume we have a numerical non-constant feature X; we can normalize X between 0 and 1 via $(X - X_{min})/(X_{max} - X_{min})$, and then compute the sample standard deviation s from the normalized samples. Theoretically, for a random variable $U \sim Uniform(0, 1)$, we have $\mu = 1/2$, and $Var(U) = \mathbb{E}[(U - \mu)^2] = \int_0^1 (x - 1/2)^2 \mathrm{d}x = 1/12$. Thus, the standard deviation of a random variable uniformly distributed is $\sqrt{1/12} \approx 0.29$. Therefore, if the sample standard deviation s approximates 0.29, we can assume that the feature X is uniformly distributed across the given cluster and is therefore not a unique property of species within the cluster. On the other hand, when the standard deviation is close to zero, this means that most sample points are close to the mean. This indicates that feature X is very discriminating for that specific cluster. To obtain a measure of variability, we need to compute the standardized standard deviation $s_s = s/0.29$.

For *categorical features*, we consider the variability measure proposed in [1]. Let X a n-dimensional categorical variable consisting of q categories labelled as $1, 2, ..., q$. The relative frequency of each category $i = 1, ..., q$ is given by $f_i = n_i/n$, where n_i is the number of samples that belongs to category i and $n = \sum_{i=1}^{q} n_i$. Let $\vec{f} = (f_1, f_2, ..., f_q)$ be the vector with all the relative frequencies. We define the variability of X as: $v_q = 1 - ||\vec{f}||_q$. Allaj [1] shows that the variability is bounded by 0 and $1 - 1/q$, where an outcome close to $1 - 1/q$ associates to high variability. We can also compute the standardized variability: $v_{q,s} = \frac{v_q}{1-1/\sqrt{q}}$, such that the variability lies between 0 and 1 for all number of categories q where again a variability close to 1 implies a high variability and hence a non-characteristic feature. A variability close to 0 indicates that the feature is highly characteristic for that sample set.

3.2 Evaluation of Clustering Methods for Profile Selection

Given a clustering method, we need to evaluate whether it is working properly with respect to the profile selection task. Unfortunately, since this task is an unsupervised problem, we do not have access to the ground truth, but we can still focus on a related problem: *Is our method stable?* The **stability** of a clustering method towards initialization of the clusters can be tested by running the algorithm multiple times with different initial cluster settings (in our case, different datapoints selected as the initial medoids), and check whether we end up with the same sets of clusters (hereby called *cluster combinations*). When the stability analysis results in two or more different cluster combinations, we want to know how similar these combinations are, and for doing this, we will introduce a measure of inter-clustering similarity.

Inter-clustering similarity is a similarity score that tells to what extent different outcomes of clustering are similar, by comparing how many elements clusters belonging to the two clustering outputs have in common with each other. This score can be computed by comparing the distribution of the elements over the clusters of two combinations. We present the process through an example:

Example: Let us consider a dataset with 20 data points. Suppose the clustering algorithm returns the following two different cluster combinations C^1 and C^2: where $C^1 = (c_1^1, c_2^1, c_3^1) = ((1, 4, 6, 7, 13, 17, 18, 20), (3, 5, 11, 12, 15, 16), (2, 8, 9, 10, 14, 19))$, and $C^2 = (c_1^2, c_2^2, c_3^2) = ((1, 4, 6, 7, 13, 17, 18, 20), (3, 11, 12, 14, 15, 16, 19), (2, 5, 8, 9, 10))$. For every cluster c_i^1 in C_1 we compute its overlap with each cluster c_j^2 in C_2. For instance, for the first cluster of C^1, $c_1^1 = (1, 4, 6, 7, 13, 17, 18, 20)$ the max overlap is 1, since c_2^1 of C^2 is exactly the same. For the second cluster c_2^1 we have:

$$\max\left(\frac{|c_2^1 \cap c_1^2|}{|c_2^1|}, \frac{|c_2^1 \cap c_2^2|}{|c_2^1|}, \frac{|c_2^1 \cap c_3^2|}{|c_2^1|}\right) = \max(0/6, 5/6, 1/6) = 0.83$$

Applying the same calculation on c_3^1 returns 0.67. The similarity score of cluster-combination C^1 with respect to cluster-combination C^2 is given by the mean over all the three maximum overlap values, which in this case is 0.83.

3.3 Profile-Based Evaluation of a Given Classifier

By means of profiles, we can assess the *historical*, *representational*, *aggregation* and *evaluation biases* (Sect. 2) of a certain ADM classification model.

Historical bias arises on the dataset used for training. We measure it using conditional demographic disparity (Def. 2); however, the resulting value may be wrong if an attribute r is not a relevant characteristic for grouping individuals. Therefore, we consider a *profile-conditioned demographic disparity*, differing from [12,18] in as much each profile label c_i acts as attribute r. In this way we capture behavioural attitudes (w.r.t. assigning or not an advantageous label) of the labeller-oracle towards elements belonging to that profile.

For **Representational bias**, by clustering the dataset around profiles, we may get insights if there are parts of the population which are overrepresented and parts which are underrepresented. A domain expert can compare the resulting distribution of profiles with the expected distribution for the population on which the decision-making algorithm is going to be used. When these two distributions differ much from each other, one can *bootstrap sampling* from profiles to get a more correct distribution in the training dataset. If no domain expertise is available, one can consider using this method during deployment. By collecting the data where the algorithm is applied on, once the dataset is sufficiently large (e.g. about the size of the training dataset), one can divide the samples of the collected dataset over the different known profiles based on distance towards the medoids associated to profiles. If the distribution of the collected dataset over the profiles relevantly differs from the distribution of the training dataset, we can conclude that there is representation bias. Alternatively, one can repeat the profile selection on the collected data, and evaluate how much they differ from the ones identified in the training dataset.

In order to evaluate **Aggregation bias**, we need a good metric to evaluate the performance of the trained model under assessment. Our goal is to evaluate the model against all profiles, i.e. to test whether the model works equally well on individuals with different profiles (i.e. individuals from different clusters). We start from the definition of equal opportunity (Def. 1), but we reformulate it in a way that equal opportunity is computed with respect to profiles instead of the protected class:

Definition 3 (Equal opportunity w.r.t. a single profile). *We say a binary predictor $\hat{y}$ satisfies equal opportunity with respect to a profile C equal to i and outcome/ground truth y iff:* $Pr\{\hat{y} = 1|C = i, y = 1\} = Pr\{\hat{y} = 1|C \neq i, y = 1\}$

Definition 4 (General equal opportunity). *A binary predictor $\hat{y}$ satisfies general equal opportunity iff it satisfies equal opportunity with respect to all profiles $C \in \{1,..,k\}$ and ground truth y. In formula:* $Pr\{\hat{y} = 1|C = 1, y = 1\} = \ldots = Pr\{\hat{y} = 1|C = k, y = 1\}$

In some cases, it might be that getting a wrong prediction for a ground truth or positive outcome label occurs more often than with a negative outcome label, and may be more valuable (e.g. in some medical disease treatment); looking at distinct values of y may give insights on the overall functioning of the model.

Evaluation bias arises when the model is evaluated on a population which differs in distribution from the data that was used for training the model. It has been shown [8] that fairness-preserving algorithms tend to be sensitive to fluctuations in dataset composition, i.e. the performance of the algorithms is affected by the train-test split. To ensure that we do not have evaluation bias, we run a Monte Carlo simulation of the decision-making algorithm. This means that we make M different train-test splits of the dataset. For each train-test split, we train the decision-making algorithm on the train set and use the test set for evaluation. For the model-evaluation, we use general equal opportunity (Def. 4). This gives us insights on which profiles are more sensitive towards train-test splitting (and thus to evaluation bias).

4 Experiments and Results

We conducted two experiments to evaluate PEBAM.[2] In the first, we considered a small artificial dataset to test the processing pipeline in a context in which we knew the ground truth. In the second, we focused on the German credit dataset, used in multiple researches on fairness [7,8,11,13]. This dataset contains 8 attributes (both numerical and categorical) of 1000 individuals, including an association of each individual to a credit risk score (good or bad). For the sake of brevity, we will limit our focus here on the German credit dataset. All Python-code that we run for obtaining the results is publicly available.[3]

Table 1. Stability analysis of the cluster combinations for all three clustering algorithms applied on the German credit data set with varying parameter settings. *For the semi-hierarchical clustering the number of clusters is not fixed, we reported the number of clusters for the cluster combination that occurs most.

Algorithm	l	k	# Clusters	# Cluster comb	Freq. most occurring comb	Running time
Simple	1	20	20	15	43	3.956 ± 0.376 s
Simple	1	30	30	50	9	8.170 ± 0.777 s
Simple	1	40	40	17	26	12.068 ± 0.875 s
Hierarchical	2	4	16	31	24	1.119 ± 0.078 s
Hierarchical	2	5	25	21	28	1.152 ± 0.077 s
Hierarchical	2	6	36	18	32	1.298 ± 0.113 s
Hierarchical	3	3	27	54	12	1.144 ± 0.047 s
Hierarchical	4	2	16	18	29	1.450 ± 0.083 s
Hierarchical	5	2	32	39	20	1.614 ± 0.071 s
Semi-hierarchical	–	–	34*	47	23	2.264 ± 0.112 s

4.1 Evaluation of Clustering for Profile Selection

We evaluate the three clustering algorithm for profile selection specified in Sect. 3.1 following the method described in Sect. 3.2. For all three clustering algorithms, we used a k-medoids clustering algorithm with the Gower distance [3,9]. The simple algorithm and the hierarchical clustering require the tuning of hyperparameters as k (number of clusters), and l (number of layers), and therefore have a fixed number of final clusters (in this context seen as profiles). Since we do not know the correct number of profiles in advance, we tried several hyperparameters. The semi-hierarchical clustering algorithm needs instead three other parameters: c_u, c_l (upper and lower bounds of variability), and n_{stop} (the threshold for cluster cardinality). For our experiments, we chose to set $c_u = 0.9$,

[2] Experimental setup: Intel Core i7-10510u, 16 GB RAM, Windows-10 64-bit.
[3] https://github.com/mcwilms/PEBAM.

$c_l = 0.1$, and n_{stop} to 5% of the size of the dataset (i.e. 50 for the German credit dataset).

As a first step, we test if clustering methods are *stable* enough. Table 1 gives a summary of the stability analysis performed for all three clustering methods on the German credit dataset, reporting (when relevant) the parameters l and c k, the number of clusters (e.g. k^l), the number of different outcomes after running the clustering algorithm 100 times with different random initializations, how often the most occurring cluster combinations occurs in these 100 runs, and the mean running time. Amongst other things, Table 1 shows that, when running the stability analysis with semi-hierarchical clustering, the German credit dataset produces 47 different cluster combinations. However, several combinations occur only once, and only one of the combination (number 0, the first one) occurs significantly more often than the other combinations, see Fig. 2a.

As a second step, we compute the *inter-clustering similarity* to test if the different cluster combinations are adequately similar. Figure 2b shows the inter-clustering similarity of the different cluster combinations we obtain on the German credit dataset via the semi-hierarchical clustering algorithm, showing only the cluster combinations that occur more than once. A dark blue tile means that two clusters are very similar (max 100%), and a white tile means that they have 50% or less of the clusters in common.

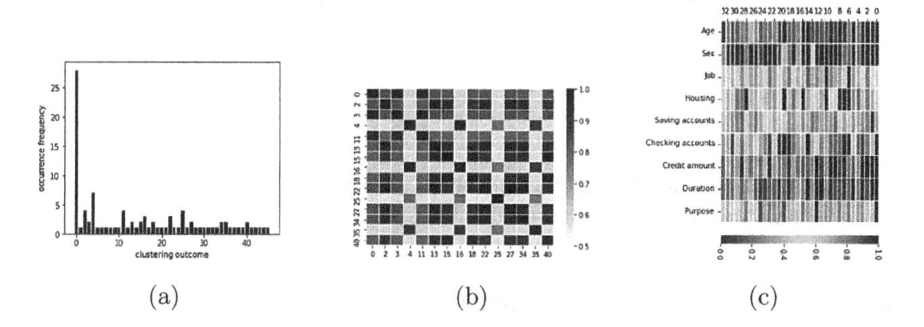

Fig. 2. Stability and variability analysis of the semi-hierarchical clustering algorithm applied on the German credit dataset: (a) frequency of each clustering outcome (or cluster combination) obtained over 100 runs; (b) inter-clustering similarity for cluster combinations that occur more than once; (c) variability of the different features for each profile in the most recurrent cluster combination.

As a confirmation that the algorithms end up on profiles which are descriptive attractors, we compute the feature *variability* of each cluster (supposedly a profile) within the most occurring clustering outcome. The variability plot of Fig. 2c shows for each profile (columns) the variability of the features (rows), where dark blue indicates a low variability and dark red a high variability. In tendency, qualitative features becomes stable, whereas numerical features show still a certain variability at the level of profiles. For each profile, however, the majority of features becomes stable.

4.2 Profile-Based Evaluation of Bias

We now apply the bias evaluation methods described in Sect. 3.3 with the profiles obtained by applying the semi-hierarchical clustering algorithm on classifiers trained on the German credit data set via three commonly used machine learning algorithms: *logistic regression classifier*, *XGboost classifier*, and *support vector machine (SVM) classifier* (e.g. [2]).[4] Following the standard practice of removing protected attributes (gender, ethnicity, etc.) as input features during training, we do not use the feature "Sex" provided by the German credit dataset for training the classifiers.

For **Representational bias**, Fig. 3 gives an overview of the presence of the identified profiles within the German credit dataset. We see that not all profiles are equally frequent; this is not necessarily an error, as long as this profile distribution is a good representation of the data on which ADM will be applied in practice. Expert knowledge or actual data collection can be used to test this assumption.

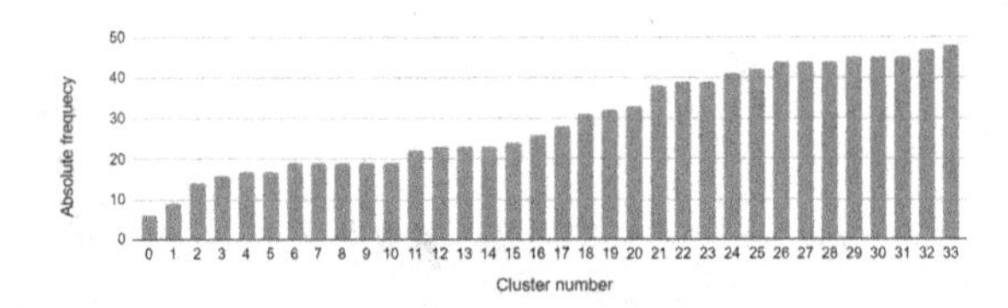

Fig. 3. Absolute frequencies of profiles obtained after performing the semi-hierarchical clustering algorithm on the German credit dataset.

To assess **Historical bias**, we first test for (general) *demographic disparity* with respect to protected attributes. A disadvantaged group with attribute x (DG_x) is the group with risk-label 'bad' and attribute x, whereas an advantaged group with attribute x (AG_x) is the group with risk-label 'good' and attribute x. Denoting with A the proportion of people from the full dataset belonging to the protected class (female) in the advantaged group over all people (female and male) in the advantaged group, and D for the proportion of people belonging to the protected class in the disadvantaged group over all people, we find:

$$D = \frac{\#DG_f}{\#DG_{f+m}} = 0.36 > 0.29 = \frac{\#AG_f}{\#AG_{f+m}} = A$$

and hence we conclude that, at aggregate level, the German credit data exhibit demographic disparity. We now will do the same computation for each profile

[4] Note however that the same approach will apply with any choice of profile selection method or of ML method used to train the classifier.

subset of the German credit dataset, to test whether there is *demographic disparity for profiles*. We write A_c for the proportion of people belonging to the protected class (female) in the advantaged group of cluster c over all people in the advantaged group of cluster c, and D_c for the proportion of people belonging to the protected class in the disadvantaged group of cluster c over all disadvantaged people in cluster c. Table 2 shows that there are 3 profiles (7, 20 and 28) that show demographic disparity. The fact that profiles show demographic disparity indicates that it might be possible that for some profiles other (not-protected) attributes correlate with the protected attribute, and so the protected attribute can indirectly be used in the training-process of the model.

In the computation of the (dis)-advantage fraction of Table 2 we still looked at the protected group *female*, however, we can also compute the measures A_c^* (D_c^*) as the fraction of (dis)advantaged individuals in a profile c over the total individuals within that profile (without distinguishing the protected class in it):

$$A_c^* = \frac{\#AG_c}{\#AG_c + \#DG_c} \qquad D_c^* = \frac{\#DG_c}{\#AG_c + \#DG_c}$$

By doing so, we get an indication of how informative a profile is for belonging to the (dis)advantaged group. Table 3 shows the fraction of advantaged and disadvantaged individuals for each profile. Note that there are profiles for which the majority of the samples is clearly advantaged (e.g. 0, 1, 2, ...), a few have some tendency towards disadvantaged outcomes (e.g. 3, 15), but in comparison could be put together with other profiles that have no clear majority (e.g. 9, 10, ...). Plausibly, for profiles exhibiting a mixed distribution of the risk label, there may be factors outside the given dataset that determine the label. Since the ADM models also do not have access to these external features, it may be relevant to evaluate performance on these profiles to evaluate this hypothesis.

Table 2. Fractions of disadvantaged (D) and advantaged (A) individuals with protected attribute *female* in each profile c.

c	D_c	A_c	c	D_c	A_c	c	D_c	A_c	c	D_c	A_c	c	D_c	A_c
0	0.00	0.00	**7**	**0.36**	**0.25**	14	0.00	0.19	21	0.00	0.03	**28**	**0.96**	**0.90**
1	0.00	0.00	8	0.00	0.00	15	1.00	1.00	22	0.00	0.00	29	0.00	0.00
2	0.00	0.00	9	0.00	0.00	16	1.00	1.00	23	0.00	0.00	30	1.00	1.00
3	0.00	0.00	10	1.00	1.00	17	1.00	0.86	24	0.00	0.00	31	1.00	1.00
4	0.00	0.07	11	0.00	0.00	18	0.00	0.00	25	0.00	0.03	32	0.00	0.00
5	1.00	1.00	12	0.00	0.00	19	0.06	0.06	26	1.00	1.00	33	0.00	0.09
6	0.00	0.08	13	0.00	0.00	**20**	**0.24**	**0.00**	27	0.00	0.00			

Table 3. Fractions of disadvantaged (D^*) and advantaged (A^*) individuals in each profile c.

c	D^*	A^*	c	D^*	A^*	c	D^*	A^*	c	D^*	A^*	c	D^*	A^*
0	0.00	1.00	**7**	**0.58**	**0.42**	14	0.09	0.91	21	0.05	0.95	**28**	**0.52**	**0.48**
1	0.11	0.89	**8**	**0.58**	**0.42**	**15**	**0.62**	**0.38**	**22**	**0.36**	**0.64**	**29**	**0.40**	**0.60**
2	0.14	0.86	**9**	**0.47**	**0.53**	**16**	**0.38**	**0.62**	23	0.15	0.85	30	0.09	0.91
3	**0.62**	**0.38**	**10**	**0.53**	**0.47**	17	0.25	0.75	24	0.37	0.63	31	0.22	0.78
4	0.12	0.88	11	0.18	0.82	18	0.19	0.81	25	0.20	0.71	32	0.13	0.87
5	0.12	0.88	12	0.35	0.65	**19**	**0.50**	**0.50**	**26**	**0.45**	**0.55**	33	0.06	0.94
6	0.32	0.68	**13**	**0.52**	**0.48**	**20**	**0.52**	**0.48**	27	0.14	0.86			

For the **Aggregation bias** we look at the blue dots in Fig. 4a, which indicate the mean performances of the algorithm over training the algorithm 100 times on different train-test splits. Looking at performance over each profile gives us a visual way to see to what extent general equal opportunity (Def. 4) is satisfied; we consider the average to provide a more robust indication. We see that the XGboost classifier performs the best of the three algorithms with respect to predicting the labels correctly, however we also observe some difference in performance depending on profile. In contrast, the SVM classifier has very low probabilities of getting an unjustified disadvantage label (Fig. 4b), while the probability of getting a correct label is not very high.

For the **Evaluation bias**, we look at the performance ranges of the different classification methods (visualized in terms of standard deviations). We see that the SVM classifier is the least sensitive towards the train-test split. The logistic regression classifier is already slightly more sensitive, however the XGboost classifier is by far the most sensitive towards the train-test split. All three algorithms are equally sensitive towards small profiles as much as larger profiles.

5 Conclusion

The paper introduced PEBAM: a new method for evaluating biases in ADM models trained on mixed datasets, focusing in particular on profiles extracted through a novel (semi-hierarchical) clustering method. Although we have proven the feasibility of the overall pipeline, several aspects need further consolidation, as for instance testing other measures of variability (e.g. to be compared with entropy-based forms of clustering, e.g. [6]), similarity scores, and distance measures. Yet, the method was already able to find biases that were not revealed by most used bias evaluation methods, since they would not test for biased decisions against groups of individuals that are regrouped by non-protected attributed values only. For instance, profile 7, exhibiting demographic disparity against women as historical bias, refers to applicants with little saving/checking accounts and renting their house, who are asking credit for cars (see Appendix for details).

Why, *ceteris paribus* (all other things being the same), men are preferred to women for access to credit for buying cars, if not in presence of a prejudice?

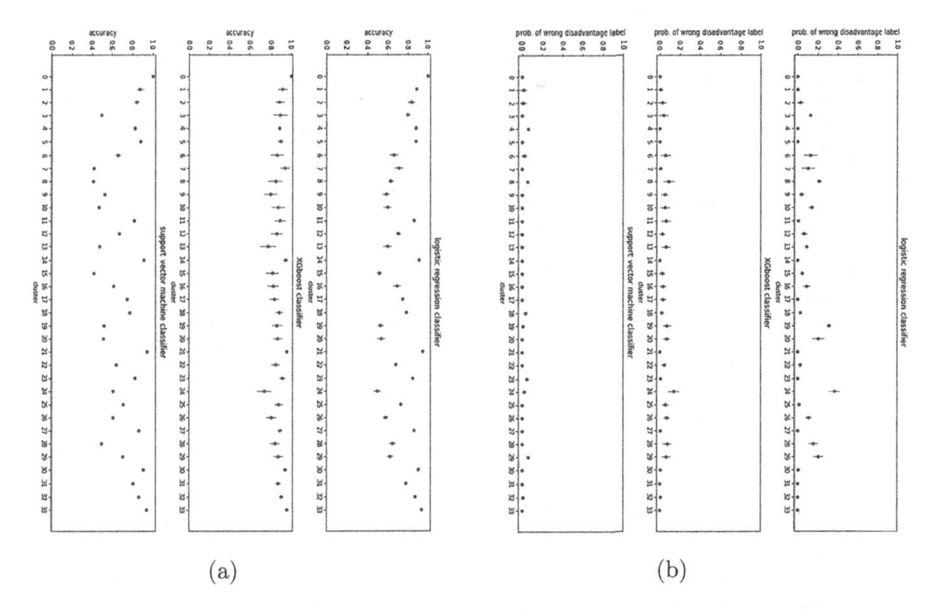

Fig. 4. (a) Probability of getting a correct prediction label. (b) Probability of getting a disadvantage ('bad') label when the true label is the advantage ('good'), for *logistic regression classifier*, *XGboost classifier*, and *SVM classifier* on the different profiles.

At a technical level, although the proposed semi-hierarchical clustering algorithms has shown a shorter running time than the baseline on the German credit dataset, the PAM algorithm does not scale well to larger datasets. Tiwari et al. propose BanditPAM as alternative for PAM [17], a method that reduces the complexity of each PAM iteration from $\mathcal{O}(n^2)$ to $\mathcal{O}(n \log n)$. When using PEBAM on large datasets one might consider using BanditPAM over PAM. This will be investigated in future work.

Acknowledgments. Giovanni Sileno was partly funded by the Dutch Research Council (NWO) for the HUMAINER AI project (KIVI.2019.006).

A Profiles on the German Credit Dataset

The following table reports the profiles selected on the German credit dataset by applying the semi-hierarchical clustering proposed in the paper, as described by their medoids:

Profile	Age	Sex	Job	Housing	Saving accounts	Checking account	Credit amount	Duration	Purpose	Sample
0	26	Male	2	Rent	Moderate	Unknown	3577	9	Car	859
1	37	Male	2	Own	Unknown	Unknown	7409	36	Business	868
2	39	Male	3	Own	Little	Unknown	6458	18	Car	106
3	26	Male	2	Own	Little	Little	4370	42	Radio/TV	639
4	31	Male	2	Own	Quite rich	Unknown	3430	24	Radio/TV	19
5	38	Female	2	Own	Unknown	Unknown	1240	12	Radio/TV	135
6	43	Male	1	Own	Little	Little	1344	12	Car	929
7	36	Male	2	Rent	Little	Little	2799	9	Car	586
8	39	Male	2	Own	Little	Little	2522	30	Radio/TV	239
9	31	Male	2	Own	Little	Moderate	1935	24	Business	169
10	33	Female	2	Own	Little	Little	1131	18	Furniture/equipment	166
11	26	Male	1	Own	Little	Moderate	625	12	Radio/TV	220
12	23	Male	2	Own	Unknown	Moderate	1444	15	Radio/TV	632
13	42	Male	2	Own	Little	Little	4153	18	Furniture/equipment	899
14	29	Male	2	Own	Unknown	Unknown	3556	15	Car	962
15	37	Female	2	Own	Little	Moderate	3612	18	Furniture/equipment	537
16	27	Female	2	Own	Little	Little	2389	18	Radio/TV	866
17	26	Female	2	Rent	Little	Unknown	1388	9	Furniture/equipment	582
18	29	Male	2	Own	Little	Unknown	2743	28	Radio/TV	426
19	53	Male	2	Free	Little	Little	4870	24	Car	4
20	36	Male	2	Own	Little	Little	1721	15	Car	461
21	38	Male	2	Own	Little	Unknown	804	12	Radio/TV	997
22	29	Male	2	Own	Little	Moderate	1103	12	Radio/TV	696
23	43	Male	2	Own	Unknown	Unknown	2197	24	Car	406
24	27	Male	2	Own	Little	Little	3552	24	Furniture/equipment	558
25	30	Male	2	Own	Little	Moderate	1056	18	Car	580
26	24	Female	2	Own	Little	Moderate	2150	30	Car	252
27	34	Male	2	Own	Little	Unknown	2759	12	Furniture/equipment	452
28	24	Female	2	Rent	Little	Little	2124	18	Furniture/equipment	761
29	34	Male	2	Own	Little	Moderate	5800	36	Car	893
30	34	Female	2	Own	Little	Unknown	1493	12	Radio/TV	638
31	30	Female	2	Own	Little	Unknown	1055	18	Car	161
32	35	Male	2	Own	Little	Unknown	2346	24	Car	654
33	35	Male	2	Own	Unknown	Unknown	1979	15	Radio/TV	625

References

1. Allaj, E.: Two simple measures of variability for categorical data. J. Appl. Stat. **45**(8), 1497–1516 (2018)
2. Belle, V., Papantonis, I.: Principles and practice of explainable machine learning. Front. Big Data **4** (2021)
3. Ben Ali, B., Massmoudi, Y.: K-means clustering based on Gower similarity coefficient: a comparative study. In: 2013 5th International Conference on Modeling, Simulation and Applied Optimization (ICMSAO), pp. 1–5. IEEE (2013)
4. Budiaji, W., Leisch, F.: Simple k-medoids partitioning algorithm for mixed variable data. Algorithms **12**(9), 177 (2019)
5. Caruso, G., Gattone, S., Fortuna, F., Di Battista, T.: Cluster analysis for mixed data: an application to credit risk evaluation. Soc.-Econ. Plan. Sci. **73**, 100850 (2021)
6. Cheng, C.H., Fu, A.W., Zhang, Y.: Entropy-based subspace clustering for mining numerical data. In: Proceedings of the 5th ACM SIGKDD Int. Conf. on Knowledge Discovery and Data Mining, KDD 2009, pp. 84–93. Association for Computing Machinery, New York (1999)
7. Feldman, M., Friedler, S., Moeller, J., Scheidegger, C., Venkatasubramanian, S.: Certifying and removing disparate impact. In: Proceedings of the 21th ACM

SIGKDD International Conference on Knowledge Discovery and Data Mining, KDD 2015, pp. 259–268. ACM (2015)
8. Friedler, S., Scheidegger, C., Venkatasubramanian, S., Choudhary, S., Hamilton, E., Roth, D.: A comparative study of fairness-enhancing interventions in machine learning. In: Proceedings of the Conference on fairness, accountability, and transparency, FAT 2019, pp. 329–338. ACM (2019)
9. Gower, J.C.: A general coefficient of similarity and some of its properties. Biometrics **27**(4), 857–871 (1971)
10. Hardt, M., Price, E., Price, E., Srebro, N.: Equality of opportunity in supervised learning. In: Advances in Neural Information Processing Systems (NIPS) (2016)
11. Kamiran, F., Calders, T.: Classifying without discriminating. In: Proceedings of 2nd IEEE International Conference on Computer, Control and Communication (2009)
12. Kamiran, F., Žliobaitė, I., Calders, T.: Quantifying explainable discrimination and removing illegal discrimination in automated decision making. Knowl. Inf. Syst. **35**(3), 613–644 (2013)
13. Kamishima, T., Akaho, S., Sakuma, J.: Fairness-aware learning through regularization approach. In: 2011 IEEE 11th International Conference on Data Mining Workshops, pp. 643–650. IEEE (2011)
14. Mehrabi, N., Morstatter, F., Saxena, N., Lerman, K., Galstyan, A.: A survey on bias and fairness in machine learning. ACM Comput. Surv. **54**(6), 1–35 (2021)
15. Pleis, J.: Mixtures of Discrete and Continuous Variables: Considerations for Dimension Reduction. Ph.D. thesis, University of Pittsburgh (2018)
16. Suresh, H., Guttag, J.: A framework for understanding sources of harm throughout the machine learning life cycle. In: Equity and Access in Algorithms, Mechanisms, and Optimization, EAAMO 2021 (2021)
17. Tiwari, M., Zhang, M.J., Mayclin, J., Thrun, S., Piech, C., Shomorony, I.: Banditpam: almost linear time k-medoids clustering via multi-armed bandits. In: Advances in Neural Information Processing Systems (NIPS) (2020)
18. Wachter, S., Mittelstadt, B., Russell, C.: Why fairness cannot be automated: bridging the gap between EU non-discrimination law and AI. Comput. Law Secur. Rev. **41**, 105567 (2021)

Author Index

Ashraf, Inaam 160

Bachlechner, Daniel 53
Barz, Michael 9
Beecks, Christian 96
Beierle, Christoph 1
Berndt, Jan Ole 131
Bhatti, Omair 9
Briesch, Martin 17
Brindise, Noel 31
Brito, Eduardo 45

Cuizon, Genesis 160
Czarnetzki, Leonhard 53

Diallo, Aïssatou 60
Dulny, Andrzej 75

Falkner, Jonas K. 160
Farahani, Aida 90
Fürnkranz, Johannes 60

Gembus, Alexander 160
Giesselbach, Sven 45
Grimm, Dominik G. 96
Gupta, Vishwani 45

Hahn, Eric 45
Haldimann, Jonas 1
Hamker, Fred H. 90
Haned, Hinda 209
Haselbeck, Florian 96
Hotho, Andreas 75
Hüwel, Jan David 96

Kainz, Fabian 53
Kollar, Daniel 1
Krause, Anna 75
Kriegel, Francesco 115
Kurchyna, Veronika 131

Lächler, Fabian 53
Laflamme, Catherine 53
Langbort, Cedric 31
Leemhuis, Mena 146
Löwens, Christian 160

Neumann, Nils 173

Özçep, Özgür L. 146

Rodermund, Stephanie 131
Rothlauf, Franz 17

Sauerwald, Kai 1
Schmidt-Thieme, Lars 160
Schwarzer, Leon 1
Sileno, Giovanni 209
Singh, Lavanya 187
Sobania, Dominik 17
Sonntag, Daniel 9
Spaderna, Heike 131

Timm, Ingo J. 131

Vitay, Julien 90

Wachsmuth, Sven 173
Wilms, Mieke 209
Wolter, Diedrich 146

Printed in the United States
by Baker & Taylor Publisher Services